ACETIC ACID BACTERIA

ACETIC ACID BACTERIA

CLASSIFICATION
AND
BIOCHEMICAL ACTIVITIES

BY

Toshinobu Asai, Ph. D.

Emeritus Professor of the
University of Tokyo

Professor of
Nihon University

UNIVERSITY OF TOKYO PRESS
Tokyo

UNIVERSITY PARK PRESS
Baltimore

© UNIVERSITY OF TOKYO PRESS, 1968
UTP No. 67199
Printed in JAPAN
All Rights Reserved

Published jointly by

University of Tokyo Press
Tokyo

and

University Park Press
Baltimore
LIBRARY OF CONGRESS CATALOG CARD NUMBER 68-24816

Dear God, what marvels there be in so small a creature.

Leeuwenhoek's draughtsman

PREFACE

Acetic acid bacteria were first recognized and isolated in 1837 by F. T. Kützing, who obtained the organism from naturally fermented vinegar and called it *Ulvina aceti*. Since then many strains have been isolated from sources such as spoiled alcoholic beverages and vinegars. These strains are responsible not only for the development of commercial vinegar production, but also for opening up new vistas to biologists and biochemists in studies of morphology, physiology, metabolism, *etc*.

The acetic acid bacteria are typical oxidative bacteria whose importance is comparable to that of the lactic acid bacteria and yeasts in anaerobic metabolism and alcoholic fermentation. Since the discovery by Kluyver and de Leeuw in 1924 of *Acetobacter suboxydans*, much has been learned about the microbial oxidation of sugars and polyols using the particular oxidative patterns of this organism as a model. Studies of cellulose biosynthesis in *Acetobacter xylinum*, to give another instance, have contributed to the understanding of polysaccharide synthesis. And, in addition to their usefulness in the study of basic biochemical processes, these bacteria produce many substances of importance to industry.

Most of the thirty years that I have spent in research have been devoted to the study of these bacteria. When I retired from the University of Tokyo six years ago I began to assemble the published data together with my own research findings for a book on their taxonomy and biochemistry. It is my hope that this book will contribute to an understanding of the acetic acid bacteria.

The following excellent reviews have been of great value in compiling this work: Dr. M.R.R. Rao's "Acetic acid bacteria" (*Ann. Rev. Microbiol.*, **11**, 1957); Prof. A. Janke's "Die Essigsäuregärung" (Handbuch der Pflanzenphysiologie, XII/1, 1960); and "Neus über den Stoffwechsel, die Systematik und Evolution der Essigsäure-Bakterien" (*Arch. f. Mikrobiol.*, **41**, 1962).

I would like to thank Associate Professor Dr. I. Suzuki of the Dept. of Microbiology, University of Manitoba, Winnipeg, Canada, and Dr. T. T. Myoda of the Rabida Institute, The University of Chicago, for their help in the preparation and proofreading of the English manuscript and for their many helpful comments, and Miss N. Ikegawa for typing the manuscript.

June, 1968   Toshinobu Asai

CONTENTS

ACETIC ACID BACTERIA

INTRODUCTION

Acetic acid bacteria of the genus *Acetobacter* are not only useful in the production of vinegar, gluconic acid, 2- and 5-ketogluconic acids and sorbose; as representative aerobic bacteria, they are also important for biochemical studies of the oxidation mechanisms of ethanol, polyalcohols and sugars.

Vinegar is described in one of the oldest and greatest books, the Bible. Man, as we may easily imagine, had obtained vinegar from ethanol-containing solutions by natural fermentation for centuries without understanding the nature of the process. Gradually it was realized that vinegar is formed by living organisms. According to Lafar[1] in *Die Essigsäure-Gärung*, the first researcher to suggest that mother of vinegar has a vegetable character was probably Boerhaave.[2]

The first biological study of the film that is occasionally formed on the surface of wine and beer was made by Persoon,[3] who proposed naming it *Mycoderma* ("viscous film"). Desmazières[4] made a further distinction between the films formed on wine (*Mycoderma vini*) and on beer (*M. cerevisiae*) and reported that the cells of the latter were oval and 8μ in length; but he considered it a member of the animal kingdom.

Kützing[5] recognized that the film formed on the surface of acidified beer or wine was actually a mass of minute organisms, $1.1-1.4\mu$ in diameter, and connected in chains. He considered it a kind of alga and named it *Ulvina aceti*. Thompson[6] later suggested changing the name to *Mycoderma aceti*.

The first systematic studies on acetic acid fermentation were carried out by Pasteur.[7] He recognized that mother of vinegar was a mass of living organisms which caused acetic acid fermentation. He also ascertained by experiment that no acetic acid fermentation could occur without *Mycoderma aceti*. Nevertheless, he was unable at the time to identify *Mycoderma* as a bacterium.

It was first recognized as such by Knieriem and Mayer[8] and later by Cohn.[9] Hansen[10] classified it into three kinds of acetic acid bacteria which included *Bacterium aceti*, *Bact. pasteurianum*, and *Bact. kützingianum*. He also recognized that acetification of beer and wine was caused by the action of these bacteria.

Since then various biologists and biochemists have carried out studies on acetic acid bacteria and acetic acid fermentation, and have succeeded in isolating a number of acetic acid bacteria and investigating their morphological and physiological characteristics and biochemical properties.

In the biochemical field, the oxidations of glucose to gluconic acid and of sorbitol to sorbose were discovered comparatively early.   Boutroux[11] first verified the formation of gluconic acid by *Mycoderma aceti* and later by *Micrococcus oblongus* (closely related to acetic acid bacteria) isolated from fruits.   Brown[12] also reported gluconic acid formation by *Bact. aceti* and *Bact. xylinum*.   From the fermentation liquor of *Vogelbeersafte* (the juice of mountain-ash berries) Bertrand[13] isolated acetic acid bacteria which were capable of oxidizing sorbitol to sorbose and which he named sorbose bacteria.

Although vinegar had been produced by natural fermentation in the Orleans region of France long before the discovery of acetic acid bateria, shortly thereafter a new technique known as the Pasteur method or quick vinegar process was adopted, and ultimately pure-cultured acetic acid bacteria were used.

Since the manufacture of vinegar is not within the scope of this book, it will not be discussed further; but the results of research on acetic acid bacteria with respect to taxonomy and biochemistry—especially oxidative metabolism of carbohydrates, alcohol, and polyalcohols—will be presented in detail.

Part I

CLASSIFICATION OF ACETIC ACID BACTERIA

Chapter 1

TAXONOMIC ALLOCATION

Generic Nomenclature.   Since their discovery, the acetic acid bacteria have been given a number of generic names.   The first, *Mycoderma*, was proposed by Persoon.[3]   *Mycoderma vini, Mycoderma cerevisiae* and *Mycoderma aceti* appeared later in the classic literature.   Kützing suggested the name *Ulvina*, though this means a kind of alga and is not acceptable for a bacterial group.   Pribram,[14] however, adopted this generic name in his *Klassifikation der Schizomyceten*.   In addition, *Umbina* was proposed by Naegeli,[15] *Termobacterium* by Zeidler,[16] *Acetobacterium* by Ludwig,[17] and *Acetimonas* by Orla-Jensen.[18] Lehmann and Neumann[19] used the generic name *Bacterium* to include acetic acid bacteria, and this name seems to have been standard for a considerable time. The name *Bacillus* has been used merely as an alternative and synonym for *Bacterium*.

The generic designation *Acetobacter*, which is universally used today, was suggested in 1900 by Beijerinck.[20]   *Acetobacter aceti* (Beijerinck)[21] first appeared as a synonym for *Bacterium aceti* Hansen. The generic name *Acetobacter* was approved by Fuhrmann[22] and others.   In 1935, Asai[23] proposed to differentiate acetic acid bacteria into two genera: *Acetobacter* and *Gluconobacter* gen. nov.   To the latter genus he assigned bacteria which have only a limited ability to oxidize ethanol and which thus do not produce much acetic acid, although their gluconic acid accumulation is very high.

In 1953, Leifson[24] observed the existence of peritrichous flagella in some commonly recognized species of *Acetobacter*, and concluded that the present *Acetobacter* genus should be divided into two genera: *Acetobacter* and *Acetomonas* gen. nov.   The re-defined *Acetobacter* thus would include only peritrichously or non-flagellated species of like physiological characteristics, while the new genus *Acetomonas* would include polarly flagellated or non-motile species of like physiological characteristics. *Acetomonas* coincides with the *Gluconobacter* genus, with respect not only to flagellation but also to physiological and biochemical properties.   In

7

1957, Tešić[25] published a simplified system of bacterial classification based largely upon the system of Krassilnikov,[26] in which the genus *Acetomonas* instead of *Acetobacter* was adopted using the description of *Acetomonas* Orla-Jensen. This, however, seems debatable in view of the fact that the original nomenclature was *Acetimonas* Orla-Jensen. In any case, Tešić's *Acetomonas* is not identical with Leifson's.

Allocation.   Several families have been proposed for the allocation of the genus *Acetobacter*. Orla-Jensen,[27] in his classification, assigned it to *Oxydobacteriaceae*. *Bergey's Manual of Determinative Bacteriology* (4th ed.)[28] and Asai[23] assigned it to the family *Nitrobacteriaceae*. These family names, however, have been rejected or have disappeared. In the 5th edition of *Bergey's Manual*,[29] *Acetobacter* was placed as a single genus in the family *Acetobacteriaceae*. Prévot[30] proposed in his *Classification of Bacteria* to set up a family *Protobacteriaceae* and within this a tribe *Acetobactereae*. The genus *Acetobacter* was included in this tribe as a single genus.

Until recently, the family *Pseudomonadaceae*, as adopted by Kluyver and van Niel[31] and Rahn,[32] was the only acceptable family in which to include the genus *Acetobacter*. Krassilnikov[33] and *Bergey's Manual* (7th ed.)[34] both supported this allocation. However, recent morphological studies, as reported separately by Leifson,[24] Asai,[35] and Shimwell,[36] have shown that typical *Acetobacter* species, such as *A. aceti* and *A. rancens*, are peritrichously flagellated. Clearly, these acetic acid bacteria cannot be assigned to the *Pseudomonadaceae*, in which the flagellum is polar. This problem will be discussed further in Chapters 6 and 7.

Chapter 2

EARLIER CLASSIFICATIONS

The first attempt to classify acetic acid bacteria was made by Hansen[37] in 1894. At that time, known species were very few, and were differentiated primarily according to morphological characteristics. His classification, based on the appearance of the pellicle in fluid media and its iodine reaction, is as follows:

    I.   Pellicle in fluid media easily separable.

        A.   Pellicle not stained by iodine.

                Bact. aceti Hansen.

        B.   Pellicle stained by iodine.

                Bact. pasteurianum Hansen.

                Bact. kützingianum Hansen.

    II.   Pellicle in fluid media thick and leathery.

                Bact. xylinum Brown.

In 1898, Beijerinck[38] proposed a classification according to natural habitat and the ability to utilize an inorganic nitrogen source.

    I.   *Bacterium aceti* Pasteur (quick vinegar bacteria).

        Found in quick vinegar acetifier. Grow in Beijerinck's nutrient solution*[39] and form pellicles on the surface. Form voluminous, transluscent viscous colonies on beer gelatin containing 10% sucrose.

                Bact. aceti Pasteur.

                Termobact. aceti Beijerinck.

                Termobact. aceti Zeidler.

    II.   *Bacterium rancens* (beer acetic acid bacteria).

        Found in beer. Do not grow in Beijerinck's nutrient solution. Formation of viscous colonies hindered on sucrose-containing beer gelatin.

---

* Beijerinck's solution : ammonium phosphate 0.05 g, potassium chloride 0.01 g, absolute alcohol 3 g, tap water 100 g.

Bact. rancens Beijerinck.
Bact. aceti Hansen.
Bact. aceti Brown.
Bact. oxydans Henneberg.
Bact. acetosum Henneberg.

III. *Bacterium pasteurianum* (beer acetic acid bacteria).
Found in beer. Do not grow in Beijerinck's nutrient solution. Pellicles stained by iodine.

Bact. pasteurianum Hansen.
Bact. kützingianum Hansen.

IV. *Bacterium xylinum.*
Form thick leathery pellicle in sugar-containing fluid media and oxidize acetic acid. Do not grow in Beijerinck's nutrient solution.

Bact. xylinum Brown.

The next year, Hoyer,[40] a follower of Beijerinck, added some species to Beijerinck's classification and included sucrose inversion among the criteria for differentiating species in the group.

I. Beer acetic acid bacteria.
Do not invert sucrose.

A. *Bacterium rancens* Beijerinck.
Pellicle not stained by iodine.

Bact. rancens Beijerinck.
Bact. aceti Hansen.
Bact. aceti Brown.
Bact. acetosum Henneberg.
Bact. oxydans Henneberg.
Bact. industrium Henneberg.
Termobact. aceti Zeidler.

B. *Bacterium pasteurianum* Hansen.
Pellicle stained by iodine.

Bact. pasteurianum Hansen.
Bact. kützingianum Hansen.

II. *Bacterium aceti* Pasteur (quick vinegar bacteria).
Invert sucrose. Grow in Beijerinck's and Pasteur's nutrient solution.*[41] Viscous growth occurs in sucrose-containing beer.

---

* Pasteur's nutrient solution : glacial acetic acid 12.75 g, absolute alcohol 22.50 ml, ammonium phosphate 0.2 g, magnesium phosphate 0.1 g, calcium phosphate 0.1 g, distilled water to make 1000 ml.

Bact. aceti Pasteur.

Bact. acetigenum Henneberg.

Bact. ascendens Henneberg.

III. *Bacterium xylinum* Brown.

Invert sucrose. Form thick leathery pellicle in liquid media. Pellicle stained blue by iodine and sulfuric acid.

Bact. xylinum Brown.

Bact. xylinum Bertrand.

Henneberg[42] in 1898 proposed classifying acetic acid bacteria according to the types of media in which they naturally occur or to which they will acclimatize themselves, ignoring morphological and physiological traits. He thus differentiated four groups:

I. Malt infusion or wort acetic acid bacteria.

Occur in breweries, distilleries and yeast factories. Cannot be used in the manufacture of vinegar. Are motile, not stained by iodine, and do not decompose acetic acid.

Bact. oxydans Henneberg.

Bact. industrium Henneberg.

II. Beer vinegar bacteria.

Found in beer. Some are used to convert beer into vinegar. The first three species are included in Hansen's classic work.

Bact. aceti Hansen.

Bact. pasteurianum Hansen.

Bact. kützingianum Hansen.

Bact. acetosum Henneberg.

Bact. rancens Beijerinck.

Termobact. aceti Zeidler.

III. Wine vinegar bacteria.

Found in wine. Some are used in the manufacture of wine vinegar.

Bact. ascendens Henneberg.

Bact. vini acetati Henneberg.

Bact. xylinoides Henneberg.

Bact. orleanense Henneberg.

Bact. xylinum Brown.

IV. Quick vinegar bacteria.

Include bacteria whose natural habitat is the shavings in acetifiers. Grow on a wide range of nutrient substances, including simple media containing only inorganic nitrogen compounds, sugar and ethanol.

Bact. acetigenum Henneberg.

Bact. schützenbachii Henneberg.

Bact. curvum Henneberg.

The difficulty of this arrangement is the fact that the relationship between each group of bacteria and its characteristic habitat is not an exclusive one. *Bact. aceti*, for example, can be isolated not only from beer but from fermented cane molasses, fruit, wine and vinegar. Hence this classification is inadequate.

In the same year, Rothenbach[43] also proposed a classification based on habitat and industrial application, as follows:

I. Quick vinegar bacteria.
    Grow in quick vinegar generator, produce high concentrations of acetic acid and form zoogloeal membranes in liquid culture media.

    Bact. schützenbachii Henneberg.

II. Useful wine acetic acid bacteria.

    Bact. aceti (Hansen) Seifert.
    Strains isolated by Rothenbach.

III. Useful beer acetic acid bacteria.

    Bact. rancens var. *zythi* Beijerinck.
    Bact. pasteurianum Hansen.
    Strains isolated by Rothenbach.

IV. Wort and malt infusion acetic acid bacteria.

    Bact. industrium Henneberg.
    Bact. oxydans Henneberg.

V. Pellicle forming acetic acid bacteria, which are able to assimilate nitrogen from inorganic ammonium sources.
    Grow well in synthetic culture media, found in quick vinegar generator, but do not produce acetic acid in large quantities.

    Bact. acetigenum Henneberg.
    Bact. aceti (Pasteur) Beijerinck.
    Bact. aceti var. *agile* Beijerinck.

VI. Harmful acetic acid bacteria.
    Found in wine, beer and quick vinegar generator. Cause cloudiness in liquid media. Often produce unpleasant odor. Are detrimental to brewing.

    Bact. ascendens Henneberg.

Termobact. aceti Zeidler.

Bact. kützingianum Hansen.

Though based in part on morphological and physiological properties —e.g., nature of the pellicle or capacity to utilize inorganic nitrogen sources —these classifications tended to rely primarily on differences in habitat, which are not considered significant in modern taxonomy.

In 1916, Janke[44] presented a new and more notable classification, which was to provide the basis for Vaughn's classification and the key adopted in *Bergey's Manual of Determinative Bacteriology* (6th ed., 1948). Janke differentiated each species primarily according to its capacity to utilize inorganic ammonium salts and acetic acid as nitrogen and carbon sources, respectively. Such factors as nature of the pellicle, motility and formation of acid from carbohydrates were used to subdivide them, as follows:

I. Haplotrophic acetic acid bacteria.

Capable of utilizing ammonium salts as a sole source of nitrogen and acetate as a sole source of carbon.*

A. *Schützenbachii* group.

Correspond to quick vinegar bacteria in Hoyer's classification.

B. *Hansenianum* group.

Produce slimy, tough pellicle in fluid media.

1. Pellicle stained yellowish by iodine.

a) Produce acids from glucose.

b) Do not produce acids from glucose.

2. Pellicle stained bluish-purple by iodine.

II. Symplotrophic acetic acid bacteria.

Cannot utilize ammonium salts as a sole source of nitrogen, or acetate as a sole source of carbon.

A. *Rancens* group.

Do not produce zoogloeal membrane in fluid media. Pellicle not stained by iodine.

1. Subgroup of *Bact. aceti* Hansen.

Nonmotile. Pellicle is separable. Rather thermophilic.

a) *Bact. aceti* Hansen type.

Produce wet immersed pellicle in beer. Medium remains transparent.

---

* Grow in Pasteur's solution or Beijerinck's solution.

(1) Produce acids from glucose.

Bact. aceti Hansen.

(2) Do not produce acids from glucose.

b) *Bact. acetosum* Henneberg type.

Produce pellicle rising on wall of flask in beer medium.

(1) Produce acids from glucose.

Bact. acetosum Henneberg.

(2) Do not produce acids from glucose.

c) Pellicle on fluid media not coherent, rather powdery, easily submerged. Medium becomes cloudy.

2. Subgroup of *Bact. aceti* Brown.

Motile. Pellicle is separable.

a) Rather psychrophilic. Pellicle does not rise on wall of flask. Do not produce acids from maltose.

b) Produce acids from maltose.

Bact. oxydans Henneberg.

Bact. industrium Henneberg.

3. Subgroup of *Bact. albuminosum* Lindner.

Cause a strong turbidity in beer. Pellicle viscous, easily precipitated as a slimy mass.

a) *Bact. aceti viscosum* Baker, Day and Hulton type.

Produce colorless or grayish-white colony on solid media. Colony wet, slimy, transparent and fluid.

(1) Include strains isolated by Janke.

(2) Produce a brown pigment.

A. melanogenum Beijerinck.

B. *Pasteurianum* group.

Pellicle generally separable, and stained by iodine under certain conditions. Easily produce involution forms* (Fig. 1).

---

* The cells of acetic acid bacteria show unusual or abnormal shapes under certain conditions, perhaps not favorable for growth. These are termed involution forms. Involution forms are very likely to occur in artificial culture and are more common with some species of *Acetobacter* than with others.

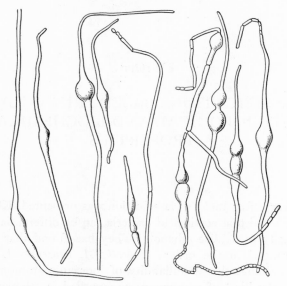

Fig. 1. Involution forms of *A. pasteurianus*
(after Hansen).

1. Subgroup of nonmotile species.
   a) *Bact. pasteurianum* Hansen type.
      Produce slimy, thin pellicle in fluid media. Often form long chains of short rods. Medium remains transparent.
   b) *Bact. kützingianum* Hansen type.
      Cause cloudiness in beer. Produce fine powdery pellicle rising on wall of flask. Do not form long chains.
   c) *Bact. pasteurianum* var. *variabile* Hoyer type.
      Staining with iodine disappears gradually, but revives after sub-inoculation into fresh medium.
2. Subgroup of motile species.
   Species not yet discovered.
C. *Xylinum* group.
   Produce thick, cartilaginous membrane on fluid media.
   1. *Bact. xylinum* Brown type.
      Do not cause cloudiness in beer. Pellicle stained by iodine and sulphuric acid.

Chapter 3

CLASSIFICATIONS BASED PRIMARILY ON PHYSIOLOGICAL AND BIOCHEMICAL PROPERTIES

Nine years after Janke's classification was presented, Kluyver and de Leeuw[45] suggested that acetic acid bacteria can be differentiated according to their capacity to oxidize ethanol. They had found that the oxidative power is arranged in the order *A. aceti* > *A. rancens* > *A. xylinum* > *A. suboxydans*, and that the assimilability of inorganic ammonium nitrogen changes with the kind of carbon source supplied, as shown in Table 1. These observations should be kept in mind in connection with Frateur's classification into four separate groups—*Peroxydans, Oxydans, Mesoxydans,* and *Suboxydans*—which will be considered later.

Table 1. Assimilability of inorganic ammonium salts with various carbon sources.

Specise	Carbon source		
	ethanol	glucose	mannitol
A. aceti	+	+	+
A. rancens	−	+	+
A. xylinum	−	+	+
A. suboxydans	−	−	+

In 1931, Hermann and Neuschul[46] noticed ketogenic oxidation by a new species of *Acetobacter*, *Bact. gluconicum*,* which produces a large amount of gluconic acid from glucose, and ketogenic compounds from polyalcohols. They examined ketogenic activities in the known species of *Acetobacter*, and found that these divided themselves into two groups which they named ketogenic and non-ketogenic acetic acid bacteria.

    I.   Ketogenic acetic acid bacteria.

---

* Hereafter species names ending in -*um* will be emended to–*us* (with the exception of *A. xylinum*) in accordance with *Bergey's Manual of Determinative Bacteriology* (7th ed.).

A. *gluconicus* Hermann.

A. *xylinoides* Henneberg.

A. *xylinum* Brown.

A. *orleanense* Henneberg.

A. *aceti* Hansen.

II.   Non-ketogenic acetic acid bacteria.

A. *pasteurianus* Hansen.

A. *acetosus* Henneberg.

A. *ascendens* (Institut für Gärungs-gewerbe, Berlin).

A. *vini acetati* Henneberg.

A. *kützingianus* Hansen.

A. *ascendens* Henneberg.

In 1925 Visser't Hooft[47] isolated a new strain, A. *peroxydans*, which he found to be catalase negative.   He accordingly divided the *Acetobacter* into catalase negative and catalase positive groups, as follows:

   I.   Catalase-negative.

     A.   Grow in Hoyer's medium.

A. *peroxydans* Visser't Hooft.

   II.   Catalase-positive.

     A.   Grow in Hoyer's medium.

A. *aceti* Beijerinck.

     B.   Do not grow in Hoyer's medium.

       1.   Form acid and carbonate on alcohol chalk plates.

A. *rancens* Beijerinck.

       2.   Form only acid on alcohol chalk plates.

         a)   Form gluconic acid on dextrose chalk agar plates, which is further oxidized to $CaCO_3$.

         (1)   Form a heavy pellicle.

A. *xylinum* Brown.

         (2)   Do not form pellicle and form brown pigment.

A. *melanogenus* Beijerinck.

         b)   Form gluconic acid on dextrose chalk agar plates, which is further oxidized to Ca 5-ketogluconate.

A. *suboxydans* Kluyver and de Leeuw.

Visser't Hooft's system was later developed into another classification by Vaughn,[48] as follows:

   I.   Oxidize acetic acid to water and carbon dioxide.

A.   Utilize ammonium salts as a sole source of nitrogen (Hoyer's medium).

   A. aceti (Kützing) Beijerinck.

B.   Do not utilize ammonium salts as a sole source of nitrogen.

   1.   Form a thick, cellulosic membrane on the surface of liquid media.

   A. xylinum (Brown) Bergey *et al.*

   2.   No cellulosic membrane formed.

   A. rancens Beijerinck.

II.   Do not oxidize acetic acid.

A.   Form pigments in glucose media.

   1.   Dark brown pigment.

   A. melanogenus Beijerinck.

   2.   Pink to rose pigment.

   A. roseus (Takahashi and Asai) nov. comb.

B.   Do not form such pigments.

   1.   Optimum temperature 30° to 35°C.

   A. suboxydans Kluyver and de Leeuw.

   2.   Optimum temperature 20°C.

   A. oxydans Henneberg.

A. peroxydans Visser't Hooft, due to its unusual catalase-negative property, was not included in the classification. Vaughn felt that further study of the species was necessary in order to determine more clearly its taxonomic position. Like Visser't Hooft, he also reduced the number of species considerably. In his opinion, *A. industrius* was a variety of *A. oxydans*, and *A. xylinoides* and *A. acetigenus* were varieties of *A. xylinum*. *A. pasteurianus*, *A. kützingianus*, and *A. zeidleri* could be assigned to the *A. aceti* or *A. rancens* group, according to their capacity to utilize ammonium salts or according to other characteristics described in the literature. *A. ascendens* he thought to be a variety of *A. aceti*, because some strains isolated by him from wine, though initially identified as *A. ascendens* Henneberg, had proved after careful observation to form acid from glucose and to grow in Hoyer's solution. *A. gluconicus*, *A. dihydroxyacetonicus*, *A. capsulatus*, and *A. viscosus* were not included.

In 1948, Vaughn's follower Shimwell[49] revised the key to include several more varieties, as follows:

I.   Oxidize acetic acid to water and carbon dioxide.

A.   Use ammonium salts as sole nitrogen source.

    A. aceti (Kützing) Beijerinck.
  Regarded as variety of *A. aceti.*
    A. ascendens (Henneberg) Bergey *et al.*
B. Do not use ammonium salts as sole nitrogen source.
  1. Form thick leathery cellulosic pellicle on liquid media.
    A. xylinum (Brown) Bergey *et al.*
  Regarded as varieties of *A. xylinum.*
    A. acetigenus (Henneberg) Bergey *et al.*
    A. xylinoides (*Bact. xylinoides* Henneberg).
  2. No leathery cellulosic pellicle formed.
    A. rancens Beijerinck.
  Regarded as varieties of *A. aceti* or *A. rancens,* according to whether cultures use ammonium salts or not.
    A. pasteurianus (Hansen) Beijerinck.
    A. kützingianus (Hansen) Bergey *et al.*
    A. zeidleri Beijerinck.
    A. acetosus (Henneberg) Bergey *et al.*
    A. turbidans Cosbie, Tošić and Walker.
    A. mobile Tošić and Walker.
II. Do not oxidize acetic acid.
 A. Pigment formed in glucose media.
  1. Dark brown (pale beer becomes darker).
    A. melanogenus Beijerinck.
  2. Pink.
    A. roseus (Takahashi and Asai) Vaughn, nov. comb.
 B. Pigment not formed.
  1. Optimum temperature 30° to 35°C.
    A. suboxydans Kluyver and de Leeuw.
  Produce ropiness in dextrin media. Regarded as possible variety of *A. suboxydans.*
    A. capsulatus Shimwell.
  2. Optimum temperature about 20°C.
    A. oxydans (Henneberg) Bergey *et al.*
  Produce ropiness in dextrin media. Regarded as varieties of *A. oxydans.*
    A. industrius (Henneberg) Bergey *et al.*

A. viscosus (Baker, Day and Hulton)
Shimwell.

Certain industrially important strains, such as *A. schützenbachii*, were excluded from the key, although Shimwell noted at the time that further comparative studies of such organisms might necessitate its expansion. He emphasized that species differentiation is largely subjective, especially in the case of acetic acid bacteria. Each attempt to define fresh criteria for the differentiation of species tends to invalidate criteria used by previous workers, without necessarily leading to a better system. Even the basic criterion that characterizes *Acetobacter*—the capacity to oxidize ethanol to acetic acid—was shaken when Stanier[50] confirmed, in 1947, that certain species of *Pseudomonas fluorescens* can also oxidize ethanol to acetic acid.

Bergey's Manual (7th ed., 1957) used Vaughn's system for its classification of *Acetobacter*. The key in the manual is nearly identical with the 6th edition's. It differs slightly from the original description in classifying *A. pasteurianus* and *A. kützingianus*, although these two species were cited as questionable in their ability to utilize ammonium salts as a sole source of nitrogen. The key is shown as follows.

I. Oxidize acetic acid to carbon dioxide and water.
    A. Utilize ammonium salts as a sole source of nitrogen (Hoyer's solution).
        1. *A. aceti.*
    B. Do not utilize ammonium salts as a sole source of nitrogen.
        1. Form a thick, zoogloeal, cellulose membrane on the surface of liquid media.
        2. *A. xylinum.*
        2. Do not form a thick, zoogloeal, cellulose membrane on the surface of liquid media.
        3. *A. rancens.*
        3a. *A. pasteurianus.*
        3b. *A. kützingianus.*
II. Do not oxidize acetic acid.
    A. Form pigments in glucose media.
        1. Dark brown to blackish pigment.
        4. *A. melanogenus.*
        2. Pink to rose pigment.
        5. *A. roseus.*
    B. Do not form pigments.

Table 4. Assignment of known species or varieties of *Acetobacter* in corresponding groups (after Frateur).

Peroxydans group :	Catalase negative.
	A. peroxydans Visser't Hooft.
	A. paradoxus Frateur.
Oxydans group :	
	A. lovaniense Frateur.
	A. ascendens Henneberg.
	A. rancens Beijerinck.
	A. rancens var. *filamentosus* Beijerinck.
	A. rancens var. *turbidans* Beijerinck.
	A. rancens var. *saccharovorans* Beijerinck.
	A. rancens var. *vini* Beijerinck.
	A. rancens var. *pasteurianus* Beijerinck.
Mesoxydans group :	
	A. aceti (Pasteur) Beijerinck.
	A. xylinum (Brown) Beijerinck.
	A. xylinum var. *xylinoides* Beijerinck.
	A. mesoxydans Frateur.
	A. mesoxydans var. *saccharovorans* Frateur.
Suboxydans group :	
	A. suboxydans Kluyver and de Leeuw.
	A. suboxydans var. *muciparum* (Kluyver and de Leeuw).
	A. suboxydans var. *biourgianum* (Kluyver and de Leeuw).
	A. melanogenus Beijerinck.
	A. melanogenus var. *maltovorans* Beijerinck.

produced in Formosa. He divided these into four types, according to the oxidation pattern of ethanol and glucose in a medium containing both substrates, as follows:

Type 1: Produces acetic acid only and does not produce gluconic acid.

Type 2: Produces gluconic acid after the complete oxidation of acetic acid to CO_2 and H_2O.

Type 3: Produces gluconic acid after the complete oxidation of ethanol to acetic acid.

Type 4: Produces acetic acid and gluconic acid simultaneously.

These patterns of selective oxidation form the basis of the following classification.

Table 5.   Arrangement of species and varieties of *Acetobacter*
by Janke after Frateur's system.

Group	Related species and varieties
I. *Peroxydans* group:	
Haplotrophic.	
A. peroxydans Visser't Hooft.	
Symplotrophic.	
A. paradoxus Frateur.	
II. *Euoxydans* group:	
Haplotrophic.	
A. lovaniense Frateur.	
Symplotrophic.	
A. ascendens Henneberg.	*Bact. aceti* Hansen (?), *Bact.*
A. rancens Beijerinck.	*aceti* Brown.
A. rancens var. *turbidans*	*A. turbidans* Cosbie, Tošić and
Beijerinck.	Walker.
A. rancens var. *vini* Beijerinck.	*A. mobile* Tošić and Walker.
A. rancens var. *saccharovorans*	*A. rancens* var. *Zythi* Hoyer.
Beijerinck.	*A. vini acetati* Henneberg.
A. rancens var. *filamentosus*	*A. acetosus* Henneberg.
Beijerinck.	*A. pasteurianus* (Hansen)
A. rancens var. *pasteurianus*	Beijerinck.
Beijerinck.	*A. kützingianus* (Hansen) Bergey
	et al.
III. *Mesoxydans* group:	
Haplotrophic.	
A. aceti (Pasteur) Beijerinck.	*A. zeidleri* Beijerinck.
A. aceti var. *muciparum*	(*Termobact. aceti* Zeidler).
Beijerinck.	*A. schützenbachii* Henneberg.
	A. curvus Henneberg.
	A. acetigenoidius Krehan.
	A. lafarinus Janke.
	A. lafarinus var. *vindovonense*
	Janke.

Table 5. *Continued.*

Symplotrophic.

A. xylinum (Brown) Beijerinck.	(= Bertrand sorbose bacteria)
A. xylinum var. *maltovorans* Beijerinck.	*A. xylinoides* Henneberg.
	A. acetigenus Henneberg.
A. xylinum var. *xylinoides* Beijerinck.	*A. orleanense* Henneberg.
A. mesoxydans Frateur.	*A. dihydroxyacetonicus* Virtanen and Bärlund.
A. mesoxydans var. *lentum* Frateur.	*A. gluconicus* Hermann and Neuschul.
A. mesoxydans var. *saccharovorans* Frateur.	*Bact. aceti* Hansen (?).
A. mesoxydans var. *lentumsaccharovorans* Frateur.	

IV.   *Suboxydans* group:

Symplotrophic.

A. suboxydans Kluyver and de Leeuw.	*A. oxydans* Henneberg.
	A. industrius Henneberg.
A. suboxydans var. *muciparum* Kluyver and de Leeuw.	*A. viscosus* Baker, Day and Hulton.
A. suboxydans var. *hoyerianum* Kluyver and de Leeuw.	*A. albuminosus* Lindner.
A. suboxydans var. *biourgianum* Kluyver and de Leeuw.	*Bact. hoshigaki* var. *rosea* Takahashi and Asai.
A. capsulatus Shimwell.	
A. roseus (Takahashi and Asai) Vaughn.	
A. melanogenus Beijerinck.	
A. melanogenus var. *maltovorans* Beijerinck.	
A. melanogenus var. *maltosaccharovorans* Beijerinck.	

Group I: Includes bacteria of Types 1, 2 and 3. These do not oxidize glucose, or oxidize it only after the oxidation of ethanol to acetic acid or to CO_2 and H_2O in the presence of both substrates. Short rods, non-motile, form pellicles creeping up sides of vessel to an extraordinary height. Comparatively tolerant to high temperatures (>60°C) and resistant to high concentrations of ethanol (10–15%) or acetic acid (2.25–5.0%). Production of acetic acid (5.9–7.8%) usually surpasses production of gluconic acid (0–5.6%).

Type 1: *A. ascendens* Henneberg var.

Type 2: Strain No. 78.*

Type 3: *A. aceti* Brown var.

A. acetosus Henneberg var.

Group II: Includes bacteria of the fourth type. These resemble group I in physiological properties, but produce acetic acid and gluconic acid simultaneously in the presence of ethanol and glucose. Short rods, motile or non-motile, form pellicles not ascending side of vessel or ascending only slightly. Death temperature (50–60°C) and resistance to ethanol (5–12%) and acetic acid (1.2–5.2%) are comparatively low. Production of gluconic acid (3.0–18.1%) usually surpasses production of acetic acid (2.3–6.5%). This group seems to be intermediate between Groups I and III.

A. Non-motile: *A. industrius* Henneberg var.

A. aceti Hansen var., Strain No. 49.

B. Motile:

1. Pellicle stained by iodine and sulphuric acid.

A. pasteurianus Hansen var.(?)

A. kützingianus Hansen var.(?)

2. Pellicle not stained by iodine and sulphuric acid.

A. curvus Henneberg var.

Group III: Includes bacteria of the fourth type which oxidize glycerol to dihydroxyacetone. Fairly long rods, form pellicles rarely ascending sides of vessel. Death temperature (45°–60°C), optimum temperature for growth (<30°C), resistance to ethanol (3–9)% and acetic acid (0.7–2.5%) with the exception of *A. xylinum*, are usually low. Gluconic acid production (11.0–19.7%) markedly exceeds acetic acid production (1.3–4.6%).

A. Non-motile:

1. Produce brown pigments from sugars. *Gluconoacetobacter liquefaciens* Asai var.

2. *A. xylinum* Brown var.

3. Strain No. 55, Strain No. 84.

B. Motile:

1. Strain No. 48, Strain No. 58.

---

* The organisms described by strain number were recognized by Tanaka to be new species or varieties.

Chapter 4

ESTABLISHMENT OF A NEW GENUS
GLUCONOBACTER ASAI

All the classifications we have discussed so far are based on taxonomic studies of bacteria isolated from alcoholic beverages or vinegar. Asai[54] in 1935 attempted for the first time to classify the oxidative bacteria living in various fruits. He isolated thirty-eight strains, most of which exhibited characteristics quite different from the usual acetic acid bacteria. Most strains produced a large amount of gluconic acid from glucose without addition of $CaCO_3$. They lacked the ability to form films in liquid media, and made the liquid turbid. Although they grew well in sugar-containing media such as malt-extract liquor, they generally grew poorly in ethanol-containing media such as saké or beer. Red or reddish-brown pigment was produced by some strains, and gelatin was liquefied by seven of thirty-eight strains investigated. A few strains were able to grow at 39° to 40°C. Most strains had maximum growth temperatures lower than 39° or 40°C, and were able to grow at temperatures as low as 7° to 9°C.

Inorganic nitrogen sources were assimilated less readily than peptones. As a source of carbon, glucose was very easily assimilated and polyalcohols such as mannitol or glycerol fairly easily. The capacity of these strains to grow in ethanol-containing media was remarkably less, however, than that of known acetic acid bacteria. Acetic acid was assimilated only by thirteen strains, but gluconic acid was assimilated easily by almost all. The optimum pH for growth was generally on the acidic side with a few exceptions near neutral. All of them grew at pH 4.2. Acids were produced by all strains from glucose, by most strains from arabinose, fructose, galactose, sucrose, and polyalcohols such as glycerol and mannitol, and by several strains from dulcitol. Acids were also produced by most strains from isobutyl alcohol and amyl alchool. Glucose and gluconic acid were oxidized to 5-ketogluconic acid by most strains; mannitol, glycerol, and isopropyl alcohol were oxidized to fructose, dihydroxyacetone, and acetone respectively. Fructose was further oxidized to kojic acid by most of the

27

strains that were able to grow at fairly low temperatures, and dulcitol was oxidized to galactose by several. On the other hand, the ability to oxidize ethanol was generally less than that of known acetic acid bacteria. Two strains were unable to oxidize ethanol at all, and only a few could oxidize acetic acid to CO_2 and H_2O.

These results led to a proposal to divide the acetic acid bacteria into two separate genera, *Acetobacter* and *Gluconobacter* gen. nov., on the basis of their capacity to oxidize ethanol and glucose. The genus *Acetobacter* was divided into two subgenera: *Euacetobacter*, which included strains capable of oxidizing ethanol but not glucose, and *Acetogluconobacter*, which included strains capable of oxidizing ethanol well* but glucose only weakly.** The new genus *Gluconobacter* was also divided into two subgenera: *Gluconoacetobacter*, which included strains capable of oxidizing glucose well but ethanol weakly, and *Eugluconobacter*, which included strains capable of oxidizing glucose but not ethanol. The description of each genus and species assignment follows:

    I.   Genus *Acetobacter*.

        Grow at 39° to 40°C, but not at 7° to 9°C. Optimum temperature for growth over 30°C. Primarily oxidize ethanol to acetic acid. Generally oxidize glucose to gluconic acid. Assimilate and oxidize acetic acid.

    A.   Subgenus *Euacetobacter*.

        Oxidize ethanol, but not glucose.

        1.   Do not invert sucrose.

            a)   Assimilate ammonium salts.

                A. ascendens Henneberg.

    B.   Subgenus *Acetogluconobacter*.

        Primarily oxidize ethanol to acetic acid. Oxidize glucose to gluconic acid weakly. Some oxidize mannitol to fructose, but do not oxidize fructose further.

        1.   Invert sucrose.

            a)   Assimilate ammonium salts.

                A. acetosus Henneberg.

                Acetogluconobacter dioxyacetonicus Asai.

        2.   Do not invert sucrose.

            a)   Assimilate ammonium salts.

---

* As indicated by the production of large amounts of acetic acid.
** As indicated by the production of small amounts of gluconic acid.

A. *ascendens* (Inst. für Gärungsgewerbe, Berlin).

A. *acetosus* (Inst. für Gärungsgewerbe, Berlin).

A. *rancens* Beijerinck.

b) Do not assimilate ammonium salts.

A. *aceti* Hansen.

A. *vini acetati* Henneberg.

A. *kützingianus* Hansen.

A. *pasteurianus* Hansen.

A. *albuminosus* Lindner.

A. *rancens* Beijerinck var. *agile* Hoyer.

II. Genus *Gluconobacter*.

Grow at 7° to 9°C, but not at 39° to 40°C. Optimum temperature for growth less than 30°C. Primarily oxidize glucose to gluconic acid. Generally oxidize ethanol to acetic acid. Neither assimilate nor oxidize acetic acid.

A. Subgenus *Eugluconobacter*.

Oxidize glucose, but not ethanol.

1. Invert sucrose.

a) Assimilate ammonium salts.

Eugluconobacter liquefaciens Asai.*

2. Do not invert sucrose.

a) Do not assimilate ammonium salts.

Eugluconobacter viscosus Asai.**

B. Subgenus *Gluconoacetobacter*.

Primarily oxidize glucose to gluconic acid. Oxidize ethanol to acetic acid weakly. Oxidize gluconic acid, fructose, and glycerol to 5-ketogluconic acid, kojic acid, and dihydroxyacetone, respectively.

1. Invert sucrose.

a) Assimilate ammonium salts.

A. *xylinum* (Brown).

---

* This strain was designated and described as *Gluconobacter liquefaciens* nov. sp. in the first publication by Asai (*J. Agr. Chem. Soc. Japan*, **11**, 689, 1935).

** This strain was wrongly assigned to *Eugluconobacter*; it probably belongs to the genus *Pseudomonas*.

Both strains were preserved too long and died, and their names were accordingly removed from the list of cultures.

>*Gluconoacetobacter cerinus* var.
>*ammoniacus* Asai.

b)   Do not assimilate ammonium salts.

>*Gluconoacetobacter liquefaciens* Asai.*
>*Gluconoacetobacter roseus* Asai.
>*Gluconoacetobacter scleroideus* Asai.
>*Gluconoacetobacter opacus* Asai.
>*Gluconoacetobacter rugosus* Asai.
>*Gluconoacetobacter cerinus* Asai.
>*Gluconoacetobacter nonoxygluconicus*
>Asai.

2.   Do not invert sucrose.

b)   Do not assimilate ammonium salts.

>*A. industrium* Henneberg (?)
>*A. oxydans* Henneberg (?)

In 1936 Yasui[55] adopted Asai's classification and assigned acetic acid bacteria that he had isolated from Japanese mother of vinegar to each subgenus, as follows:

Strain belonging to *Euacetobacter*.

>*A. ascendens* Henneberg.

Strains belonging to *Acetogluconobacter*.

>*A. orleanense* Henneberg var. I.
>*A. acetosus* Henneberg var.
>*A. rancens* Beijerinck var.
>*A. aceti* Pasteur var.
>*A. ascendens* Henneberg var. II, III, and IV.
>*A. aceti* Brown var.
>*A. aceti* Hansen var.
>*A. schützenbachii* Henneberg var.

Strains belonging to *Gluconoacetobacter*.

>*A. vini acetati* Henneberg var.(?)
>*A. xylinum* Brown var.
>*A. xylinoides* Henneberg var.
>*A. industrius* Henneberg var.

---

* The subgeneric names of *Eugluconobacter*, *Gluconoacetobacter*, *Euacetobacter* and *Acetogluconobacter* were withdrawn in a later publication by Asai and Shoda (*J. Gen. Appl. Microbiol.*, **4**, 289, 1958). Hence this strain is identical with the strain *Gluconobacter liquefaciens* G–1 which appeared in that publication.

A. oxydans Henneberg var.

No strain corresponding to subgenus *Eugluconobacter* was found, probably because acetic acid-containing material was used for the isolation of bacteria.

Five years later, Uemura and Kondo[56] tried isolating oxidative bacteria from fruits and various flowers (without the petals). Nectar yeasts were known, but investigations on flowerborne oxidative bacteria had not been undertaken previously.

They isolated 32 strains and arranged them according to Asai's classification. Five strains were assigned to subgenus *Eugluconobacter*, and the remaining 27 strains to subgenus *Gluconoacetobacter*. The latter were identified as *G. liquefaciens*, *G. roseus*, and *G. cerinus* respectively. Typical *Acetobacter* were not found, presumably owing to the fact that sugar-containing material was used for the isolation of bacteria.

Uemura *et al.*[57] also compared Asai's classification with Tanaka's suggested differentiation of acetic acid bacteria, as shown in Table 6. *Eugluconobacter* Asai may correspond to a presumptive Group IV of *Acetobacter* (Tanaka) even though the group does not appear in Tanaka's classification.

Table 6.   Comparison of the classifications of Asai and Tanaka.

	Asai's classification	Tanaka's classification
Acetobacter	*Euacetobacter*	Type 1 of Group I of *Acetobacter*
	Acetogluconobacter	Type 2 or 3 of Group I of *Acetobacter*
Gluconobacter	*Gluconoacetobacter*	Type 4 of Group II or Group III of *Acetobacter*
	Eugluconobacter	Group IV of *Acetobacter*

Uemura *et al.* supported the existence of the *Eugluconobacter* group on the basis of experiments by Warburg's manometric method on the Qo_2 for glucose and ethanol. According to these experiments, the Qo_2 of the *Eugluconobacter* group for ethanol is very low (6.7–48.4) or almost negligible, but that for glucose is very high (over 200). The RQ for glucose (0.4–1.1) and ethanol (0.7), on the other hand, is comparatively large, whereas the RQ for gluconic acid (0.3) is small. Oxidizable substances are few and resistance to free gluconic acid is weak compared with the *Gluconoacetobacter* group. Moreover, the *Eugluconobacter* were proved to assimilate gluconate but not acetate, and to grow in bouillon.

In media containing ethanol, such as beer and saké, however, growth was impeded.

The next question was whether these two subgenera *Eugluconobacter* and *Gluconoacetobacter* had anything in common. It was confirmed that both *Eugluconobacter* and *Gluconoacetobacter* show a high order of Qo_2 for glucose at a level greater than 100. The glucose dehydrogenase of both subgenera can reduce 2,6-dichlorophenolindophenol better than methylene blue, and the cells of both, when cultivated in sugar-containing media, show stronger dehydrogenase activity than cells harvested from bouillon cultures.

With respect to the relation between *Gluconoacetobacter* and *Acetogluconobacter*, Uemura *et al.* noted a problem for further study. They observed that some strains corresponding closely to Asai's *Gluconoacetobacter* persistently showed Qo_2 values for glucose that were less than the values for ethanol. On this basis they felt that these *Gluconoacetobacter* strains should be reassigned to the *Acetogluconobacter* group, there being insufficient differentiation between the two groups to support the assignment to separate subgenera.

Chapter 5

A CLASSIFICATION PROPOSED BY KONDO
AND AMEYAMA

In 1958 Kondo and Ameyama[58] proposed a new system of classification for *Acetobacter*. The classification generally followed Asai's, especially with respect to oxidizability of alcohols or glucose. According to Asai, species isolated from substances containing ethanol or acetic acid have strong oxidative activity for alcohols, but weak oxidative activity for sugars; they oxidize glucose but not gluconic acid, and lack oxidizability for sugar alcohols such as sorbitol, mannitol and glycerol. On the other hand, species isolated from sugar-containing materials such as fruits or flowers have strong oxidative activity for sugars and oxidize glucose and gluconic acid well, but are weak for ethanol; they can also oxidize sugar-alcohols. Kondo and Ameyama did, in fact, find two species, one incapable of oxidizing ethanol (strain No. 20, identified as *Acetobacter suboxydans* α nov. var., isolated from *Fragaria chiloensis* Duch.) and the other incapable of oxidizing glucose (*Acetobacter ascendens* Henneberg NRRL).

These two species could to be assigned to *Eugluconobacter* and *Euacetobacter* respectively after Asai's classification, but the authors felt that they lie rather at two opposite poles—one having the least and the other the greatest oxidizability for ethanol. Both, therefore, should preferably be retained in the same genus of *Acetobacter*.

Kondo and Ameyama pointed out that *Acetobacter* species can be classified into two groups according to their oxidizability for carbohydrates.

> Group I: Do not oxidize gluconic acid, sorbitol, and mannitol. Oxidize acetic acid.
>
> Group II: Oxidize gluconic acid, sorbitol, and mannitol. Do not oxidize acetic acid.

To differentiate them several other properties were taken into consideration, namely, whether they (1) produce pigment in the media, (2) do not oxidize ketohexoses, (3) oxidize only fructose among the ketohexoses,

33

Table 7. Relation of glucose oxidation to general characteristics (after Kondo and Ameyama).

Group		I			II					
Type		1	2	3	4	5	5	6	7	8
Species No.		1	2	3-5	9-11	6,7	8	12-17	18,19	20
Oxidative product from glucose	GA[a]	−	+	+	+	+	+	+	+	+
	5-KGA	−	−	−	−	−	−	+	+	+
	2-KGA	−	−	−	+	+	+	+	+	−
	2,5-KGA	−	−	−	±	+	+	−	−	−
Oxidative product from mannitol	Fructose	−	−	−	+	+	+	+	+	+
	Kojic acid	−	−	−	−	+	+	+	+	+
Kojic acid from sorbose		−	−	−	−	−	−	−	+	+
Pigment production in media		−	−	−	−	+	+	−	−	−
Oxidation of acetate		+	+	+	−	−	−	−	−	−
Acetoin from lactate		+	+	+	−	−	−	−	−	−
Temp. for growth	8°C	−	−	−	+	+	+	+	+	+
	40°C	+	+	+	−	−	−	−	−	−
pH for growth	3	−	−	−	+	+	+	+	+	+
	8.5	+	+	+	−	−	−	−	−	−

Resistance against	1.5% Acetic acid	+	+	–	–	–	–
	5% GA	–	–	+	+	+	+
Oxidative capacity for	Alcohol	→————————————→					
	Glucose						
Oxidation pattern of glucose and ethanol in a medium containing both substances		I	II	III	IV	IV'	V

Ordinate: acidity; abscissa: time. Solid line: acetic acid (volatile acid); dashed line: gluconic acid (non-volatile acid).

a) GA, gluconic acid; KGA, ketogluconic or diketogluconic acid.

Species No. 1. *A. ascendens* Henneberg NRRL.
      2. *A. aceti* Beijerinck ATCC.
      3. *A. pasteurianus* Hansen ATCC.
      4. *A. kützingianus* Hansen ATCC.
      5. *A. rancens* Beijerinck NRRL.
      6. *A. melanogenus* Beijerinck NRRL.
      7. *A. rubiginosus* nov. sp.
      8. *A. aurantius* nov. sp.
      9. *A. acetosus* Henneberg NRRL.
    10. *A. dioxyacetonicus* Asai.
    11. *A. orleanense* Henneberg.
    12. *A. cerinus* Asai.
    13. *A. gluconicus* Hermann ATCC.
    14. *A. xylinum* Brown ATCC.
    15. *A. industrius* Henneberg NRRL.
    16. *A. oxydans* Henneberg NRRL.
    17. *A. roseus* Asai.
    18. *A. albidus* nov. sp.
    19. *A. suboxydans* Kluyver and de Leeuw NRRL.
    20. *A. suboxydans* α nov. var.

and (4) oxidize fructose and sorbose.

The relation between the two groups is shown in Table 7. The key proposed by Kondo and Ameyama[59] is described below.

Key to species of genus *Acetobacter*

   I. Grow well at 40° but not at 9°C.

      Oxidize acetic acid to CO_2 and H_2O.

      Oxidize glucose to gluconate, but not to ketogluconate.

     A. Do not oxidize glucose.

               A. ascendens Henneberg (NRRL).

     B. Oxidize glucose.

        1. Oxidize glucose after peroxidation of alcohol.

             A. aceti Beijerinck (ATCC).

        2. Oxidize glucose after oxidation of alcohol.

             A. rancens Beijerinck (NRRL).

             A. pasteurianus Hansen (ATCC).

             A. kützingianus Hansen (ATCC).

  II. Grow well at 9° but not at 40°C.

      Do not oxidize acetic acid.

      Oxidize glucose to ketogluconate via gluconate.

      Oxidize mannitol to fructose, sorbitol to sorbose, and glycerol to dihydroxyacetone.

     A. Form pigment in glucose media.

        Oxidize gluconate to 2,5-diketogluconate *via* 2-ketogluconate.

        1. Dark reddish brown pigment.

          a) Acid from maltose.

             A. rubiginosus nov. sp.

          b) No acid from maltose.

             A. melanogenus Beijerinck (NRRL).

        2. Brown pigment.

          Strongly oxidize pentose to pentonic acid.

             A. aurantius nov. sp.

     B. Do not form pigment.

        1. Oxidize gluconate to 2-ketogluconate primarily.

          Oxidize mannitol to fructose, but not to kojic acid.

          a) Utilize ammonium salt.

             A. acetosus Henneberg (NRRL).

          b) Do not utilize ammonium salt.

             A. dioxyacetonicus Asai.

A. orleanens Henneberg.

2. Oxidize gluconate to 5-ketogluconate and 2-keto-gluconate.

Oxidize fructose to kojic acid, but do not oxidize sorbose.

a) Utilize ammonium salt.

A. gluconicus Hermann (ATCC).

A. cerinus Asai.

A. roseus Asai.

b) Do not utilize ammonium salt.

(1) Form a thick zoogloeal membrane.

A. xylinum Brown (ATCC).

(2) Do not form a thick zoogloeal membrane.

A. oxydans Henneberg (NRRL).

A. industrius Henneberg (NRRL).

3. Oxidize gluconate to 5-ketogluconate primarily.

Oxidize fructose and sorbose to kojic acid.

a) Acid from maltose.

A. albidus nov. sp.

b) No acid from maltose.

(1) Oxidize ethyl and amyl alcohol.

A. suboxydans Kluyver and de Leeuw (NRRL).

(2) Do not oxidize ethyl and amyl alcohol.

A. suboxydans var. α nov. var.

Chapter 6

ESTABLISHMENT OF A NEW GENUS
ACETOMONAS LEIFSON

For a long time it was generally accepted that all motile species of *Acetobacter* have polar monotrichous flagella. In *Bergey's Manual* (7th ed., 1957) the motile species are described as polarly flagellated, and the genus *Acetobacter* is therefore assigned to the family *Pseudomonadaceae*. In 1953, however, Leifson[24] discovered that thirty strains of the genus representing commonly recognized species failed to show any polar monotrichous flagella under direct microscopic observation. Two other types of flagellation were found—polar multitrichous and peritrichous. The latter type had not been recognized until Leifson's discovery. It was accordingly proposed to divide the existing *Acetobacter* into two separate genera, *Acetobacter* and *Acetomonas* gen. nov. The biochemical properties of each genus were confirmed to be distinctly different with respect to acetate and lactate oxidizability; the *Acetomonas* gen. nov. include only polar multitrichous species and non-flagellated species which are unable to oxidize acetate and lactate to CO_2 and H_2O, while the redefined *Acetobacter* include only peritrichously flagellated species and non-flagellated species which can oxidize acetate and lactate to CO_2 and H_2O. *Acetomonas suboxydans* is the type species for the genus *Acetomonas* and *Acetobacter aceti* is the type species for the genus *Acetobacter*. The flagellation andp hysiological reactions of species belonging to both genera are listed in Tables 8 and 9.

In Leifson's opinion, it would be most appropriate to place the genus *Acetomonas* in the family *Pseudomonadaceae*, but to fit the redefined *Acetobacter* genus into another family. He also felt that the capacity to grow in a medium in which ammonium salts are the sole source of nitrogen (Hoyer's solution), which was used to differentiate *A. aceti* from *A. xylinum* and *A. rancens* in the key adopted in *Bergey's Manual*, was a useless criterion, as shown in Table 8. Since this view is shared by Janke and others, it might be worthwhile to discuss this capacity in relation to the

38

Table 8. Some comparative characteristics of species of the genus *Acetobacter* (after Liefson).

Species	Reaction in			Growth in Hoyer's medium	Catalase	Flagellation	Leathery pellicle
	peptone	acetate	lactate				
A. aceti B-746 and B-1036	sl. alk.	v. alk.	v. alk.	−	−	peritrichous	−
A. aceti F-4	"	"	"	+	+	peritrichous	−
A. aceti var. muciparum F-6	"	"	"	+	+	none	−
A. orleanense B-55	"	"	"	?	+	peritrichous	−
A. acetum var. nariobiense B-679	"	"	"	?	+	none	−
A. acetum-mucosum B-145	"	"	"	−	+	none	−
A. aceti B-578	"	"	"	−	+	none	−
A. rancens F-2	"	"	"	−	+	none	−
A. rancens var. turbidans F-5	"	"	"	−	+	peritrichous	−
A. turbidans B-1025	"	"	"	−	+	none	−
A. rancens H-277	"	"	"	−	+	none	−
A. oxydans H-263	"	"	"	−	+	none	−
A. rancens var. saccharovorans F-12	neutral	sl. alk.	f. alk.	−	+	none	−
A. xylinum F-3	sl. alk.	f. alk.	f. alk.	+	+	none	+
A. xylinum var. maltovarans F-9	"	"	"	+	−	none	+
A. xylinum B-43	"	"	"	+	+	none	+
A. aceti B-581	"	"	"	+	+	none	+
A. mesoxydans vini F-10	neutral	neutral	sl. alk.	−	−	none	−
A. mesoxydans var. saccharovarans F-14	"	?	neutral	−	−	none	−

sl. = slightly ; f. = fairly ; v. = very ; alk. = alkaline ; neutral = no change of reaction.

Table 9. Some comparative characteristics

Species	Reaction in		
	peptone	acetate	lactate
Am. suboxydans var. *biourgianum* F-1	sl. acid	sl. acid	sl. acid
Am. suboxydans F-16	,,	,,	,,
Am. suboxydans var. *muciparum* F-15	,,	,,	,,
Am. suboxydans B-72	,,	,,	,,
Am. capsulatus B-1225	,,	,,	,,
Am. viscosus B-1226	,,	,,	,,
Am. melanogenus var. *maltovorans* F-8	,,	,,	,,
Am. melanogenus var. *maltosaccharovorans* F-13	,,	,,	,,
Am. melanogenus B-58	,,	,,	,,
Am. melanogenus B-63	,,	,,	,,

Am. = *Acetomonas*

general nutritional requirements of the species — not only for carbon sources, but also for vitamins and growth factors.

Leifson's discovery of *Acetobacter* strains with peritrichous flagella attracted a great deal of attention among taxonomists. Why the flagellation had been misjudged is not clear, but it may have been due to the difficulty of flagella staining in acetic acid bacteria. The flagella are easily detached in the process of staining, and this may have given the bacteria the appearance of having polar monotrichous flagella. Leifson stated that the acetic acid bacteria are more difficult subjects for flagella staining than most bacteria, because of the high acidity of the medium, the presence of capsular material round the cell, and the tendency of the cells to clump when washed in the centrifuge.

Shimwell[60] further confirmed the peritrichous flagellation of *Acetobacter* strains with a motile celluloseless mutant from a cider strain of *A. xylinum* and with *A. rancens* and *A. aceti*. He was unable, however, to confirm Leifson's finding that the peritrichous flagella of *Acetobacter* are always of orthodox type with a wavelength averaging 2.9μ, or that the number of flagella per organism tends to be few. One strain of *A. aceti* was found to have numerous flagella, and the wavelength (1.4 μ) was in the range considered by Leifson to be confined to the polar flagella of *Acetomonas*. He also felt that Leifson's definition of *Acetomonas* gen. nov. was too narrow, and suggested substituting " species with one or more polar flagella " for " polar multitrichous species." He was able to find a of the genus *Acetomonas* (after Leifson).

Growth in Hoyer's medium	Catalase	Flagellation	Pigment
−	+	polar, multitrichous	pink
−	−	,,	pink
−	+	,,	−
−	+	,,	−
−	+	,,	−
−	+	,,	−
−	−	,,	brown
−	+	,,	brown
−	+	,,	brown
−	+	none	brown

strain with a single polar flagellum in *A. melanogenus*, though polar mono-trichous strains are relatively rare. He proposed reviving the family *Acetobacteriaceae* which had been adopted in *Bergey's Manual* (5th ed., 1939), and placing the genus *Acetobacter* in this family to distinguish it from the genus *Acetomonas*.

Chapter 7

RECLASSIFICATION OF ACETIC ACID BACTERIA INTO *ACETOBACTER* AND *GLUCONOBACTER*

Twenty-three years after their first published classification,[54] Asai and Shoda[61] attempted to reclassify the acetic acid bacteria, taking into account Leifson's discovery. They found that peritrichous flagella always occur in the motile and acetate-oxidizing strains of *Acetobacter*, whereas polar flagella only occur in the motile strains of *Gluconobacter* Asai. The latter genus, when established by Asai in 1935, was characterized by its inability to oxidize acetate. Thus a correspondence between *Gluconobacter* Asai and *Acetomonas* Leifson was verified.

Experiments showed that the turbidity in liquid culture, representing the degree of growth in the liquid phase, was generally heavier in *Gluconobacter* than in *Acetobacter*. Pellicle formation in liquid media was scarcely recognized in strains of *Gluconobacter*, and the optimum temperature for growth was lower than that of *Acetobacter*. These characteristics fully agreed with Asai's earlier experiments.[54] It was also reconfirmed that in *Acetobacter*, the capacity to oxidize ethanol surpassed or at least remained on the same level as the capacity to oxidize glucose, while the reverse was true in *Gluconobacter*. A high level of accumulation of gluconic acid, surpassing that of acetic acid, was seen in all strains of *Gluconobacter*, and among these *G. suboxydans* H-15 and *G. cerinus* strain 24 were found to be scarcely able to produce acetic acid from ethanol at all. On the other hand, acetic acid was accumulated to a greater extent by strains of *Acetobacter* than by those of *Gluconobacter*. Some *Acetobacter* species, *viz.*, *A. acetosus* ATCC 6438 and *A. pasteurianus* ATCC 6033, had a rather high level of gluconic acid accumulation and poor acetic acid accumulation. This does not mean, however, that these species are weak in producing acetic acid; over-oxidation of acetic acid occurs. It was also confirmed in this experiment that *A. ascendens* NRRL B-56 fails to form acid from glucose. This characteristic had been observed by Asai[54] in a strain of *A. ascendens* Henneberg, by Tošić and Walker[62] in *A. ascendens* (Lister

Institute Reference number 4937) and by Tanaka[63] in the same species *A. ascendens* Henneberg. The observation was of particular interest because there exist some strains of *Gluconobacter*, as mentioned above, which are virtually unable to produce acetic acid from ethanol.

Most remarkable, however, is the ability of *Gluconobacter* to form certain substances from glucose, galactose, sorbose and fructose, giving a reddish violet color reaction with ferric chloride; this ability is completely lacking in *Acetobacter*. The formation from fructose of kojic acid and isokojic acid (later identified as 5-oxymaltol by Terada *et al.*) was reported by Takahashi and Asai,[64] Aida *et al.*,[65] Sakaguchi *et al.*[66] and Ikeda.[67] Formation of comenic acid from glucose was also reported by Aida *et al.*[68] and from galactose by Takahashi and Asai[69]; the formation from glucose of new γ-pyrone compounds called rubiginol and rubiginic acid was also reported for some strains of *Gluconobacter* by Aida.[70] The substances that give a positive ferric chloride reaction are most probably the γ-pyrone compounds mentioned above. This characteristic was used to distinguish between *Gluconobacter* and *Acetobacter*. It was also reconfirmed that the genus *Gluconobacter* exhibits strong ketogenic activity towards polyalcohols such as mannitol, sorbitol, and glycerol, whereas the strains of *Acetobacter* generally do not.

On the basis of these characteristics, Asai and Shoda again proposed to divide the acetic acid bacteria into two separate genera—one, *Acetobacter*, primarily oxidizing ethanol to acetic acid, and the other, *Gluconobacter*, primarily oxidizing glucose to gluconic acid. The subgenera proposed in the original classification—*Eugluconobacter* and *Gluconoacetobacter* in genus *Gluconobacter*, and *Euacetobacter* and *Acetogluconobacter* in genus *Acetobacter*—were, however, withdrawn in order to simplify the new classification.

In their report,[61] Asai and Shoda described three strains, *G. liquefaciens* G-1, *G. melanogenus* AC-8 and *G. melanogenus* U-4, as anomalous types of *Gluconobacter*. These three strains are almost identical to *G. melanogenus* Beijerinck, especially in the reddish brown color of the cells and in the pigmentation in culture broth containing glucose and $CaCO_3$, except that they can oxidize acetate to CO_2 and H_2O. The flagellation was determined to be polar and they were therefore placed in the genus *Gluconobacter*. In later experiments by Asai *et al.*,[71] however, the three strains were ascertained to possess peritrichous flagella, as was pointed out by Shimwell and Carr,[72] by Stouthamer[73] and by Hodgkiss *et al.*,[74] though the cells did not appear to be homogeneously peritrichous but in some cases exhibited mixed-type flagellation with polar flagella.

Table 10.   Criteria for the classification of *Acetobacter*

| Genus | Flagellation | Production of | |
		acetic acid	gluconic acid
Acetobacter	peritrichous or none	strong	weak to negligible[a]
Gluconobacter	polar or none	weak to negligible[a]	strong

a), *b*) ; The original descriptions "weak or none" and " − or rarely + " are emended here

Consequently these strains were removed from the *Gluconobacter* and temporarily placed as " intermediate " strains belonging neither to the *Gluconobacter* nor to the *Acetobacter*.   A detailed discussion of these interesting strains will be presented later.

In any case the differentiation of acetic acid bacteria into the two genera *Acetobacter* and *Gluconobacter* was again verified.   The remarkable differences between the genera are cited in Table 10; micrographs of flagella staining are shown in Figures 2, 3 and 4.

The genus *Acetobacter* includes organisms that exhibit only peritrichous flagellation if motile, oxidize ethanol strongly,* oxidize glucose weakly or scarcely,** and oxidize acetate and lactate completely.   Ketogenic oxidative activity towards glycerol is negative or rarely positive.   No substances that react positively to ferric chloride are formed from fructose. The type (and sole ?) species is *A. aceti* Beijerinck.

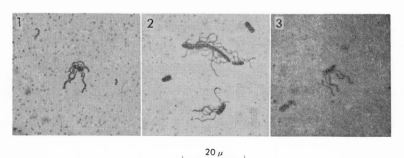

20 μ

Fig. 2.   Micrographs of *Acetobacter* (after Asai *et al.*).

1.   *A. aceti* strain 4.   2.   *A. aceti* strain 5.   3.   *A. albuminosus* IAM 1807.
Each strain was cultured on Koji extract-ethanol-CaCO₃ agar slant for 18-24 hr at 25°C.

---

 * Produce a large quantity of acetic acid.
** Produce a small or negligible quantity of gluconic acid.

and *Gluconobacter* (after Asai and Shoda).

Oxidation of		Ketogenic activity	Formation of FeCl₃-reaction positive substances from fructose
acetic acid	lactic acid		
+	+	weak or none	—
—[b]	—[b]	strong	+

to "weak to negligible" and " — " respectively.

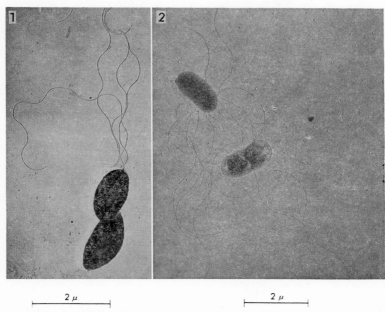

                2 μ   2 μ

Fig. 3. Electron micrographs of *Gluconobacter* and *Acetobacter* (after Asai *et al.*).
1. *G. suboxydans* IFO 3172. Cells grown on Koji extract-CaCO₃ agar slant for 18 hr at 20°C. 2. *A. aceti* strain 4. Cells grown on Koji extract-ethanol-CaCO₃ agar slant for 18 hr at 25°C.

The genus *Gluconobacter* includes organisms that exhibit only polar flagellation if motile, oxidize glucose strongly, oxidize ethanol weakly or scarcely, and do not oxidize acetate and lactate to carbonate. Ketogenic oxidative activity towards glycerol is positive. Substances that react positively to ferric chloride are formed from fructose. The type (and sole) species is *G. oxydans* (Henneberg) Asai.

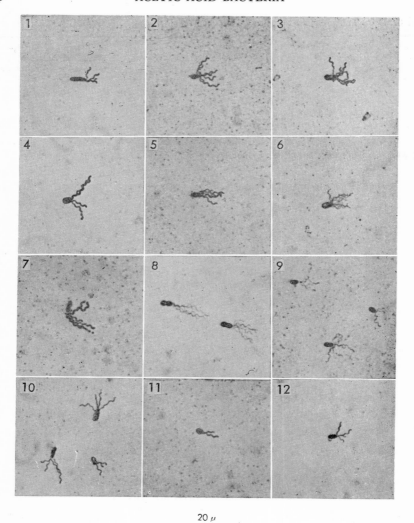

20 μ

Fig. 4.   Micrographs of *Gluconobacter* (after Asai *et al.*).

1.   *G. roseus* IAM 1841.   2.   *G. roseus* IAM 1838.   3.   *G. suboxydans* IFO 3172.
4.   *G. suboxydans* var. α IFO 3254.   5.   *G. cerinus* IAM 1833.   6.   *G. cerinus* IFO
3267.   7.   *G. gluconicus* IFO 3171.   8.   *G. albidus* IFO 3250.   9.   *G. rubiginosus*
IFO 3244.   10.   *G. melanogenus* IAM 1818.   11.   *G. oxydans* IFO 3189.   12.
G. capsulatus IFO 3462.   Each strain was cultured on mannitol agar slant for 18–24 hr
at 20°C.

Other existing species were to be considered as varieties of single species in each genus, owing to reasons presented later (see Chapter 12). Thirty-one strains belonging to the *Gluconobacter* were divided into three groups by Asai *et al.*[71] according to the pigmentation of colonies.

Group I. Strains forming pink colonies on glucose-yeast extract-CaCO$_3$ agar:

G. roseus IAM 1838, IAM 1839, IAM 1840, and IAM 1841. *G. suboxydans* NRRL B-755, *G. cerinus* IAM 1833. *G. scleroideus* IAM 1842, *G. nonoxygluconicus* IFO 3275.

Group II. Strains forming brown colonies with water-soluble brown pigment on glucose-yeast extract-CaCO$_3$ agar:

G. melanogenus IFO 3292, IFO 3293, IFO 3294 and IAM 1818. *G. rubiginosus* IFO 3243 and IFO 3244.

Group III. Strains not forming pigmented colonies on glucose-yeast extract-CaCO$_3$ agar:

G. suboxydans H-15, IFO 3172, IFO 3130 and IFO 3254, *G. cerinus* IAM 1832, IFO 3262 and IFO 3267, *G. gluconicus* IFO 3171, *G. industrius* IFO 3260, *G. capsulatus* IFO 3462, *G. suboxydans* IAM 1828, *G. dioxyacetonicus* IFO 3271 and IFO 3272, *G. albidus* IFO 3250 and IFO 3251, *G. oxydans* IFO 3189, *G.*(?) *turbidans* IFO 3225.*

Gluconobacter was placed in the family *Pseudomonadaceae*. *Acetobacter*, since it shows no close relationship to any existing family, was placed in the family *Acetobacteriaceae*, which appeared previously in *Bergey's Manual* (5th ed., 1939) and which we propose to revive.

---

* This strain was the only one found unable to oxidize glycerol to dihydroxyacetone and to grow on mannitol agar. All other strains of *Gluconobacter* can grow on mannitol agar and oxidize glycerol to dihydroxyacetone.

Chapter 8

OTHER PROPOSALS ON THE SYSTEMATIZATION
OF ACETIC ACID BACTERIA

A systematization based upon nutritional requirements was presented in 1953 by Rainbow and Mitson,[75] and Brown and Rainbow,[76] who separated the bacteria into two well-defined groups. One group with a predominant lactate metabolism was termed " lactaphilic "; the other, with a predominant glucose metabolism, was termed " glycophilic." Their nutritional requirements are shown below.

I. Lactaphilic group. Grew rapidly when lactate was provided as the only source of energy and carbon. When other organic acids were substituted for lactate, it was necessary to add glucose for even partial effectiveness. Typical lactaphilic strains did not require exogenous supplies of growth factors, utilized ammonium sulphate as the sole source of nitrogen, and gave rise to new ninhydrin-reacting substances when cell suspensions were incubated in solutions of proline, glutamate and asparatate. The species belonging to this group are *A. aceti*, *A. mobile*, *A. acetosus*, *A. ascendens*, *A. orleanense*, *A. oxydans*, *A. rancens*, *A. suboxydans* NCTC 6430, *A. suboxydans* NCTC 7113, *A. suboxydans* NCTC 7099 and *A. acidum-mucosum*.

II. Glycophilic group. Grew well when glucose, but not lactate, was supplied as the chief source of energy and carbon; utilized ammonium sulphate as the sole source of nitrogen slowly and to a limited extent in glucose medium, the degree of growth being increased by the presence of certain organic acids. Characteristically, the strains required certain sugars or sugar alcohols for growth, but were only slightly or not at all stimulated by lactate. Glycophilic strains yielded no new ninhydrin-reacting substances under the previously stated conditions and required added nicotinate, pantothenate and *p*-aminobenzoate in a defined medium. The species belonging to this group are *A. capsulatus*, *A. gluconicus*, *A. turbidans* and *A. viscosus*.

This systematization may be compared to Vaughn's, in which groups

I (oxidize acetic acid) and II (do not oxidize acetic acid) correspond to the lactaphilic and glycophilic groups, respectively. Rainbow and Mitson, referring to the utilization of ammonia as a nitrogen source, concluded that Vaughn's subdivision based on this criterion was unsatisfactory, since all strains tested by them utilized ammonia when the other growth requirements were supplied. They also pointed out that *A. suboxydans* should, according to Vaughn's description, fit into their glycophilic groups, whereas the three strains they examined were unquestionably lactaphilic.

Recently Stouthamer[73, 77] stated that organisms belonging to the genus *Acetobacter* have a tendency to convert from one species to another by mutation, and proposed the following modification of Frateur's system in order to obviate the effects of mutation. 1) Strains of the *Mesoxydans* group should oxidize at least two polyalcohols. 2) Only strains which are catalase negative and which do not attack glucose should be classified as *A. peroxydans*. 3) Strains that do not attack glucose and are catalase positive and strains which attack glucose and are catalase positive or negative should be combined. Of the latter, those that do not oxidize polyalcohols should be classified as *A. rancens*, and those that oxidize at least two polyalcohols as *A. mesoxydans*. The classification scheme is shown in Figure 5.

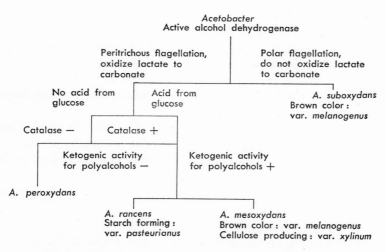

Fig. 5. Scheme for the classification of *Acetobacter* (after Stouthamer).

The description of each group was presented as follows.

1. *A. peroxydans* group.
    a) Do not attack glucose
    b) Catalase negative.
    c) Give irisation on Ca lactate.
    d) Usually oxidize molecular hydrogen.
    e) Do not usually oxidize gluconate, ketogluconate, polyalcohols and ketoses.
    f) Most strains grow in Hoyer's medium.
    a), b) and c) are specific characters for this group.

2. *A. rancens* group.
    a) Give irisation on Ca lactate.
    b) Oxydograms are negative in at least three polyalcohols.
    c) Usually catalase positive.
    d) Most strains produce acid in glucose-$CaCO_3$ medium. Some strains give irisation on glucose-$CaCO_3$ agar. One strain produces 2-ketogluconate from glucose.
    e) Do not generally oxidize fructose and mannitol. Most strains oxidize glycerol and dihydroxyacetone after phosphorylation.
    f) Most strains produce acetoin in lactate-containing medium.
    g) Do not oxidize molecular hydrogen.
    h) Motile strains are peritrichously flagellated.
    a) and b) are specific characters for this group.

3. *A. mesoxydans* group.
    a) Give irisation on Ca lactate.
    b) Oxydograms are positive in at least two polyalcohols.
    c) Usually catalase positive.
    d) Most strains produce acid in glucose-$CaCO_3$ medium. Some produce Ca 5-ketogluconate. *A. mesoxydans* var. *melanogenus* turns the medium brown.
    e) Most strains oxidize fructose and dihydroxyacetone. Oxidize glycerol in two ways, with or without phosphorylation.
    f) Most strains do not produce acetoin in lactate–containing medium.
    g) Motile strains are peritrichously flagellated.
    a) and b) are specific characters for this group.

4. *A. suboxydans* group.
    a) No irisation on Ca lactate.
    b) Oxydograms are positive in at least two polyalcohols.
    c) Usually catalase positive.
    d) Produce acid in glucose-$CaCO_3$ medium. Rapidly produce Ca

5-ketogluconate. *A. suboxydans* var. *melanogenus* turns the medium brown.

e) Most strains oxidize fructose, sorbose and erythrulose. Dihydroxyacetone is weakly attacked by only one strain.

f) Most strains do not produce acetoin in lactate-containing medium.

g) Motile strains are polarly flagellated.

a) and b) are specific characters for this group.

Stouthamer confirmed the presence of the TCA cycle in *Acetobacter;* its presence in *Acetobacter* and absence in *Gluconobacter* (*Acetomonas*) has also been confirmed by other workers, and will be discussed in a separate chapter.

Shimwell and Carr[78] proposed including this characteristic in differentiating acetic acid bacteria. But Joubert *et al.*[79] found that the acetate-oxidizing strains are able to deaminate and oxidize a number of amino acids while the non-acetate-oxidizers are unable to oxidize and deaminate any amino acids supplied. Carr and Shimwell[80] adopted this trait to differentiate *Acetobacter* and *Acetomonas;* their comparison is shown in Table 11.

Table 11. Differentiation of *Acetobacter* and *Acetomonas*
(after Carr and Shimwell).

	Acetobacter	*Acetomonas*
Ethanol oxidized to $CO_2 + H_2O$	+	−
Lactate oxidized to carbonate	+	−
Nutrition type	Lactaphilic	Glycophilic
Deaminate and oxidize various amino acids	+	−
TCA cycle	+	−
Flagella (if motile)	Peritrichous	Polar

From comprehensive studies on the carbohydrate metabolism of the acetic acid bacteria, De Ley[81] pointed out that the investigated strains can be arranged in two main lines, and that within each line strains can be arranged according to a stepwise decreasing gradation of oxidative properties. The *Oxydans* group, contrary to what one might expect from the name, had only limited oxidative powers, and the *Peroxydans* even less. De Ley proposed separating the acetic acid bacteria into two biotypes—*Gluconobacter oxydans* and *Acetobacter aceti*—and considering the existing species as varieties within these two types. He classified them into three groups:

I. Do not attack glucose at all. Neither glucose oxidase nor glucose dehydrogenase detected.

    A. ascendens strain A, *A. paradoxus* 30, *A. peroxydans.*

II. Oxidize glucose only, with an uptake of 0.5 mole O_2/mole substrate. Contain glucose oxidase or glucose dehydrogenase or both, but no hexokinase or other mechanism to introduce glucose into the hexose monophosphate oxidative cycle.

    A. rancens 15, 23 and " Davis," *A. pasteurianus* 11 and 8856, *A. kutzingianus* 3924, *A. rancens* var. *turbidans* 8619 and *A. ascendens* 8163.

Table 12. Oxidation of some carbon substrates by acetic acid bacteria (after De Ley).

Substrate	*Peroxydans* (1 strain)	*Oxydans* (22 strains)	*Mesoxydans* (14 strains)	*Gluconobacter* (9 strains, excluding the strains which are unable to oxidize ethanol)
Glucose	—	+[a]	+	+
Na gluconate	—	— (most strains)	+	+
Na 2-ketogluconate	—	—	+ or —	+ or —
Na 5-ketogluconate	—	—	+ or —	+ or —
Mannitol	—	— (most strains)	+ (nearly all)	+ (nearly all)
Fructose	—	— (most strains)	+ (nearly all)	+ (nearly all)
Mannose	—	+[a] (nearly all)	+[a]	+[a]
Galactose	—	+[a] (nearly all)	+[a]	+[a]
Xylose	—	+[a] (nearly all)	+[a]	+[a]
Glycerol	—	+	+	+
i-Erythritol	—	—	+	+
Na D-lactate	+	+	+	+
Ethanol	+	+	+	+
Na acetate	+	+	+	—

a) Can oxidize only to the corresponding sugar acids. *Gluconobacter liquefaciens* 20 was able to oxidize acetate and *G. cerinus* 21, although it did not oxidize acetate, nevertheless oxidized D-lactate and ethanol nearly to completion. *G. cerinus* 21 and *G. liquefaciens* 20 oxidized galactose with the uptake of 1.0 or more mole O_2/mole substrate. *G. cerinus* 21 oxidized xylose with the uptake of 0.95 mole O_2/mole substrate.

ERRATA

Page	Line	For	Read
24	31	*A. acetigenoidius* Krehan	*A. acetigenoideus* Krehan
38	20	andp hysiological	and physiological
77	29	as *Acetobacter* dose	as *Acetobacter* does
219	15	epidehydroshikimate	epidihydroshikimate
332 (left)	31	Acetobacter	*Acetobacter*
332 (left)	34	for alde	for alde-
334 (left)	14	lysozyme Lysate	lysozyme lysate
342 (right)	8	D-Sorbitol oxidation of	D-Sorbitol, oxidation of

III. Oxidize glucose with an uptake of more than 0.5 mole O_2/mole substrate (most strains took up 1–3 mole O_2/mole substrate).

A. estunensis, A. turbidans 6424, A. aceti Ch 31, A. vini acetati 4939, A. mesoxydans 8622, A. mesoxydans 4, A. xylinum 8737, G. capsulatus 4943, G. suboxydans 26, G. melanogenus 8086, G. cerinus 21, G. liquefaciens 20.

Note: In most of the strains of the Mesoxydans group and Gluconobacter, ketohexonates are formed.

De Ley studied the oxidative behavior of each group further and obtained the results shown in Table 12.

The oxidation of i-erythritol seemed an especially promising factor to use in classification, since no strain of the Oxydans group of Acetobacter can oxidize it, while all the strains of the Mesoxydans group and of Gluconobacter can.

Cooksey and Rainbow[82] drew attention to the fact that lactaphiles and glycophiles are not only nutritionally different but also metabolically distinct. The capacity of cell extracts to effect reversible transamination from glutamate to aspartate is well developed in lactaphiles, but only poorly or not at all in glycophiles. This was ascertained by using A. acidum-mucosum, A. mobile, A. rancens, A. ascendens and A. oxydans as lactaphiles and A. capsulatus, A. gluconicus, and A. viscosus as glycophiles. Cell suspensions of lactaphiles possessed greater ability to oxidize intermediates of the TCA cycle than did those of glycophiles, and also had citrate synthase activity, which was not detected in glycophiles. The lactaphiles were able to convert L-aspartate into α-alanine, possibly by β-decarboxylation rather than transamination. These facts indicate that lactaphiles are enzymatically better equipped than glycophiles, and that the TCA cycle which operates in lactaphiles may not occur in glycophiles.

Since the initial work of Riddle et al.,[83] spectrophotometric observation for the identification of bacteria has been extended to devise a method for obtaining and comparing reproducible spectra. Recently Scopes[84] examined the infrared spectra of representative strains of acetic acid bacteria to see whether the method could give some information on the interrelationships between Acetobacter and Acetomonas (Gluconobacter). Twenty-two strains of Acetobacter showed spectra significantly different from those of nine strains of Acetomonas; the only exeptions were Acetomonas melanogenus NCIB 8080 and 8084, which showed an Acetobacter type spectrum. These strains, however, were polarly flagellated and did not oxidize ethanol to CO_2 and H_2O, and should therefore be regarded as *Acetomonas* strains despite their *Acetobacter* type spectra; they seem situated, essentially, on both sides of the generic boundary. Within each genus, no startling differences which could be correlated with species boundaries, or peculiarities which could be traced to the presence or absence of particular biochemical activities, were seen. It was concluded from these results that the division of acetic acid bacteria into two genera *Acetobacter* and *Acetomonas* is to be supported, but that the existence of different species within each genus is unlikely, as was suggested by Shimwell[85] and De Ley.[81]

Shimwell and McIntosch[86] recently reported on an interesting serological investigation of *Acetomonas* and *Acetobacter*, in which strains of *Acetomonas* (syn. *Acetobacter suboxydans* or *Gluconobacter suboxydans*) gave a good definite serological group distinguishable from that of *Acetobacter*. They used 25 strains of *Acetomonas suboxydans* and *Acetobacter rancens* and presented a scheme for the classification of both genera on the basis of flagella arrangement, biochemical activities and serological properties.

The base composition of DNA in the cells of 28 strains of acetic acid bacteria was determined by De Ley and Schell.[87] It was shown that most strains of *Gluconobacter* (*Acetomonas*) clustered closely together at 60.6–63.4 per cent GC (guanine+cytosine) content to total base and all strains of *Acetobacter* lay within the range of 55.4–64.0 per cent GC content, suggesting a close relationship and possible common phylogenetic origin for *Gluconobacter* and *Acetobacter*. The strains with greater biochemical activity showed on the whole a higher percentage of GC content in their DNA, though it was stressed that the existence of independent species in each genus cannot be accepted, as was demonstrated in a previous report by De Ley.[81]

An attempt to prepare DNA hybrids of acetic acid bacteria was made by De Ley and Friedman.[88] Among the tested strains, deuterated ^{15}N-labeled DNA from *A. aceti* (*mesoxydans*) 4 formed hybrids with ordinary DNA obtained from other species of the genus, viz., *A. xylinum*, *A. pasteurianus*, *A. estunensis* and possibly *A. xylinoides*, when the guanine plus cytosine base composition did not vary by more than 1 to 2 per cent. No hybrids were formed beyond this limit (*A. aceti* Ch 31 and *A. aceti* var. *muciparus* 5) in spite of great morphological, physiological, and biochemical similarities; hence the authors supported the concept of genetic separation rather than division of genus into sharply separated species based on small phenotypic differences. The results of the experiment are shown in Table 13.

A recent investigation by De Ley *et al.*[89] on DNA homology in the

Table 13. Hybrid formation between deuterated ^{15}N-DNA from *A. aceti* (*mesoxydans*) 4 and ordinary DNA from other organisms (after De Ley and Friedman).

Strain	% Hybrid formed	Mean % G+C	Variance of the compositional distribution σ	Compositional distribution of DNA molecules expressed as % G+C ±2σ
A. aceti (*xylinoides*) 4940	ca. 10	62.4	1.5	59.5–65.4
A. aceti (*estunensis*) E	ca. 25	62.2	1.1	60.0–64.4
A. aceti (*xylinum*) 8747	ca. 20	60.7	2.25	56.2–65.2
A. aceti (*mesoxydans*) 4	complete	60.6	1.6	57.4–63.8
A. aceti (*pasteurianus*) 11	ca. 15	59.8	1.75	56.3–63.3
A. aceti Ch 31	0	59.6	1.6	56.4–62.8
A. aceti (var. *muciparus*) 5	0	59.5	1.5	56.5–62.5
Gluconobacter oxydans (*melanogenus*) 116	0	60.6	0.87	58.9–62.3

Pseudomonas-Xanthomonas group has shown that *Gluconobacter* and *Acetobacter* have some 20–30 per cent DNA similarity with *Pseudomonas* from the DNA hybridization experiments.

Williams and Rainbow[90] reported on the enzyme activities involved in the TCA cycle in extracts of five lactaphilic and five glycophilic strains of acetic acid bacteria. The enzyme activities present in all strains are citrate synthase, aconitate hydratase, fumarate hydratase, succinate dehydrogenase, L-malate dehydrogenase, α-ketoglutarate dehydrogenase, and oxaloacetate decarboxylase. Isocitrate dehydrogenase and isocitrate lyase were detected in lactaphilic strains but not in glycophilic strains. Malate synthase activity was not detected in extracts of either lactaphiles or glycophiles, although an extract prepared from a strain of *A. aceti* grown on Hoyer's medium (with ethanol as the sole carbon source and ammonium sulfate as the sole nitrogen source) showed some activity. In general, enzymatic activities were greater in the lactaphilic strains. Extracts of glycophiles possessed only weak citrate synthase, aconitate hydratase, fumarate hydratase, and L-malate dehydrogenase activity and only α-ketoglutarate and succinate dehydrogenase activity was comparable to that in extracts of lactaphiles. The activity of oxaloacetate decarboxylase was also greater in lactaphiles than in glycophiles. NADP-linked L-malate dehydrogenase was revealed only in extracts of *A. acidum-mucosum*. In other strains, another L-malate oxidizing enzyme, which required no added cofactors and was considered to be cytochrome-linked, was detected.

Williams and Rainbow felt that the TCA cycle may make a quantitatively greater contribution to the metabolism of lactaphilic than glycophilic organisms, but that members of the *Suboxydans* (glycophilic) group, which had been reported as lacking the TCA cycle, have rather a severely restricted ability to carry out its reactions. Table 14 shows the enzymatic activities expressed relative to that of the most active extract (taken as one hundred).

Table 14.   Comparative activities of the TCA cycle enzymes in extracts of acetic acid bacteria (after Williams and Rainbow).

	C	A	ID/NAD	ID/NADP	KD	SD	F	MD
Lactaphiles :								
A. acidum-mucosum	48	100	7	93	58	45	37	23
A. ascendens	100	80	7	100	92	100	56	21
A. mobile	16	8	4	10	15	64	16	3
A. oxydans	79	98	100	38	63	36	100	100
A. rancens	51	72	11	44	68	74	45	8
Glycophiles :								
A. capsulatus	<1	1	0	0	31	48	2	5
A. gluconicus	2	2	0	0	19	37	2	<1
A. suboxydans	2	2	0	0	49	23	5	2
A. turbidans	<1	2	0	0	74	10	4	3
A. viscosus	1	2	0	0	100	11	4	5

Abbreviations :   C, citrate synthase ; A, aconitate hydratase ; ID/NAD and ID/NADP, isocitrate dehydrogenase (NAD- and NADP-linked) ; KD, α-ketoglutarate dehydrogenase ; SD, succinate dehydrogenase ; F, fumarate hydratase ; MD, L-malate dehydrogenase (dichlorophenolindophenol-linked).

Leisinger[91] presented a paper dealing with the taxonomy and physiology of the acetic acid bacteria.   In a study using nineteen strains of bacteria it was concluded that the criteria for Frateur's classification, as well as nutritional characteristics, remained stable over a period of two years.   The presence of a glyoxylate bypath was supported only in strains able to grow on Hoyer's medium, while the Entner-Doudoroff pathway was demonstrated only in *A. xylinum.*

Leisinger divided his nineteen strains into the three following groups according to their capacity to utilize ethanol or glucose as the sole carbon source.

I. Utilize ethanol as the sole carbon source. Glyoxylate bypath present, TCA cycle present. *A. peroxydans* 109, *A. aceti* 3, 114 and *A. liquefaciens* 101 (=*G. liquefaciens* G-1 Asai).

II. Not homogeneous, comprise strains of Frateur's *Mesoxydans* and *Oxydans* group. Many strains utilize glucose as the sole carbon source. Glyoxylate bypath absent, TCA cycle present. *A. rancens* 98, 97, 31, 14, 32, *A. ascendens* 194, *A. mesoxydans* 96, *A. xylinum* 27, 95, and 113.

III. Do not utilize either glucose or ethanol as the sole carbon source. Glyoxylate bypath absent, TCA cycle absent. *A. rancens* 112, *Acetomonas* (*Gluconobacter*) *oxydans* 111, 100, 90, and *Acetomonas melanogenus* 93.

When grown on Hoyer's medium the group I strains most probably utilize ethanol by the reactions of the glyoxylate bypath. Experiments with ethanol-^{14}C in the presence of fluoroacetate indicated that the group II strains can transform amino acids into compounds which permit operation of the TCA cycle and hence utilization of ethanol. The group III strains that correspond to *Acetomonas* (*Gluconobacter*) are unable to incorporate the carbon in ethanol into cellular material, probably due to lack of the TCA cycle.

Ameyama *et al.*[92] studied the oxidative activities of acetic acid bacteria for glucose, K gluconate, Ca 2-ketogluconate, sucrose, L-arabinose, mannitol, fructose, sorbose, Na succinate, Na acetate, ethanol, and amyl alcohol; from their results they divided the sixty strains they had used into five groups (shown below). Thus the existence of the intermediate strains reported by Asai *et al.*[71] was confirmed independently, though the authors differ slightly on *A. aurantius*.

Acetobacter group. Oxidize ethanol and succinate strongly.
Typical species: *A. aceti.*

Pigment-producing group I (intermediate type). Oxidize succinate.
A. aurantius, G. liquefaciens.

Pigment-producing group II. Do not oxidize succinate.
Typical species: *A. melanogenus.*

Gluconobacter group I. Oxidize glucose strongly and ethanol weakly; do not oxidize succinate; produce both 2- and 5-ketogluconate.
Typical species: *G. oxydans.*

Gluconobacter group II. Oxidize ethanol very weakly; most strains do not oxidize succinate; produce mainly 5-ketogluconate.
Typical species: *G. suboxydans* var. α.

Ameyama and his co-workers also observed that variations with respect to oxidizability of carbon sources and to pigment production were induced in some species or strains during the preservation of stock cultures in the laboratory (Table 15).

Table 15.   Some characters of the variant strains (after Ameyama *et al.*).

Strain	Oxidative activity[a]		Brown pigment production	Presumable variation
	Succinate	Acetate		
A. pasteurianus IFO 3223	12	6	+	*Acetobacter* → Pigment-prod. I
A. acetigenus IFO 3278	1	0	—	*Acetobacter* → *Gluconobacter* I
A. melanogenus IFO 3190	11	4	—	Pigment-prod. II → Pigment-prod. I (→ No pigment)
A. industrius IFO 3261	15	2	—	*Gluconobacter* I → *Acetobacter*

The figures are expressed as relative activities of Qo_2 towards those for glucose which are calculated as 10.

Chapter 9

MUTATION OF ACETIC ACID BACTERIA

The variability of strains of acetic acid bacteria was first reported by Tošić and Walker.[62] They found that certain characteristics of *A. aceti*, *A. pasteurianus*, and *A. kützingianus* differed from the original descriptions of these species and suggested the following reasons for the discrepancies: The experimental conditions under which these organisms were described by former workers were not identical to theirs; the details of earlier experiments were too meagre to repeat the early work; and the three species were the first acetic acid bacteria isolated in pure culture and, consequently, were maintained under the artificial conditions of the laboratory for much longer periods than any of the others: therefore, the possibility of mutation was greater in these three species than in others.

Walker and Kulka[93] reported that *A. viscosus* lost the ability to form mucilaginous matter in wort after long periods of artificial cultivation. Biochemical variations were also observed. A strain of *A. suboxydans* which at first yielded gluconic acid and 5-ketogluconic acid when grown on a glucose yeast-water medium in the presence of $CaCO_3$, later yielded gluconic acid and a little 2-ketogluconic acid. A similar phenomenon was observed by Asai (unpublished) with a strain of the same species which after a long period of cultivation lost the ability to form Ca 5-ketogluconate crystal on malt extract-$CaCO_3$ agar slopes, and the colonies turned brown. Schramm and Hestrin[94] observed the change of *A. xylinum* to a non-cellulosic mutant. Creedy *et al.*[95] found that the addition of sodium arsenite converted *A. xylinum* to a non-cellulosic mutant. Kulka and Walker[96] reported that *A. ascendens*, which originally did not form gluconic acid, acquired the ability to do so after long periods of preservation.

The occurrence of celluloseless mutants of *A. acetigenus* NCIB 8132, *A. acetigenus* NCIB 5346 and *A. xylinum* var. *africanum* NCIB 7029 by repeated subculture was reported by Steel and Walker,[97] who later presented an interesting paper[98] on the physiological and biochemical properties of the celluloseless mutant. The twenty-six mutants they obtained, unlike the parent strains, did not oxidize ethanol to acetic acid, showed optimum growth at alkaline pH values (7.5–8.5), were able to tolerate high concentrations of ethanol (10–12 per cent by volume), liquefied gelatin, grew at a high temperature (40°C) and in yeast extract medium.

The biochemical activities of the mutants compared to the parent strains are shown in Table 16.

Table 16. Comparison between cellulose-forming parent
strains and cellulose-less mutants.

Biochemical activities	Cultures	
	Parent	Mutant
Catalase present	3/3	26/26
Oxidation of calcium lactate	3/3	0/26
Oxidation of ethanol and acetate	3/3	0/26
Oxidation of glycerol to dihydroxyacetone	3/3	14/26
Oxidation of glucose to gluconic acid	3/3	0/26

Steel and Walker pointed out that the mutants, apart from minor differences, may possibly fit into the genus *Pseudomonas* and are very similar to *Pseudomonas geniculata* in *Bergey's Manual* (6th ed.). However, their attempts to isolate from vinegar samples bacteria that exhibited properties similar to those shown by the mutant organisms were unsuccessful.

Shimwell and Carr[99] repeated these experiments using the same NCIB parent strains. They, too, failed to isolate a mutant possessing even one of the properties found among those of Steel and Walker. However, they obtained from a cider strain of *A. xylinum* a motile celluloseless mutant biochemically equivalent to *A. mesoxydans* and possessing peritrichous flagella.

Gromet *et al.*[100] came to the following conclusions on the biochemical evolution of *Acetobacter*. In their opinion, *A. suboxydans* and *A. xylinum* have the same series of successive intermediary steps in carbohydrate oxidation, but the former lacks (1) the ability to convert a P-ester essential in the pentose cycle into cellulose, and (2) the ability to oxidize acetate to carbon dioxide.

These two functions are situated terminally in the metabolic scheme and neither ability is necessary for the metabolic utilization of carbohydrates as a source of energy or carbon. Shimwell,[101] in fact, found a mutant strain of *A. xylinum* indistinguishable from the established type species of noncellulose forming *Acetobacter*.

Biochemically speaking, if a loss mutation occurred consecutively for the deletion of the non-essential enzyme systems mentioned above, *A. xylinum* could possibly be the phylogenetic ancestor of *A. suboxydans*. Shimwell, however, emphasized that evolution proceeds from a simple to a more complex state, and that the ancestor of *Acetobacter* was equipped with a mutative potential so broad as to comprise within its range two enzyme systems (for cellulose synthesis and acetate oxidation) which elsewhere have assumed cardinal roles in cellular survival but are facultative in *Acetobacter*.

De Ley[102] posed a related problem with respect to the phylogenetic ancestor of *A. peroxydans*, which is characterized by its catalase-negative property. In his opinion, *A. peroxydans* in the course of evolution originated from the *Mesoxydans* or *Oxydans* group by mutations characterized by the loss of catalase, kinases and permeases, so that polysaccharides, C_6 and C_5 sugars and their derivatives could not be introduced into the intermediary metabolism. One point supporting this hypothesis is an experiment by Frateur and Simonart,[103] in which they succeeded in isolating catalase-negative strains of *Acetobacter* biochemically indistinguishable from *A. rancens* and *A. mesoxydans*. De Ley and Stouthamer[104] observed a strain of *A. melanogenus* which had lost the capacity to make brown pigment and now produced considerable amounts of Ca 5-ketogluconate, suggesting a possible transformation from *A. melanogenus* to *A. suboxydans*. Shimwell[105] published a detailed study on these phenomena of mutability. He inoculated a quick vinegar generator with *A. mesoxydans* and after five days plated a portion of mash on a beer-gelatin medium. He was able to isolate from the plate colonies strains identified as *A. xylinum*, *A. xylinoides* and *A. orleanense*. In another experiment he isolated strains identified as *A. rancens* from a generator inoculated with *A. mesoxydans*. He further succeeded[101] in discovering that, with the exception of *A. lovaniense* NCIB 8620 and *A. suboxydans* NCIB 3734, all the strains studied already contained " morphological mutants," some of which gave rise under laboratory cultivation to further mutants still more diverse in form and characteristics.

From these results, he concluded that there may be only one or two species of *Acetobacter*, and that the numerous species or varieties described in the literature may be merely mutants of one or two parent types. Members of the *Suboxydans* group seem to be of a different phylogenetic origin, and consequently should not be included in *Acetobacter* and should be excluded from the evolutionary pattern suggested for *Acetobacter* proper. Shimwell's experimental results showing transfor-

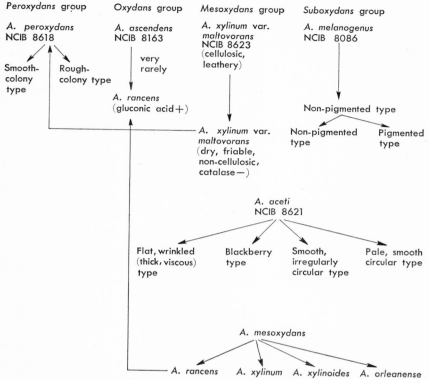

Fig. 6.   Transformation between the species of *Acetobacter*
(after Shimwell).

mation between the species are shown in Figure 6.

In Frateur's system, the species of *Acetobacter* proper can be divided into three groups according to their biochemical properties, as seen in Table 17.

Starting from the left, it may be observed that there is a gradual but stepwise diminution in the number of positive biochemical properties for each species, and a variation in the combination of these properties. Since there is one species (*A. paradoxus*) which does not possess any of the positive criteria, while others possess respectively one, two, three or four, there is no reason why all possible combinations of the five criteria should not occur in nature. In this case, the number of possible species would be 2^5 or thirty-two species. Carr,[106] in fact, has reported on a cellulose-forming variety of *A. aceti* and has also isolated from apple juice and cider

Table 17. Biochemical properties of species of *Acetobacter* proper (after Shimwell).

Character	*Mesoxydans* group			*Oxydans* group			*Peroxydans* group	
	A. aceti	*A. xylinum*	*A. mesoxydans*	*A. lovaniense*	*A. rancens*	*A. ascendens*	*A. peroxydans*	*A. paradoxus*
Catalase	+	+	+	+	+	+	−	−
Growth in Hoyer's medium[a]	+	−	−	+	−	−	+	−
Formation of — dihydroxyacetone	+	+	+	−	−	−	−	−
Formation of — gluconic acid	+	+	+	+	+	−	−	−
Formation of — cellulose	−	+	−	−	−	−	−	−
Biochemical activity (+ number)	4	4	3	3	2	1	1	0

Shimwell theorized that the members of the *Suboxydans* group are probably not phylogenetically related to any of the other species and are probably descended from some *Pseudomonas* type.

a) The Hoyer's medium used in this experiment is the one improved and modified by Frateur (*La Cellule*, **53**, 333, 1950). Composition: $(NH_4)_2SO_4$ 1 g., K_2HPO_4 0.1 g., KH_2PO_4 0.9 g., $MgSO_4$ 0.25 g., ethanol (95%, w/v) 30 ml, distilled water to make 1 liter, add 0.5 ml 1% ferric chloride solution.

an organism with the same characteristics as Frateur's *A. lovaniense*, plus cellulose-forming ability. The latter was named *A. estunense* nov. sp.,[107] but whether it is a true species or not is still open to question. The variations shown in Table 17 are limited to the five criteria used by Frateur in his classification; the addition of other criteria would presumably affect the number of species similarly. Experiments by Shimwell and others have, in fact, indicated the following variations.

1) Loss of cellulose-forming ability (*A. xylinum→A. mesoxydans*)
2) Acquirement of cellulose-forming ability (*A. mesoxydans→A. xylinum*)
3) Loss of dihydroxyacetone-forming ability from glycerol (*A. mesoxydans→A. rancens*)
4) Loss of catalase (catalase-negative *A. xylinum* var. *maltovorans*)
5) Acquirement of gluconic acid-forming ability (*A. ascendans→ A. rancens*)
6) Conversion of 2-ketogluconic acid-forming ability to 5-ketogluconic acid-forming ability.

7) Conversion of 5-ketogluconic acid-forming ability to 2-ketogluconic acid-forming ability.

These variations are the result only of the conversion of biochemical properties; if morphological and cultural properties were also considered, a tremendous number of " species " might be derived from a single so-called species. Shimwell[105] was able to isolate four different types of colonies from plates of *A. aceti* and twenty colonies from *A. mesoxydans* with different cultural characteristics depending on the rate of aeration.

As for the non-motile strains of the *Suboxydans* group (*Acetomonas* Leifson), he stated that though they do not oxidize acetic acid and in the course of handling many cultures of acetate-oxidizing *Acetobacter* strains no mutant had been obtained which had lost this property, the possibility could not be ruled out that non-motile strains of *A. suboxydans* are derived from non-motile strains of *A. mesoxydans* by the loss of their ability to oxidize acetic acid. He doubted whether all the strains of the *Suboxydans* group should be included in the genus *Acetomonas* and suggested that it may be a heterogenous group consisting of two types, phylogenetically different, converting on the one hand from *Acetobacter* (by the loss of acetate-oxidizing power), and on the other from *Pseudomonas* (by the gain of acid tolerance). In discussing transformation among species, Shimwell[85] stated that among the biochemical properties ketogenic ability and the ability to form starch or cellulose were most readily lost. Growth on Hoyer's medium, on the other hand, was quite stable and was never lost. Also, when a mutant was obtained from pure culture, it almost invariably gave rise to further mutants, and these in turn to still others. He concluded, finally, that *Acetobacter* strains cannot be classified at all, no matter what criteria are used.

Shimwell and Carr[72] also carried out various experiments on species of *Acetomonas* (Frateur's *Suboxydans*) which are characterized biochemically by the inability to overoxidize ethanol to CO_2 and H_2O. They collected fifty strains of *A. suboxydans* and *A. melanogenus* and their varieties, and tried to differentiate them by such criteria as the production of acid, of pigment (water-soluble), and of mucilage.

Surprisingly, one strain of *A. suboxydans* (Frateur's original strain) produced a brown pigment from glucose, which would class it as *A. melanogenus* or its variety. The strain also produced acid from glucose, maltose and fructose, just as *A. melanogenus* did. It differed from *A. capsulatus* in two characteristics: it freely produced hypertrophied cells, and it did not produce viscidity in dextrin media. These two characteristics had been used by Henneberg to separate *A. oxydans* from *A. industrius*.

Shimwell theorized that *A. suboxydans* and *A. capsulatus* might be derived from *A. oxydans* and *A. industrius* respectively, and that the capacity to produce a water-soluble brown pigment from glucose, used by Beijerinck[108] as a characteristic for *A. melanogenus*, was inadequate for differentiating one species from another. *A. suboxydans* can produce such a pigment from fructose or maltose. According to Shimwell's observations, loss of pigmentation occurs even within a single strain, and a colorless mutant may regain pigment production. Hence Shimwell felt that the distinction between *Acetomonas* and *Acetobacter* should be accepted, but that the specific distinction between *A. suboxydans* and *A. melanogenus* within the genus *Acetomonas* is so slight that they should be considered varieties instead. He amended the description of genus *Acetomonas* Leifson as follows: oxidize ethanol to free acetic acid, but not to carbon dioxide and water; motile or non-motile; if motile the cells possess one or more polar flagella. Type (and sole) species: *Acetomonas oxydans* (Henneberg) Shimwell and Carr.

Recently Shimwell and Carr,[109] while studying mutation in acetic acid bacteria (" mutation " being used in a broad sense to mean all inheritable

Table 18. Biochemical activities of colony mutants of *Acetobacter rancens* (after Shimwell and Carr).

	Parent strain	Mutants								
		1	2	3	4	5	6	7	8	9
Alcohol to acetic acid	+	O	+	O	O	+	+	+	O	O
Acetate to carbonate	+	+	+	+	+	+	+	+	+	+
Lactate to carbonate	+	+	+	+	+	+	+	+	+	+
Catalase	+	+	+	+	+	+	+	+	+	+
Growth on Hoyer's medium	O	O	O	O	O	O	O	O	O	O
Acid from glucose	+	+	+	+	+	+	+	+	+	O
Dihydroxyacetone	O	O	O	O	O	O	O	O	O	O
Cellulose	O	O	O	O	O	O	O	O	O	+
Starch	O	O	O	O	O	O	O	O	O	O
Colony form	N	ST	SO	R	RH	SO	Z	R	R	SO
Colony texture	B	C	B	C	B	B	B	B	C	V
Motility	O	O	O	O	O	O	O	O	O	+ (peritrichous)

+ = positive, O = negative, N = numerous, ST = smooth transluscent, SO = smooth opaque, R = rough (various types), RH = rhizoid, Z = zoned, B = buttery, C = creamy, V = viscid.

altered traits, whether due to changes in " nuclear genes " or in " plasmagenes " or to some other cause), were able to isolate a strain lacking the ability to oxidize ethanol to acetic acid from a species of *A. rancens*. This strain retained all other characteristics of the acetic acid bacteria except colony shape, which differed from that of the parent strain. The biochemical characteristics of the colony mutants are shown in Table 18.

Shimwell and Carr also isolated from cider a strong starch-producing strain that did not produce acetic acid from ethanol, although it oxidized acetate to carbonate. This strain gave rise to mutants indistinguishable from two classic *Acetobacter* species, one corresponding to *A. rancens* and the other to *A. pasteurianus* in biochemical properties. Both strains regained the capacity to produce acetic acid from ethanol, and also were able to produce acid from glucose, which the parent strain could not. Although the relationship is not clear, this phenomenon was associated with the presence, in the parent strain, of many swollen filaments and large bodies. Therefore the investigators suggested that these mutations may be correlated with the production of such bodies, and may indicate some heterodox form of reproduction other than simple fission. They termed their non-acetifying bacteria " quasi-acetobacters," and concluded that their advent seemed essentially to demolish the genus *Acetobacter*, but that it would be impossible to indicate an alternative under the existing rigid taxonomic conventions.

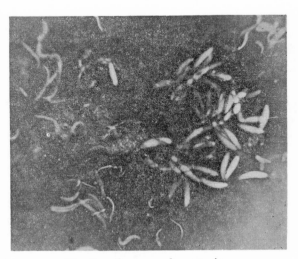

Fig. 7.   *A. mesoxydans* strain.

Large bodies and small rods (courtesy of Shimwell).

Shimwell[85] also described another interesting occurrence supporting the possibility of reproduction by means other than simple fission in *Acetobacter*. When certain strains of *A. mesoxydans* were incubated on beer agar slope for about 48 hours, large swollen cells containing a mass of small spherical to rod-shaped bodies were occasionally observed. In the immediate vicinity of these large cells there were almost invariably numbers of small rods similar to those within, suggesting that these had been liberated by the rupture of the large cells. The small rods were observed only in negative-staining preparations (without fixation); positive-staining by various methods and using various dyes failed to bring them out. A photograph is shown in Figure 7.

While Shimwell and his coworkers supported the theory of facile mutability in *Acetobacter*, Schell and De Ley[110] had some doubts. They suggested (a) that the " mutants " reported by Shimwell and others might have been contaminants; (b) that some " mutant " colonies might have been merely juvenile forms of the parent culture; and (c) that other " mutant " colonies were probably merely colony variants. Shimwell and Carr,[111] however, denied these allegations and reaffirmed their view in a later publication.

Similar variations in character have also been observed by Ameyama et al.[92] with strains of *Acetobacter* and *Gluconobacter*.

Chapter 10

THE EXISTENCE OF STRAINS PRODUCING
LITTLE ACETIC ACID FROM ETHANOL

Carr and Shimwell[80] rejected the name *Gluconobacter* because its description included "weak or no acetic acid production from ethanol," and bacteria with no acetic acid production would be *Pseudomonas* strains which should be excluded from *Gluconobacter* by definition.

Asai *et al.*,[71] however, have in fact encountered strains that are unable to produce any significant amounts of acetic acid from ethanol, and yet cannot justifiably be assigned to *Pseudomonas*. These are *G. cerinus* IAM 1832, *G. suboxydans* IAM 1828 (=*G. suboxydans* YG-21), and *G. suboxydans* H-15 (see Table 19).

Table 19. Ethanol and glucose oxidation by three strains
of *Gluconobacter*.

Strain	Production of acetic acid on ethanol +CaCO₃	Acid production from ethanol		Acid production from glucose	
		Stationary	Shaking	Stationary	Shaking
G. cerinus IAM 1832	−	2.6 (12)	4.7 (7)	79.1 (12)	88.7 (1)
G. suboxydans IAM 1828	+ (Very weak)	2.2 (12)	4.1 (7)	47.0 (12)	91.7 (1)
G. suboxydans H-15	+ (Very weak)	2.5 (12)	12.2 (7)	46.3 (12)	88.4 (1)

The figures on the left represent mg of acetic or gluconic acid produced per 5 ml of culture fluid containing 2% of the appropriate substrate. Figures in parentheses show the day of maximum acid production after inoculation.

It shoud be noted that the Qo_2 (N) value for ethanol with washed cells of *G. cerinus* IAM 1832 was only 289 against 5421 for glucose. The three strains did not oxidize lactate to carbonate but did produce dihydroxyacetone from glycerol, two principal characteristics of the *Gluconobacter*. Remarkable ketogenic activities towards mannitol and sorbitol were also confirmed. *G. cerinus* IAM 1832 (syn. *G. cerinus* nov. sp. Asai, strain 24) was isolated from apples in 1935 by Asai[112]; at that time the strain was capable of oxidizing ethanol. When re-examined twenty-three years later, it had lost this capacity, although all other characteristics were still unchanged, and satisfied the conditions for generic assignment to *Gluconobacter* as described in Table 20.

Table 20. Comparative properties of *G. cerinus* IAM 1832 when isolated and 23 years after isolation.

	When isolated (1935)	23 years after isolation (1958)
Motility	—	—
Catalase	Not tested	+
Oxidation of acetate	—	—
Production of		
acetic acid	+	—
gluconic acid	+	+
Pigment production	—	—
Ketogenic activity to		
mannitol	+	+
sorbitol	Not tested	+
glycerol	+	+
Production of 5-ketogluconate	+	+
Opt. pH for growth	3.4–5.5	6.0
Opt. temp. for growth	20–25°C	25°C
Kojic acid from fructose	+	+

Recent experiments on the growth of *G. cerinus* IAM 1832 at different pH's in liquid media showed that although it grew poorly in a mannitol medium or in a yeast extract-glucose-ethanol medium supplemented with ammonium sulfate in stationary culture at a pH range of 4.5 to 7.0, it grew well when each medium was on the acidic side (with an initial pH of 4.0) and when it was incubated in a shaken culture. These characteristics contrast with those of *Pseudomonas*, while they satisfy the criteria for *Acetomonas* (*Gluconobacter*) proposed by Shimwell *et al.*[113]

We concluded, therefore, that *G. cerinus* IAM 1832 still belongs to the genus *Gluconobacter* and should not be placed in the *Pseudomonas*. The capacity to oxidize ethanol to acetic acid is not necessarily an essential characteristic of the *Gluconobacter;* the case is similar to that of organisms of the genus *Acetobacter* which cannot oxidize glucose to gluconic acid, as reported with strains of *A. ascendens*,[61, 62] since the capacity to oxidize glucose is not essential for the *Acetobacter*. Even in the *Gluconobacter*, organisms which are unable or scarcely able to oxidize ethanol can exist by mutation or in nature. The important question is whether or not their other characteristics satisfy the criteria for inclusion in the *Pseudomonas* or the *Gluconobacter*. Kondo and Ameyama[58] pointed out the existence of one organism, *A. suboxydans* var. α (isolated from *Fragaria chiloensis* Duch., the Dutch strawberry) which was very similar to *G. suboxydans* but showed only a weak capacity to oxidize ethanol, and in stationary culture did not oxidize it at all. The " quasi-acetobacters " of Shimwell and Carr[109] should also be kept in mind.

Chapter 11

THE EXISTENCE OF " INTERMEDIATE " STRAINS

Two types of brown pigment-producing bacteria have been reported. One is characterized by polar flagellation when motile, cannot oxidize acetate, and is represented by *G. melanogenus* Beijerinck; the other is characterized by peritrichous flagellation when motile, can oxidize acetate to CO_2 and H_2O, and is represented by strains *G. liquefaciens* IAM 1834 (*G. liquefaciens* G-1), *G. melanogenus* IAM 1835 (*G. melanogenus* AC-8), and *G. melanogenus* IAM 1836 (*G. melanogenus* U-4) (Asai and Shoda[61]), all of which were once pooled as a specific type in the genus *Gluconobacter*. In addition, Kondo and Ameyama[58] isolated and identified two strains of *Acetobacter aurantius* (IFO 3245 and IFO 3246) from the flower of *Lilium*

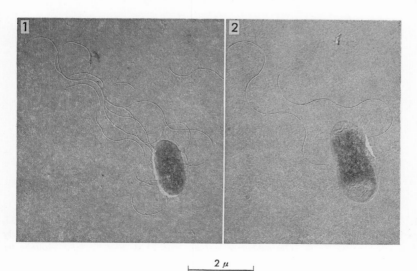

$\vdash$ ——— 2 μ ——— $\dashv$

Fig. 8.   Electron micrographs of " intermediate " strains (after Asia *et al.*).
1.   *Gluconobacter liquefaciens* IAM 1834.   Cells grown on mannitol agar for 18 hr at 20°C.
2.   *Gluconobacter melanogenus* IAM 1835.   Cells grown on mannitol agar for 18 hr at 20°C.

aurantium Lindl., which, according to their description, produce brown pigment and oxidize ethanol to acetic acid but do not oxidize acetic acid. They are motile, but the flagellation was not described. Asai *et al.*[71] re-examined the strains to study their flagellation and taxonomy. The results are shown in Tables 21 and 22; see also Figures 8 and 9.

Shimwell and Carr[72] and Stouthamer[73] had confirmed the peritrichous flagellation of *G. liquefaciens* G-1, and Hodgkiss *et al.*[74] had done the same

Table 21. Morphological characteristics of the five anomalous strains.

Strain	Form	Size (μ)	Motility	Flagellation
G. liquefaciens IAM 1834	Rod	0.6×1.8	+	Peritrichous
G. melanogenus IAM 1835	Rod	$0.6 \times 1.0 - 1.4$	+	Peritrichous
G. melanogenus IAM 1836	Rod	$0.6 \times 1.0 - 1.4$	+	Peritrichous
A. aurantius IFO 3245	Rod	$0.4 - 0.6 \times 1.2 - 1.8$	+	Polar
A. aurantius IFO 3246	Rod	$0.4 - 0.6 \times 1.2 - 1.6$	+	Polar

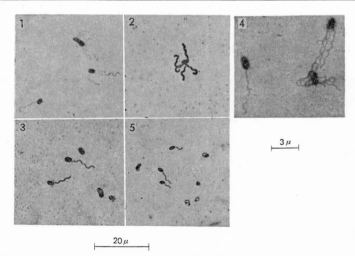

Fig. 9. Micrographs of "intermediate" strains (after Asai *et al.*).

1. *Gluconobacter liquefaciens* IAM 1834. 2. *Gluconobacter melanogenus* IAM 1835, peritrichously flagellated cell. 3. *Gluconobacter melanogenus* IAM 1835, polarly flagellated cells. 4. *Gluconobacter melanogenus* IAM 1836. 5. *Acetobacter aurantius* IFO 3245.

G. liquefaciens IAM 1834, *G. melanogenus* IAM 1835 and *A. aurantius* IFO 3245 were cultured on mannitol agar slant for 18–24 hr at 20°C. *G. melanogenus* IAM 1836 was cultured for 18 hr at 25°C on agar slant consisting of : peptone, 3.0 g ; yeast extract, 2.0 g ; sodium acetate, 2.0 g ; glucose, 2.0 g ; agar, 20.0 g ; and distilled water, 1000 ml (pH 6.0).

for *G. melanogenus* AC-8 and *G. melanogenus* U-4. Figures 8 and 9 show the same results. Surprisingly, however, the two strains of *A. aurantius* were shown to possess single polar flagella. Even in the former three strains the flagellation is not perfectly homogeneous, as seen in the photographs (Figure 9), and often shows a mixed type of lateral and polar flagellation—which may be related to their phylogenetic line. All five strains, as shown in Table 22, grew well on glutamate agar* while the strains of *Gluconobacter* and *Acetobacter* did not. Since even the polarly flagellated melanogenic *Gluconobacter* strains (*G. melanogenus* Beijerinck) were unable to grow on this medium, this growth ability might help to differentiate these five strains from the genus *Gluconobacter*.

Komagata and Iizuka[114, 115] reported that the *Pseudomonas*, especially the fluorescent group, like these strains grew well on glutamate agar. All strains grew well on mannitol agar (containing mannitol, yeast extract, and peptone) like the *Gluconobacter* strains, and produced some substance or substances giving a reddish violet color with ferric chloride from glucose as was the case with the polarly flagellated *G. melanogenus* Beijerinck. All strains produced a water-soluble brown pigment in glucose-$CaCO_3$ agar and in glutamate agar cultures, and the colonies were also brown. Two strains of *A. aurantius* produced a water-soluble yellow pigment in mannitol agar whereas the peritrichously flagellated strains did not. On the other hand, all strains oxidized acetate and lactate. With respect to this behavior only, these strains, at least the peritrichously flagellated ones, could be assigned to the genus *Acetobacter*.

Ketogenic activity towards glycerol, one of the criteria adopted by Shimwell *et al.*[113] for the differentiation of *Acetomonas* from *Pseudomonas*, was seen only in peritrichously flagellated melanogenic strains. Two strains of *A. aurantius* did not produce dihydroxyacetone. Three peritrichously flagellated strains assimilated ammonium nitrogen when ethanol or mannitol was supplied as the carbon source. *A. aurantius* IFO 3245 assimilated ammonium nitrogen when glucose or mannitol was supplied, while *A. aurantius* IFO 3246 was unable to assimilate it regardless of the kind of carbon source supplied. An acid formation test presented a clear-cut difference. The peritrichous group did not produce acids from galactose, fructose, and melibiose while the polar group did.

G. liquefaciens IAM 1834 (*G. liquefaciens* G-1) was isolated by Asai in 1935[116] from the dried fruit of *Diospyros kaki* Thunb. (the Japanese per-

---

* 10 g glucose, 5 g sodium glutamate, 1 g KH_2PO_4, 0.2 g $MgSO_4 \cdot 7H_2O$, 0.1 g KCl, 20 g agar and 1000 ml distilled water (pH 7.2).

Table 22. Physiological and biochemical

Subject	IAM 1834	IAM 1835
Pigmentation	Brown	Brown
Growth on		
glutamate agar	+ (Brown)	+ (Brown)
mannitol agar	+ (Pale brown)	+ (Pale brown)
Soluble pigment	Brown	Brown
FeCl$_3$ reaction	+	+
(glucose broth)		
Reaction in		
acetate	Alk.	Alk.
lactate	Alk.	Alk.
Production of acetic acid on ethanol+CaCO$_3$	+	+
Oxidation of lactate to carbonate	+	+
DHA from glycerol	+	+
Growth at 37°C	−	−
Acid from		
galactose	−	−
fructose	−	−
melibiose	−	−

simmon) and designated as *Gluconoacetobacter liquefaciens* Asai at that time.   A reexamination of this strain was undertaken by Asai and Shoda[61] and again by Asai *et al.*[71]   The results are compared in Table 23.

It is not clear whether the strain was originally mistaken for non-motile (the flagella may have been dormant or in a " paralyzed " state), or whether it aquired motility during the long preservation of the culture, but it is evident that a rare mutation occurred, with transformation from a strain which does not oxidize acetate into one which does—a change of the principal characteristic that divides the *Gluconobacter* and *Acetobacter*. Other biochemical and physiological characteristics are almost identical with the polarly flagellated *G. melanogenus* strains, except for gelatin liquefaction, capacity to oxidize acetate, and flagellation.   From this standpoint, we dis-agreed with the opinion of Carr and Shimwell[117] that this strain is merely a pigment-producing mutant of *A. aceti,* or that of Kimmitt and Williams[118] characteristics of the five anomalous strains.

IAM 1836	IFO 3245	IFO 3246	G. melanogenus IAM 1818 (Control)
Brown	Brown	Brown	Brown
+ (Brown)	+ (Brown)	+ (Brown)	–
+ (Pale brown)	+ (Pale brownish yellow)	+ (Pale brownish yellow)	+ (Pale brown)
Brown	Yellow to brown	Yellow to brown	Brown
+	+	+	+
Alk.	Alk.	Alk.	Acid
Alk.	Alk.	Alk.	Acid
+	+	+	+
+	+	+	–
+	+	+	+
–	+	+	–
–	+	+	+
–	+	+	+
–	+	+	+

supporting the generic transfer of the strain to genus *Acetobacter*. We feel that this unusual strain has appeared in the course of an evolutionary trend from non-acetate-oxidizing *G. melanogenus* into acetate-oxidizing *Acetobacter* strains. Therefore we support the view that the ascendant of the genus *Acetobacter* is most probably the genus *Gluconobacter*, and that the three peritrichous strains mentioned are " intermediate " strains which can be assigned neither to *Gluconobacter* nor to *Acetobacter*.

The flagellation of these strains must also be taken into consideration. The cells often appear as a mixed type, with lateral and polar as well as peritrichous flagellation, as shown in the photographs. A similar case was reported by Leifson and Hugh,[119] who observed that the organism K-57, which can be classified as *Xanthomonas* or *Flavobacterium*, exhibited two different types of flagellation, one the typical polar flagellation and the other lateral.

Table 23. Changes in the characteristics of *G. liquefaciens* IAM 1834 during preservation in the laboratory.

Characteristic	Original description (1935)	19 years after isolation (1954)	28 years after isolation (1963)
Motility	Non motile[a]	Motile	Motile
Flagellation	Not observed	Positive (wrongly reported as polar)	Peritrichous
Oxidation of acetate to CO_2	Negative	Positive	Positive
Oxidation of lactate to CO_2	Not observed	Positive	Positive
Production of acetic acid from ethanol	Positive[b]	Positive	Positive
Production of gluconic acid from glucose	Positive[c]	Positive	Positive
Production of DHA from glycerol	Positive	Positive	Positive
Brown pigment production	Positive	Positive	Positive
$FeCl_3$ reaction-positive substance from fructose	Positive	Positive (also from glucose)	Positive (also from glucose)
Gelatin liquefaction	Positive	Negative	Not observed
Opt. temp. for growth	25–30°C	25–30°C	No growth at 37°C
Opt. pH for growth	5.4–6.3	6.0	Not observed

a) Motile strains with similar properties were also isolated at the time but died during the long preservation.
b) 0.37 g acetic acid per 100 ml was produced in medium containing 2% ethanol after 14 days' incubation.
c) 17.2 g gluconic acid per 100 ml was produced in medium containing 20% glucose.

On the other hand, the brown pigment-producing *A. aurantius* strains are distinctly polarly flagellated and are very similar in character to *Pseudomonas*. Recently Shimwell *et al.*[113] suggested differentiating *Pseudomonas* from *Acetomonas* (*Gluconobacter*) by the following criteria: (1) negative acetic acid production, (2) positive oxidation of lactate to carbonate, (3) negative production of dihydroxyacetone from glycerol, (4) no growth at pH 4.5. According to these criteria, the *A. aurantius* strains cannot be assigned to the *Pseudomonas*, since they were confirmed to grow well at pH 3.5 in a yeast extract-ethanol-glucose medium and produced a noticeable amount of acid from ethanol. Nor can they be assigned to the *Gluconobacter*, since they oxidized lactate completely and did not produce dihydroxyacetone from glycerol. Hence these *A. aurantius* strains, though closely related to *Pseudomonas*, are probably in a transitional position in the evolutionary course of *Pseudomonas* to *Gluconobacter*, and until a reasonable generic assignment can be determined for them, may be termed simply " intermediate."

The existence of " intermediate " strains has been also discussed in detail by Ameyama and Kondo.[120] Eleven strains comprising *G. liquefaciens* IMA 1834, two *G. melanogenus* strains (IAM 1835 and 1836), and eight *A. aurantius* strains were classed by them as " intermediate." These strains produce a brown pigment but differ from non-acetate-oxidizing and brown pigment-producing *Gluconobacter* strains in biochemical characteristics. They showed the lactaphilic character of Rainbow and Mitson in casamino acid-containing medium; but in the medium containing glutamate as the nitrogen source, they showed a glycophilic character, in the order of *G. melanogenus* < *G. liquefaciens* < *A. aurantius*. Ketogenic activities toward polyalcohols were seen in the strains of *G. melanogenus* and *G. liquefaciens* but not in *A. aurantius*. They did not require any vitamin for growth and oxidized succinate, acetate and lactate as *Acetobacter* dose, but were closely related to *Gluconobacter* in carbohydrate availability, in that they grew well in many carbohydrates tested. Strong oxidation of L-serine was characteristic of these strains. Details of their activities are summarized in Table 24.

It should be emphasized, however, that the flagellation of these strains is quite different from that observed by other investigators, especially in the case of *G. melanogenus* and *G. liquefaciens*. These strains were shown to possess a single polar (lateral ?) flagellum but not multi-peritrichous ones, while the strains of *A. aurantius* were shown to be polarly and singly or rarely multi-flagellated.

Table 24. Characteristics of the "intermediate" strains compared with *Acetobacter* and *Gluconobacter* (after Ameyama and Kondo).

	Acetobacter	"Intermediate" strains			Gluconobacter
		G. melanogenus	G. liquefaciens	A. aurantius	
Growth					
with casamino acid	Lactate > Glucose	Lactate > Glucose	Lactate > Glucose	Lactate > Glucose	Glucose > Lactate
with glutamate	Lactate > Glucose	Lactate > Glucose	Lactate > Glucose	Glucose > Lactate	Glucose > Lactate
Production of reducing substances from					
Glucose	−	+	.	+	+
Gluconate	−	+	+	+	+
Glycerol	−	+	+	−	+
Mannitol	−	+	+	+	+
Sucrose	−	−	−	+	−
D-Xylose	−	+	−	+	−
Oxidation of acetate and succinate	+	+	+	+	−
Oxidation and deamination of amino acids	+	+	+	−	−
Vitamin requirement	−	−	−	−	+
Flagellation	not tested	single lateral	single lateral	polar	polar

Chapter 12

CAN ONE RELIABLY ESTABLISH " SPECIES " DIFFERENTIATION IN *GLUCONOBACTER* AND *ACETOBACTER*?

Asai *et al.*[71] had attempted, as stated in the preceding chapter, to divide the genus *Gluconobacter* into three groups, *viz.*, (a) organisms forming pink colonies, (b) organisms forming brown colonies and producing water-soluble brown pigment, and (c) organisms producing no noticeable pigment. These characteristics seemed, at first, applicable for species differentiation in *Gluconobacter*. Later, however, it was reported that pigmentation often varied. Shimwell and Carr[72] found pigment-producing *Acetomonas suboxydans* and *Acetomonas capsulatus* NCIB 4943. Fewster[121] reported the appearance of pink colonies with *A. suboxydans* ATCC 621 when grown in the presence of $CaCO_3$. De Ley and Stouthamer[104] observed that a strain of *A. melanogenus* had lost the capacity to make brown pigment and had begun to produce considerable amounts of Ca 5-ketogluconate, suggesting the possibility of transformation from *A. melanogenus* to *A. suboxydans*. A similar loss of pigmentation was reported by Shimwell,[101] who found that a strain of *A. melanogenus* gave rise to two colony forms, one pigment-producing and the other not. The colorless mutant later reverted and again produced pigment. Biochemical variations were reported by Walker and Kulka,[93] who isolated a strain of *A. suboxydans* which at first produced gluconic acid and 5-ketogluconic acid, and later produced gluconic acid and a little 2-ketogluconic acid. The occurrence of a non-ketogenic mutant of *A. suboxydans* NCIB 3734 was observed by Shimwell and Carr.[72] The four strains used by Asai and his co-workers,[71] *G. suboxydans* NRRL B-755, *G. cerinus* IAM 1833, *G. scleroideus* IAM 1842, and *G. nonoxygluconicus* IFO 3275, had originally produced no pigments, but later gave rise to pink colonies. In view of these observations and those reported by other workers, they concluded that pigmentation alone was inadequate for species characterization.

What other properties might be singled out to validate the species concept in *Gluconobacter*? Strains that produced brown pigment also produced certain substances (probably γ-pyrone compounds) from glucose, giving a reddish violet color with ferric chloride. But the observation that *G. melanogenus* can mutate into a *G. suboxydans*-like strain prevents the acceptance of this characteristic for species differentiation. Transformations of cellulose-forming *A. xylinum* into cellulose-less *A. mesoxydans* and *A. rancens*, and of catalase-positive *A. ascendens* into catalase-negative *A. paradoxus*, as well as the loss of the dihydroxyacetone-forming ability in *A. mesoxydans* have all been verified. Acquisition of gluconic acid-forming ability in *A. ascendens* was demonstrated by Kulka and Walker.[96] Shimwell[85] reported that the existing species of *Acetobacter* are unstable and mutate spontaneously into other species even in the laboratory, and concluded, as mentioned earlier, that *Acetobacter* strains cannot be classified at all, no matter what criteria are used.

De Ley[81] proposed abolishing specific names for all acetic acid bacteria, since there are no abrupt changes in the sequence of strains, and it is difficult to draw the line in the gradation of properties defining the limits of species.

Asai *et al.*[71] had observed a strain of *A. xylinum* A-24 that exhibits ketogenic activity towards glycerol and lacks the capacity to oxidize lactate to carbonate, and had stated that this strain should be transferred into the genus *Gluconobacter* although it was non-motile. The capacity to assimilate ammonium salts in Hoyer's solution had been recognized earlier as a criterion differentiating *A. aceti* from *A. xylinum* and *A. rancens*. Although this criterion was adopted in *Bergey's Manual of Determinative Bacteriology* (6th ed.), new findings by Leifson[24] showed that four *A. xylinum* strains were able to grow in Hoyer's solution also, suggesting that it is impossible to differentiate between *A. xylinum* and *A. aceti* on this basis only.

Thus we concluded, as Shimwell and Carr and De Ley had, that species characterization of acetic acid bacteria is unreliable. Recent studies by Scopes[84] on spectrophotometric identification and the serological studies on *Acetomonas* and *Acetobacter* of Shimwell and McIntosch[86] also supported generic division only of acetic acid bacteria. With the exception of *A. melanogenus* Beijerinck, no remarkable differences were observed within the species belonging to each genus.

We have proposed the adoption of *A. aceti* Beijerinck as the type (and sole ?) species of genus *Acetobacter*, and *G. oxydans* (Henneberg) Asai as the type (and sole) species of genus *Gluconobacter* because of their historical position in the literature. Other existing species should be considered varieties of the type species for each genus. This differentiation is almost identical with Leifson's and De Ley's classifications, though the latter used the designation *biotype* instead of *genus* for *Acetobacter* and *Gluconobacter*.

Chapter 13

THE PHYLOGENETIC RELATIONSHIP
OF ACETIC ACID BACTERIA

The acetic acid bacteria or *Acetobacter* are defined as aerobic bacteria that oxidize ethanol to acetic acid, and accumulate the acid to a greater or lesser degree. These organisms were initially isolated from materials containing vinegar or ethanol, such as beer or wine. *A. aceti, A. pasteurianus, A. kützingianus* and *A. xylinum* were all isolated from such sources.

Later, in the course of investigations, carbohydrate-containing materials were also used to isolate acetic acid bacteria. Among those isolated were found organisms which oxidized ethanol weakly, produced little acetic acid but showed a high capacity for oxidizing* glucose with the accumulation of a large amount of gluconic acid. Bertrand's sorbose bacteria, Hermann's *Bact. gluconicum*, and a large number of acetic acid bacteria isolated by Asai from fruits all fit into this category.

Stanier,[50] on the other hand, reported the presence in *Pseudomonas fluorescens* of strains which could oxidize ethanol to acetic acid, demonstrating that the production of acetic acid is not a property limited to acetic acid bacteria. It is well known also that pseudomonads generally oxidize glucose to gluconic acid and in this respect are identical to the strains of acetic acid bacteria which accumulate gluconic acid from glucose.

Asai found, among the bacteria isolated from fruits, many strains that formed little acetic acid and lacked the capacity to assimilate and oxidize it. He proposed, as mentioned earlier, separating this group from the *Acetobacter* and giving it the new genus name *Gluconobacter*, because of the bacteria's ability to accumulate a large amount of gluconic acid. Later, Ikeda,[122] studying the relationship among *Acetobacter, Gluconobacter,* and *Pseudomonas,* proposed calling the latter "pseudooxidative bacteria," because of their high oxygen requirement in sugar

---

* Accumulating the first oxidation products; not oxidizing completely to CO_2 and water without accumulation of intermediate products.

oxidation (shaken culture or culture with aeration and agitation), optimum growth in basic conditions, weak resistance to acid, and ability to utilize inorganic nitrogen sources readily. He also proposed calling *Acetobacter* and *Gluconobacter* " homo-oxidative bacteria." A comparison of these homo- and pseudo-oxidative bacteria is given in Table 25.

Vaughn[123] recognized the similarities between *Acetobacter* and *Pseudomonas* (see Table 26), but questioned the validity of including *Acetobacter* in the family *Pseudomonadaceae* and suggested, provisionally, including it in the family *Acetobacteriaceae*. Leifson[24] discovered that acetic acid-producing and acetate-oxidizing strains of *Acetobacter*, *i.e.* typical acetic acid bacteria, are not polarly but peritrichously flagellated when motile; thus he too concluded that they should not be classed in the *Pseudomonadaceae*. Only non-acetate oxidizers are polarly flagellated when motile, and these he named *Acetomonas*. Shimwell,[85] after studying mutation in *Acetobacter* (see the preceding chapter) had agreed to the distinction between *Acetomonas* and *Acetobacter*, but felt that strains of *Acetomonas* could not be taxonomically distinguished. He described *Acetomonas oxydans* (Henneberg) Shimwell and Carr as the type (and only) species, confining both *Acetobacter* gen. nov. and *Acetomonas* gen. nov. to one species respectively.

There exists, as was mentioned earlier, a non-acetifying strain of *Gluconobacter* (*Acetomonas*), *G. cerinus* IAM 1832 (*G. cerinus* 24); there are also two other strains which produce very little acetic acid. In contrast, some strains of *A. ascendens* (Henneberg) fail to produce a significant amount of gluconic acid or almost entirely lack the ability to oxidize glucose.[61, 62, 63] *Gluconobacter* (or *Acetomonas*) is polarly flagellated when motile, as Leifson pointed out; it does not oxidize acetate. Thus, *G. melanogenus* Beijerinck produces brown pigment, is polarly flagellated, and does not oxidize acetate. We encountered, nevertheless, a mutant* (provisionally named " intermediate " strain IAM 1834) that is peritrichously flagellated and oxidizes acetate; the other characteristics are quite similar to those of *G. melanogenus* Beijerinck. Analogous strains (" intermediate " strains IAM 1835 and 1836) were also found (Asai *et al.*[71]). The flagellation of these strains, however, is not strictly homogeneous and the cells often have lateral or even polar flagella. There also exists an " intermediate " strain (IFO 3245) which produces brown pigment and oxidizes acetate but has polar flagella.

---

* Very recently the strain has been assigned by Asai to the genus *Acetobacter* on the basis of its peritrichous flagellation and acetate oxidizability, and was named *Acetobacter intermedius* nov. sp.

Table 25.   Comparative characteristics of the oxidative bacteria (after Ikeda).

	Homo-oxidative bacteria		Pseudo-oxidative bacteria
	Acetobacter	*Gluconobacter*	*Pseudomonas*
Optimum pH for growth	Acidic	Acidic	Basic
Oxygen requirement	Aerobic	Aerobic	Facultative aerobic
Tolerance to acid and ethanol	Strong	Strong	Weak
Assimilation of inorganic N	Difficult	Difficult	Easy
Habitat	Ethanol or acetic acid-containing materials	Fruits or flowers, sugar-containing materials	Soil, sewage, *etc.*
Flagellation (if motile)	Peritrichous[a]	Polar	Polar

a) Confirmed later by Leifson, Shimwell and Asai.

Table 26.   Similarities between *Acetobacter* and *Pseudomonas* (after Vaughn).

Character	*Acetobacter*	*Pseudomonas*
Morphology	Rods (ovoid to long) Involution forms very common	Rods (ovoid to long)
Motility	Polar flagellation[a] (if motile)	Polar flagellation (if motile)
Endospore formation	None	None
Pigmentation	Two species	Numerous species (yellow or blue-green fluorescent)
Oxygen requirements	Aerobic	Aerobic
Catalase	Positive[b]	Positive
Oxidation products of glucose	Gluconic acid, 5-ketogluconic acid[c]	Gluconic acid, 2-ketogluconic acid
Habitat	Acid media	Neutral or alkaline media

a)   Later emended to peritrichous by Leifson for acetic acid-oxidizing acetobacters.
b)   Exceptions are *A. peroxydans* and *A. paradoxus* which are catalase negative.
c)   Some strains of *Acetobacter* produce 2-ketogluconic acid.

What are the phylogenetic relationships of these acetic acid bacteria, non-acetic acid or non-gluconic acid producers included? Here is one possibility. Bacteria which had originally inhabited the soil began to inhabit fruits or flowers. In adapting themselves to their surroundings they acquired the ability to oxidize glucose or fructose in order to obtain energy for growth. They also acquired the ability to grow well in acid conditions, adapting themselves to the organic acids formed by sugar oxidation. But fruit is also a good habitat for yeasts, and often contains ethanol produced by these yeasts in the natural fermentation of sugars. Hence the sugar-oxidizing bacteria had to adapt themselves to ethanol and acquired the ability to tolerate and to oxidize ethanol to acetic acid. Eventually they tended to lose the ability to oxidize glucose, and the production of gluconic acid was shifted to that of acetic acid. These "terminal" bacteria are probably the so-called acetic acid bacteria. The original bacteria, therefore, were probably pseudomonads. Some strains of pseudomonads are reported to oxidize ethanol to acetic acid (Stanier[50]); others have acquired, by adaptation, the ability to oxidize sorbitol ketogenically (Sebek and Randles[124]). These facts substantiate the relationship between pseudomonads and acetic acid bacteria.

It can easily be imagined that *Pseudomonas*, adapting to its environment, was transformed into *Gluconobacter* (*Acetomonas*) and further into *Acetobacter*. The strains lacking or with a very low capacity to oxidize ethanol (Asai's non-acetifier and other strains, excepting Shimwell's quasi-acetobacter) can perhaps be situated in the evolutionary line from *Pseudomonas* to *Gluconobacter*. The strains lacking the capacity to oxidize glucose (*A. ascendens* or *A. peroxydans*) may occupy the extreme point, perhaps in the genus *Acetobacter*.

Shimwell's quasi-acetobacters are probably back-mutants of acetobacters that still have the capacity to oxidize glucose. Shimwell emphasized that he could not demonstrate experimentally mutation of *Acetomonas* (*Gluconobacter*) into *Acetobacter*, and concluded that these two genera are phylogenetically different groups, although he did not exclude the possibility that certain *Acetomonas* might have developed from *Pseudomonas*.

Phylogenetically speaking, bacterial flagella are believed to have evolved from polar to peritrichous flagella (Bisset[125]). It would not be impossible, therefore, to suppose that the polarly flagellated *Gluconobacter* were transformed into an intermediate type with mixed flagellation, and then into the peritrichously flagellated *Acetobacter*. If we accept this assumption, the "intermediate" strain IAM 1834 and other analogous strains can be placed as intermediates in the transformation of *Gluconobacter* into *Acetobacter*. The polarly flagellated "intermediate" strain IFO 3245 and its analogous strain can then be placed in the evolutionary line *Pseudomonas→Gluconobacter*, or can perhaps be seen as the immediate predecessor of the "intermediate" strain IAM 1834. Figure 10 shows this hypothetical relationship, as presented by Asai et al.[71]

Pseudomonas is the common ancestor. Through the "intermediate" strain IFO 3245 two independent lines have resulted. One, having acquired peritrichous flagella, developed into the genus *Acetobacter via* the "intermediate" strain IAM 1834. The other remained polar in flagellation, lost the ability to oxidize acetate and lactate to carbonate, acquired ketogenesis for glycerol and developed into the genus *Gluconobacter*.

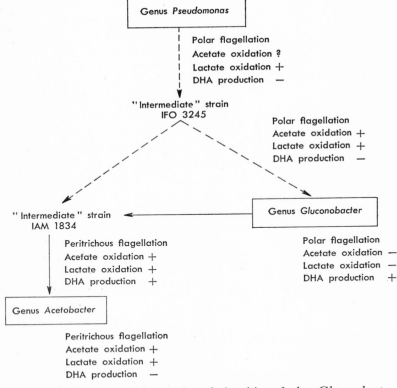

Fig. 10. Presumed phylogenetic relationship of the *Gluconobacter* and *Acetobacter* with reference to "intermediate" strains (after Asai et al.).

Apart from these, one other route—from *Gluconobacter* to *Acetobacter via* the "intermediate" strain IAM 1834, with the acquirement of peritrichous flagella and the capacity to oxidize acetate and lactate—was suggested in the investigation of IAM 1834. In this scheme, of course, the acquirement of biochemical activities is included.

The scheme does not necessarily satisfy the concept of regressive evolution; however, we did in fact encounter a case of transformation of a non-acetate oxidizing *Gluconobacter* into an acetate oxidizing one, as seen with the "intermediate" strain IAM 1834. In addition, several experiments on mutation carried out by other workers supported the possibility of such a trend in natural mutation. Walker and Kulka[93] reported that *A. ascendens*, which had not originally produced gluconic acid, acquired the ability to do so after long periods of preservation. The transformation of a non-cellulosic *A. mesoxydans* strain into a cellulosic *A. xylinum* strain was also reported by Shimwell.[85] We noted a case in which a strain of *G. suboxydans*, after a long period of cultivation, gained the ability to form a large amount of 2-ketogluconate from glucose, and the colonies turned light brown on malt extract-$CaCO_3$ agar.

De Ley[81] viewed the natural relationship and phylogeny of the acetic acid bacteria from the standpoint of comparative carbohydrate metabolism. His theory, based on regressive evolution, formed the basis for his arrangement of bacterial strains in a sequence of decreasing biochemical activities. The *Acetobacter* and *Gluconobacter* have many features in common —the enzymatic mechanism for oxidation of D-lactate and ethanol to acetate, the enzymes of the hexose monophosphate pathway, especially the oxidosome–linked nature of the hexose and pentose oxidations, the formation of 2-ketogluconate, ketogenic characters, acid resistance, *etc.* There is no question, then, regarding their close phylogenetic relationship. Many of their characteristics are found, among other bacteria, only in the genus *Pseudomonas*. One can arrange the species in these two genera in a smooth gradation from the most enzymatically complex to the simplest, *i.e.* from *G. liquefaciens* 20 ("intermediate" strain IAM 1834) to *G. viscosus* and from *A. xylinum* to *A. peroxydans*.

Following precisely this theory, De Ley presented the scheme of phylogenetic relations shown in Figure 11. The position of *G. liquefaciens* should be noted. De Ley considered this acetate-oxidizing *Gluconobacter*(?) species (an "intermediate" strain according to Asai) a representative of species closest to the ancestral pool of acetic acid bacteria, and thought that the present acetic acid bacteria might have evolved from these species in the course of time.

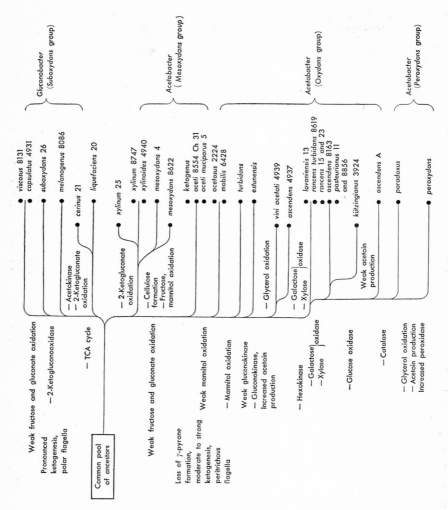

Fig. 11.   Possible phylogenetic relationship of the acetic acid bacteria, based primarily on comparative carbohydrate metabolism (after De Ley).

Concerning the phylogenetic relationship between the catalase-negative *A. peroxydans* and the catalase-positive acetic acid bacteria, De Ley and Vervloet[126] argued that even though this organism lacks catalase, it should not be set apart from the *Acetobacter*, since (1) its enzymatic mechanism for lactate and ethanol oxidation is nearly identical with other strains of *Acetobacter*; (2) the peroxidase mechanism, while not essential, is universal if weak in other acetic acid bacteria ; and (3) its carbohydrate metabolism is nearly indistinguishable from that of *A. ascendens*. They felt that *A. peroxydans* is undoubtedly a true acetic acid bacterium, but simply has the poorest enzyme equipment.

If the peroxidases in *Acetobacter* have the same prosthetic group as the enzymes from other origins, a mutant might, in the course of evolution, have arisen from a strain of *A. ascendens*, without the capacity to synthesize apocatalase and with more protohematin IX linked to the apoperoxidase protein, resulting in higher peroxidase activity. In this way a strain of *A. paradoxus* would develop, which, by further loss of acetoin forming enzymes, would give rise to *A. peroxydans*. The hydrogenase activity of this organism, however, was not detected in other strains of *Acetobacter*, hence it is not known whether a gain mutation occurred or whether the same sequence hypothesized for peroxidase may be applied. The new catalase-negative strains described by Wiame et al.[127] under the name of *A. acidophilus* were considered by De Ley[128] as the most specialized of all, since they can oxidize only primary alcohol, grow only in the pH range of 2.7–4.0, and seem to require CO_2. De Ley also suggested, at this time, that *Acetobacter* might well be related phylogenetically to *Achromobacter*. Table 27 shows the distribution of some enzymes in *Acetobacter*, in decreasing sequence.

Janke[129] published another phylogenetic scheme based on the groups represented in Frateur's classification. According to his theory, the *Suboxydans* group (*Gluconobacter* or *Acetomonas*) is the oldest, in evolutionary terms, among the acetic acid bacteria, because of its (1) polar flagellation, (2) phosphate-independent dehydrogenation and linkage to pentose cycle oxidation, and (3) absence of DCA cycle, TCA cycle, and complex cytochrome-cytochrome oxidase system. Thus it cannot oxidize acetic acid to CO_2 and H_2O.

The *Oxydans* (*Euoxydans*) group and *Mesoxydans* group cannot reasonably be separated from each other because of their ready inter-transformation and their many overlapping properties (as reported by Shimwell). Janke placed these groups in a later position phylogenetically than the *Suboxydans* group. The *Peroxydans* group, by contrast, lacks catalase and is believed to be capable of utilizing H_2O_2 as a hydrogen acceptor and of activating molecular hydrogen; consequently, Janke concluded that this group, although it might have the same origin, occupies a different position in phylogenetic development.

The species of acetic acid bacteria, and their varieties, that have appeared in the literature up to the present are listed in Table 28.

Table 27. The distribution of some enzymes in *Acetobacter* (after De Ley).

Organism	Fructose and gluconate oxidation	Mannitol dehydrogenase	Gluconokinase	Hexokinase	Galactose and xylose oxidase	Glucose oxidase	Catalase	Lactate and ethanol oxidation
A. mesoxydans 4	+	+	+	+	+	+	+	+
A. aceti Ch 31	Weak	+	+	+	+	+	+	+
A. mobilis 6428	−	Weak	+	+	+	+	+	+
A. turbidans 6424	−	−	+	+	+	+	+	+
A. estunensis E	−	−	Weak	+	+	+	+	+
A. lovaniensis 13	−	−	−	−	+	+	+	+
A. rancens 15	−	−	−	−	−	+	+	+
A. ascendens A	−	−	−	−	−	−	+	+
A. paradoxus 30	−	−	−	−	−	−	−	+

Table 28. List of the species or varieties of acetic acid bacteria.

Species	Author	Habitat	Reference
A. aceti (Kützing) Beijerinck (*Ulvina aceti.* Kützing)	Kützing	vinegar	F.T. Kützing: *J. prakt. Chem.*, **11**, 385 (1837).
A. aceti Pasteur (*Mycoderma aceti non visqueux*)	Pasteur	quick-vinegar	L. Pasteur: Etudes sur le vinaigre, 60, 82 (1868).
Mycoderma aceti gélatineux (*A. xylinum*)	Pasteur	vinegar	L. Pasteur: *ibid.*, 67, 95 (1868).
A. aceti Hansen	Hansen	beer	E.C. Hansen: *Comptes rendus de Carlsberg*, **1**, 49, 96(1879), **3**, 182 (1894), **5**,39 (1900).
A. pasteurianum Hansen	Hansen	beer	E.C. Hansen: *ibid.*

Table 28 (Continued).

Species	Author	Habitat	Reference
A. xylinum Brown	Brown	beer, alcohol mash, vinegar	A.J. Brown: *J. Chem. Soc. London, Transactions*, **49**, 432 (1886).
A. aceti Brown	Brown	beer	A.J. Brown: *ibid.*, 172 (1886).
A. xylinum Wermischeff	Wermischeff	red wine	Wermischeff: *Ann. Inst. Pasteur*, **7**, 213 (1893).
A. kützingianum Hansen	Hansen	beer	E.C. Hansen: *Comptes Rendus de Carlsberg*, **3**, 182 (1894), **5**, 39 (1900).
A. friabile Lindner (*A. aceti* var. *friabile* Lindner)	Lindner Zeidler	beer	P. Lindner: Mikroskop. Betriebskontrolle in den Gärungsgewerben, *etc.*, 1 Aufl., Paul Parey, Berlin (1895), 4 Aufl., (1905). A. Zeidler: *Wochenschr. f. Brauerei*, **7**, 1213 (1890).
A. albuminosum Lindner (*A. aceti* var. *albuminosum* Lindner)	Lindner	beer	P. Lindner: *ibid.*
A. zeidleri Beijerinck (*Termobact. aceti* Zeidler, *A. lindneri*)	Zeidler	beer	A. Zeidler: *Cent. Bakt.*, 2 Abt., **2**, 729 (1896), M.W. Beijerink: Kral's Sammlung v. Mikroorg., Prague, **7**, (1898).
A. oxydans Henneberg	Henneberg	beer	W. Henneberg: *Cent. Bakt.*, 2 Abt., **3**, 223 (1897).
A. acetosum Henneberg	Henneberg	beer	W. Henneberg: *ibid.*
A. aceti Seifert	Seifert	vinegar	W. Seifert: *Cent. Bakt.*, 2 Abt., **3**, 337 (1897).
A. xylinum Seifert	Seifert	white wine	W. Seifert: *ibid.*
A. pasteurianum var. colorium Beijerinck	Beijerinck	beer	M.W. Beijerinck: *Cent. Bakt.*, 2 Abt., **4**, 209 (1898).
A. acetigenum Henneberg	Henneberg	vinegar	W. Henneberg: *Cent. Bakt.*, 2 Abt., **4**, 14 (1898).
A. industrium Henneberg	Henneberg	yeast, wort	W. Henneberg: *ibid.*, 933 (1898).
A. ascendens Henneberg	Henneberg	vinegar	W. Henneberg: *ibid.*

Table 28 (Continued).

Species	Author	Habitat	Reference
A. rancens Beijerinck	Beijerinck	beer, vinegar	M.W. Beijerinck: *Cent. Bakt.*, 2 Abt., **4**, 209 (1898).
A. pasteurianum var. *variabile* Hoyer	Hoyer	beer	D.P. Hoyer: Bijdrage tot de Kennis van de Azijnbakteriën. Leidner Dissert., Delft (1898), *Die deutsche Essigind.*, **3**, Nr. 1–25 (1899).
A. pasteurianum var. *agile* Hoyer	Hoyer	beer	D.P. Hoyer: *ibid.*
A. rancens var. *zythi* Hoyer	Hoyer	beer	D.P. Hoyer: *ibid.*
A. rancens var. *celiae* Hoyer	Hoyer	beer	D.P. Hoyer: *ibid.*
A. rancens var. *agile* Hoyer	Hoyer	beer	D.P. Hoyer: *ibid.*
A. rancens var. *muciparum* Hoyer	Hoyer	beer	D.P. Hoyer: *ibid.*
A. aceti var. *agile* Beijerinck	Beijerinck	quick-vinegar	M.W. Beijerinck: *Cent. Bakt.*, 2 Abt., **4**, 209 (1898).
A. acidi oxalici Zopf	Zopf	acid, viscid liquid from oak (barrel)	F. Banning: *Cent. Bakt.*, 2 Abt., **8**, 395 (1902).
A. plicatum Fuhrmann	Fuhrmann	wine	F. Fuhrmann: *Cent. Bakt.*, 2 Abt., **15**, 377 (1906). *Beihefte z. Botan. Centralbl.*, **19**, Abt. 1, Heft 1 (1905).
A. vini acetati Henneberg	Henneberg	wine-vinegar	W. Henneberg: *Die deutsche Essigind.*, **10**, 89 (1906).
A. xylinoides Henneberg	Henneberg	wine-vinegar	W. Henneberg: *ibid.*
A. orleanense Henneberg	Henneberg	quick-vinegar	W. Henneberg: *ibid.*
A. schützenbachii Henneberg	Henneberg	quick-vinegar	W. Henneberg: *ibid.*
A. curvum Henneberg	Henneberg	quick-vinegar	W. Henneberg: *ibid.*
A. kützingianum var.	Takahashi	hiochi-saké (spoiled saké)	T. Takahashi: *Bull. Coll. Agr. Tokyo Imp. Univ.*, **7**, 531 (1907).

Table 28 (Continued).

Species	Author	Habitat	Reference
A. aceti Brown var. *Tanezu* Takahashi	Takahashi	tanezu (mother of Japanese vinegar)	T. Takahashi: *J. Tokyo Chem. Soc.*, **29**, 891 (1908).
A. ascendens Henneberg var. *Tanezu* Takahashi	Takahashi	tanezu	T. Takahashi: *ibid.*
A. acetosum Henneberg var. *Tanezu* Takahashi	Takahashi	tanezu	T. Takahashi: *ibid.*
A. aceti Pasteur var. *Tanezu* Takahashi	Takahashi	tanezu	T. Takahashi: *ibid.*
A. xylinoides Hanneberg var. *Tanezu* Takahashi	Takahashi	tanezu	T. Takahashi: *ibid.*
A. rancens Beijerinck var.	Takahashi	hiochi-saké	T. Takahashi: *J. Coll. Agr. Imp. Univ. Tokyo*, **1**, 103 (1909).
A. aceticum rosaceum Holm	Holm	cacao nut	A. Jörgensen: Die Mikroorganismen der Gärungsindustrie, 5 Aufl., Paul Parey, Berlin (1909), s. 109.
A. melanogenum Beijerinck	Beijerinck	beer	M.W. Beijerinck: *Cent. Bakt.*, 2 Abt., **29**, 169 (1911).
A. aceti viscosum Baker, Day and Hulton	Baker, Day and Hulton	viscous beer	J.L. Baker, F.E. Day and H.F.E. Hulton: *Cent. Bakt.*, 2 Abt., **36**, 433 (1913).
A. aceti Hansen var.	Janke	beer	A. Janke: *Cent. Bakt.*, 2 Abt., **45**, 48 (1916).
A. acetosum var.	Janke	beer	A. Janke: *ibid.*
A. aceti Brown var.	Janke	beer	A. Janke: *ibid.*
A. aceti viscosum var.	Janke	beer	A. Janke: *ibid.*
A. pasteurianum var.	Janke	beer	A. Janke: *ibid.*
A. xylinum var.	Janke	beer	A. Janke: *ibid.*
Acetobacter sorbose (?)	Fred, Peterson and Anderson		E.B. Fred *et al.*: *J. Bacteriol.*, **8**, 277 (1923).
A. industrium var.	Yagi and Hashitani	strawberry	K. Yagi and G. Hashitani: *J. Soc. Chem. Ind. Japan*, **26**, 1265 (1923).

Table 28 (Continued).

Species	Author	Habitat	Reference
A. aceti Hansen var.	Miyaji	tanezu	K. Miyaji : *J. Sci. Agr. Soc. Japan*, No. 256, 157, No. 263, 630 (1924).
A. acetigenum var.	Miyaji	tanezu	K. Miyaji : *ibid.*
A. schützenbachii var.	Miyaji	tanezu	K. Miyaji : *ibid.*
A. vini acetati var.	Miyaji	tanezu	K. Miyaji : *ibid.*
A. xylinum Brown var.	Miyaji	tanezu	K. Miyaji : *ibid.*
A. suboxydans Kluyver and de Leeuw	Kluyver and de Leeuw	beer	A.J. Kluyver and F.J. de Leeuw : *Tijdschr. Vergelik. Geneesk.*, **10**, 170 (1924).
A. peroxydans Visser't Hooft	Visser't Hooft	hydrogen peroxide solution	Visser't Hooft : Inaug. Dissert., Delft (1925).
A. dihydroxy- acetonicum	Virtanen and Bärlund	beet juice	A.I. Virtanen and B. Bärlund : *Biochem. Z.*, **169**, 170 (1926).
A. aceticum var.	Vande- caveye	cider	S.C. Vandecaveye : *J. Bacteriol.*, **14**, 1 (1927).
A. gluconicum Hermann	Hermann	kombucha (alga-tea)	S. Hermann : *Biochem. Z.*, **205**, 297 (1929).
A. acetigenoideum Krehan	Krehan	fruit	M. Krehan : *Arch. Microbiol.*, **1**, 493 (1930).
A. ascendens var.	Tanaka	saké-mash	S. Tanaka : *J Agr. Chem. Soc. Japan*, **8**, 962 (1932).
A. aceti Brown var.	Tanaka	fermented cane molasses	S. Tanaka : *ibid.*, **9**, 593 (1933).
A. curvum var.	Tanaka	apple	S. Tanaka : *ibid.*, **10**, 793 (1933).
A. industrium var.	Tanaka	native Formosan wine	S. Tanaka : *ibid.*
Gluconobacter liquefaciens Asai (*Gluconoacetobac- ter liquefaciens* Asai)	Asai	persimmon, fig, banana, strawberry, longan	T. Asai : *J Agr. Chem. Soc. Japan*, **11**, 610, 611– 614 (1935).
G. liquefaciens Asai var.	Tanaka	cane juice	S. Tanaka : *J. Agr. Chem. Soc. Japan*, **12**, 33 (1936).
A. aceti Brown var.	Asai	almond	T. Asai : *J .Agr. Chem. Soc. Japan*, **11**, 507 (1935).

Table 28 (Continued).

Species	Author	Habitat	Reference
A. albuminosum Lindner var.	Asai	tomato, plum, loquat	T. Asai : *ibid.*, 505, 507 (1935).
A. aceti Hansen var.	Asai	grape, pineapple	T. Asai : *ibid.*, 506, 508 (1935).
A. rancens Beijerinck var.	Asai	"new summer orange"	T. Asai : *ibid.*, 509 (1935).
A. rancens var. *agile*	Asai	fig	T. Asai : *ibid.*
A. dioxyacetonicus Asai (*Acetogluco- nobacter dioxy- acetonicus* Asai)	Asai	"sanpôkan," (a kind of citrus fruit) persimmon, pear	T. Asai : *ibid.*, 510, 511–513 (1935).
G. cerinus Asai	Asai	apple, dried date, cherry, "ponkan" (a kind of grapefruit)	T. Asai : *ibid.*, 614, 615–617 (1935).
G. rugosus Asai	Asai	apricot	T. Asai : *ibid.* 617 (1935).
G. cerinus var. *ammoniacus* Asai	Asai	persimmon, mandarin orange, plum, strawberry, longan	T. Asai : *ibid.*, 618– 620, 674–676 (1935).
G. roseus Asai (*A. roseus* Asai)	Asai	persimmon	T. Asai : *ibid.*, 676 (1935).
G. nonoxygluconicus Asai	Asai	pear	T. Asai : *ibid.*, 677 (1935).
G. opacus Asai	Asai	peach	T. Asai : *ibid.*, 667 –678 (1935).
G. scleroideus Asai	Asai	persimmon	T. Asai : *ibid.*,679 (1935).
A. capsulatum Shimwell	Shimwell	ropy beer	J.L. Shimwell : *J. Inst. Brew.*, **42**, 585 (1936).
A. viscosum Shimwell	Shimwell	beer	J.L. Shimwell : *ibid.*, 586 (1936).
A. ketogenum	Utkin	kombucha	L.M. Utkin : Micro- biology, USSR, **6**, 421 (1937).
A. turbidans Cosbie, Tošić and Walker	Cosbie, Tošić and Walker	beer	A.J.C. Cosbie, J. Tošić and T.K. Walker : *J. Inst. Brew.*, **48**, 82 (1942).
G. 2-ketogluconicum	Kondo, Narita	persimmon	H. Kondo and Z. Narita : *J. Pharm. Soc. Japan*, **63**, 289, 301 (1943).

Table 28 (Continued).

Species	Author	Habitat	Reference
G. cerinus Asai var.	Nehira	cherry, dried persimmon	T. Nehira : J. Ferment. Technol. (Japan), 21, 124 (1943).
G. liquefaciens Asai var.	Nehira	strawberry	T. Nehira : ibid.
A. acidum-polymyxa Tošić and Walker	Tošić and Walker	yeast	J. Tošić and T.K. Walker : J. Inst. Brew., 50, 294 (1944).
A. mobile Tošić and Walker	Tošić and Walker	beer	J. Tošić and T.K. Walker : ibid., 295 (1944).
G. liquefaciens var.	Uemura and Kondo	peach	T. Uemura and K. Kondo : J. Agr. Chem. Soc. Japan, 18, 20 (1942).
G. roseus var.	Uemura and Kondo	morning glory flower	T. Uemura and K. Kondo : ibid.
G. cerinus var.	Uemura and Kondo	flowers and fruits	T. Uemura and K. Kondo : ibid.
A. lafarinum Janke	Janke		A. Janke : Arch. Mikrobiol., 15, 116 (1950).
A. lafarinum var. vindovonense Janke	Janke		A. Janke : ibid.
A. paradoxum Frateur	Frateur	wine	J. Frateur : La Cellule, 53, separate volume 3, 335 (1950).
A. lovaniense Frateur.	Frateur	(collection)	J. Frateur : ibid., 336 (1950).
A. rancens var. filamentosum Beijerinck	Frateur	beer	J. Frateur : ibid., 342 (1950).
A. rancens var. turbidans Beijerinck	Frateur	(collection)	J. Frateur : ibid., 346 (1950).
A. rancens var. saccharovorans Beijerinck	Frateur	(collection as A. acetosum)	J. Frateur : ibid., 351 (1950).
A. rancens var. vini Beijerinck	Frateur	(collection)	J. Frateur : ibid., 352 (1950).
A. rancens var. pasteurianum Beijerinck	Frateur	(collection as A. pasteurianum)	J. Frateur : ibid., 354 (1950).
A. aceti var. muciparum Beijerinck	Frateur	(collection)	J. Frateur : ibid., 359 (1950).

Table 28 (Continued).

Species	Author	Habitat	Reference
A. xylinum var. *xylinoides* Beijerinck	Frateur	(collection as *A. xylinum* or *A. xylinoides*)	J. Frateur : *ibid.*, 362 (1950).
A. xylinum var. *maltovorans* Beijerinck	Frateur	(collection as *A. xylinum*)	J. Frateur : *ibid.*, 364 (1950).
A. mesoxydans Frateur	Frateur	(collection)	J. Frateur : *ibid.*, 364 (1950).
A. mesoxydans var. *saccharovorans* Frateur	Frateur	beet juice (collection as *A. dihydroxy-acetonicum*)	J. Frateur : *ibid.*, 366 (1950).
A. mesoxydans var. *lentum* Frateur	Frateur	(collection as *A. orleanense*)	J. Frateur : *ibid.*, 367 (1950).
A. mesoxydans var. *lentum-saccharo-vorans* Frateur	Frateur	(collection as *A. aceti* Hansen)	J. Frateur : *ibia.*, 368 (1950).
A. suboxydans var. *muciparum* Kluyver and de Leeuw	Frateur	beer (collection as *A. suboxy-dans*)	J. Frateur : *ibid.*, 371 (1950).
A. melanogenum var. *maltovorans* Beijerinck	Frateur	(collection as *A. melano-genum*)	J. Frateur : *ibid.*, 372 (1950).
A. melanogenum var. *maltosaccharo-vorans* Beijerinck	Frateur, Kluyver, Simonart, Verhoven, Weisman and Coulon		J. Frateur : *ibid.*, 377 (1950).
A. suboxydans var. *biourgianum* Kluyver and de Leeuw	Frateur	yeast	J. Frateur : *ibid.*, 377 (1950).
A. suboxydans var. *hoyerianum* Kluyver and de Leeuw	Frateur	beer	J. Frateur : *ibid.*, 378 (1950).
A. acidum-mucosum Tošić and Walker	Tošić and Walker	brewery yeast	J. Tošić and T.K. Walker : *J. Gen. Micro-biol.*, 4, 192, (1950).
A. operans	Shimwell	vinegar acetifier	J.L. Shimwell : British Pat., 781, 584 (1957).

Table 28 (Continued).

Species	Author	Habitat	Reference
A. acidophilum Wiame, Dothey and Harpigny	Wiame, Dothey and Harpigny	experimental acetifier	J.M. Wiame, R. Dothey and R. Harpigny: Presented at the International Vinegar Congress, The Hague (1957).
A. alcoholophilus Kozulis and Parsons (*A. rancens*)	Kozulis and Parsons	beer (ale)	J.A. Kozulis and R.H. Parsons: *J. Inst. Brew.*, **64**, 47 (1958). J.G. Carr: *ibid.*, **65**, 338 (1959).
A. estunense Carr	Carr	apple juice, cider	J.G. Carr: *Antonie v. Leeuwenhoek*, **24**, 157 (1958).
A. aurantium Kondo and Ameyama	Kondo and Ameyama	golden-banded lily	K. Kondo and M. Ameyama: *Bull. Agr. Chem. Soc. Japan*, **22**, 370 (1958).
A. albidus Kondo and Ameyama	Kondo and Ameyama	dahlia	K. Kondo and M. Ameyama: *ibid.*
A. suboxydans var. α Kondo and Ameyama	Kondo and Ameyama	strawberry	K. Kondo and M. Ameyama: *ibid.*
A. rubiginosus Kondo and Ameyama	Kondo and Ameyama	loquat	K. Kondo and M. Ameyama: *ibid.* 369 (1958).
A. acetosum var. *nairobiense*	Kulka, Singh, Nattrass, Hall and Walker	selected in East Africa	D. Kulka *et al.*: *J. Sci. Food Agr.*, **8**, 487 (1958).
A. xylinum var. *africanum*	Kulka, Singh, Nattrass, Hall and Walker	selected in East Africa	D. Kulka *et al.*: *ibid.*
A. rancens Beijerinck	Suomalainen	shavings of quick vinegar process	H. Suomalainen: *Brauwissenschaft*, **15**, 356 (1962).

The generic and specific names were given after the original description.

The following organisms are also included in the "Index Bergeyana" (The Williams & Wilkins Company, Baltimore, 1966): *A. oxydans* var. *africanum* Kulka, Nattrass and Walker, *A. nikitinsky* Gamova-Kayukova, *A. acetum* Locke and Main, *A. ketogenum* Walker and Thomas., and *A. lermae* Goncalves de Lima.

REFERENCES

1. F. Lafar: Die Essigsäure-Gärung, Sonderabdruck aus " Handbuch der Technischen Mykologie," Bd. V. Verlag von Gustav Fischer, Jena (1913), p. 1.

2. H. Boerhaave: *Elementa chemiae, Luduni Batavorum.*, **2**, 179, 207 (1732).

3. C. H. Persoon: *Mycologia europaea*, **1**, 96 (1822).

4. L. Desmazières: *Annales des Sciences nat.*, **10**, 42 (1826).

5. F. T. Kützing: *J. prakt. Chem.*, **11**, 385 (1837).

6. R. T. Thompson: *Liebigs Ann.*, **83**, 89 (1852).

7. L. Pasteur: Etudes sur le vinaigre, Paris (1868). In this book one can find his " Mémoire sur la fermentation acétique " abstracted from "Annales scientifiques de l'École Normale supérieure " T. I (1864).

8. W. von Knieriem and A. Mayer: *Landw. Versuchsstation*, **16**, 305 (1873).

9. F. Cohn: Beitr. z. Biol. d. Pflanz., **1**, Heft 2, 127 (1872).

10. E. C. Hansen: *Compt. Rend. de Carlsberg*, **1**, 49, 96 (1879). A. Klöcker: Gesammelte theoretische Abhandlungen über Gärungsorganismen, Verlag von Gustav Fischer, Jena (1911), p. 511.

11. L. Boutroux: *Compt. Rend. Séan. Acad. Sci.*, **91**, 236 (1880).

12. A. J. Brown: *J. Chem. Soc.*, **49**, 172 (1885), **50**, 463 (1886), **51**, 638 (1887).

13. G. Bertrand: *Compt. Rend. Séan. Acad. Sci.*, **122**, 900 (1896), *Ann. Inst. Pasteur*, **12**, 385 (1898).

14. E. Pribram: Klassifikation der Schizomyceten, Franz Deuticke, Leipzig und Wien (1933), p. 75.

15. C. Naegeli: Bot. Zeitung, p. 760 (1857).

16. A. Zeidler: *Cent. Bakt.*, 2 Abt., **2**, 739 (1896).

17. F. Ludwig: *Cent. Bakt.*, 2 Abt., **4**, 867 (1898).

18. S. Orla-Jensen: *Cent. Bakt.*, 2 Abt., **22**, 312 (1909).

19. K. B. Lehmann and R. O. Neumann: Altas und Grundriss der Bakteriologie, I und II, Munchen (1896).

20. M. W. Beijerinck: Proc. Acad. v. Wetenshapp., Amsterdam **2**, 495 (1900).

21. M. W. Beijerinck: Kral's Sammlung v. Mikroorg., **4**, Prague (1898).

22. F. Fuhrmann: *Beihefte Bot. Cent. Orig.*, **19**, Abt. 1, 8 (1905).

23. T. Asai: *J. Agr. Chem. Soc. Japan*, **11**, 686 (1935).

24. E. Leifson: *Antonie v. Leeuwenhoek*, **20**, 102 (1954).

25. Ž. P. Tešić: Int. Bull. Bacteriol. Nomenclature and Taxonomy, **7**, 117 (1957).

26. N. Krassilnikov: Manual of Determining the Bacteria and the Acinomycetes (in Russian), Moscow (1949).

27. S. Orla-Jensen: *Cent. Bakt.*, 2 Abt., **22**, 305 (1909).

28. D. H. Bergey: Bergey's Manual of Determinative Bacteriology, 4th edition. The Williams and Wilkins Company, Baltimore (1934), p. 37.

29. H. D. Bergey, R. S. Breed, E. G. D. Murray and A. P. Hitchens: Bergey's Manual of Determinative Bacteriology, 5th edition. The Williams and Wilkins Company, Baltimore (1939), p. 222.

30. A. R. Prévot: Traité de Systématique Bacterienne, **2**, Dunod, Paris (1961).

31. A. J. Kluyver and C. B. van Niel: *Cent. Bakt.*, 2 Abt., **94**, 369 (1936).

32. O. Rahn: *Cent. Bakt.*, 2 Abt., **96**, 273 (1937).

33. N. A. Krassilnikov: Diagnostik der Bakterien und Aktinomyceten, Gustav Fischer, Jena (1959), p. 411.

34. R. S. Breed, E. G. D. Murray and N. R. Smith: Bergey's Manual of Determinative Bacteriology, 7th edition. The Williams and Wilkins Company, Baltimore (1957), p. 183.

35. T. Asai: *J. Gen. Appl. Microbiol.*, **4**, 289 (1958).

36. J. L. Shimwell: *Antonie v. Leeuwenhoek*, **25**, 66 (1959).

37. E. C. Hansen: *Res. du Meddel. fra Carlsberg Lab.*, **3**, Heft 3, 265 (1894).

38. M. W. Beijerinck: *Cent. Bakt.*, 2 Abt., **4**, 209 (1898).

39. M. W. Beijerinck: *Cent. Bakt.*, 2 Abt., **4**, 214 (1898).

40. D. P. Hoyer: *Deutsch. Essigind.*, **3**, Nr. 1–25 (1899). Bijdrage tot de Kennis van de Azijnbakteriën, Leidener Dissert., Univ. Leiden, Waltman, Delft (1898).

41. L. Pasteur: Etude sur le vinaigre, Paris (1868).

42. W. Henneberg: *Deutsch. Essiging.*, Nr. 14 (1898), Handbuch der Gärungsbakteriologie, 2 Aufl., Bd. 2, Paul Parey, Berlin (1926), p. 191.

43. F. Rothenbach: *Deutsch. Essigind.*, Nr. 35 (1898), Die Untersuchumgsmethoden und Organismen des Gärungessigs und seiner Rohstoffe, Paul Parey, Berlin (1907), p. 207.

44. A. Janke: *Cent. Bakt.*, 2 Abt., **45**, 48 (1916).

45. A. J. Kluyver and F. J. G. de Leeuw: *Deutsch. Essigind.*, Nr. **22**, 175 (1925).

46. S. Hermann and P. Neuschul: *Biochem. Z.*, **233**, 130 (1931).

47. F. Visser't Hooft: Biochemische onderzoekingen over het geslacht Acetobacter, Dissert., Tech. Univ. (Hoogeschool), Meinema, Delft (1925).

48. R. H. Vaughn: *Wallerstein Lab. Commun.*, **5**, 20 (1942).

49. J. L. Shimwell: *Wallerstein Lab. Commun.*, **11**, 27 (1948).

50. R. Y. Stanier: *J. Bacteriol.*, **54**, 191 (1947).

51. A. Janke: *Cent. Bakt.*, 2 Abt., **110**, 728 (1957).

52. J. Frateur: *La Cellule*, **53**, separate volume 3, 288 (1950).

53. S. Tanaka: *J. Agr. Chem. Soc. Japan*, **13**, 844 (1937).

54. T. Asai: *J. Agr. Chem. Soc. Japan*, **10**, 621, 731, 932, 1124 (1934), **11**, 50, 331, 377, 499, 610, 674 (1935).

55. Y. Yasui: *J. Brewing* (Japan), **14**, 680 (1935).

56. T. Uemura and K. Kondo: *J. Agr. Chem. Soc. Japan*, **17**, 747, 825, 919, 1019 (1941), **18**, 15 (1942).

57. T. Uemura, K. Kondo and R. Kodama: *J. Agr. Chem. Soc. Japan*, **18**, 201 (1942).

58. K. Kondo and M. Ameyama: *Bull. Agr. Chem. Soc. Japan*, **22**, 269 (1958).

59. K. Kondo and M. Ameyama: *The Annual Repts. of the Dept. of Agr., Shizuoka Univ.* (Japan), **7**, 149, 153 (1957).

60. J. L. Shimwell: *Antonie v. Leeuwenhoek*, **24**, 187 (1958).

61. T. Asai and K. Shoda: *J. Gen. Appl. Microbiol.*, **4**, 289 (1958).

62. J. Tošić and T. K. Walker: *J. Soc. Chem. Ind.*, **115**, 180 (1946).

63. K. Tanaka: *J. Sci. Hiroshima Univ.* (Japan), Series B, Div. 2, Vol. 3, 94 (1938).

64. T. Takahashi and T. Asai: *J. Agr. Chem. Soc. Japan*, **9**, 55, 369 (1933), *Cent. Bakt.*, 2 Abt., **38**, 286 (1933).

65. K. Aida, M. Fujii and T. Asai: *Proc. Japan Acad.*, **32**, 600 (1956).

66. K. Sakaguchi, T. Asai and Y. Ikeda: *J. Agr. Chem. Soc. Japan*, **20**, 155 (1944).

67. Y. Ikeda: *J. Gen. Appl. Microbiol.*, **1**, 152 (1955).

68. K. Aida, T. Kojima and T. Asai: *J. Gen Appl. Microbiol.*, **1**, 23 (1955).
69. T. Takahashi and T. Asai: *Cent. Bakt.*, 2 Abt., **93**, 248 (1936).
70. K. Aida: *J. Agr. Chem. Soc. Japan*, **28**, 523 (1954), *Bull. Agr. Chem. Soc. Japan*, **19**, 97 (1955).
71. T. Asai, H. Iizuka and K. Komagata: *J. Gen. Appl. Microbiol.*, **10**, 95 (1964).
72. J. L. Shimwell and J. G. Carr: *Antonie v. Leeuwenhoek*, **25**, 353 (1959).
73. A. H. Stouthamer: Koolhydraatstofwisseling van de azijnzuurbacteriën, Thesis, Utrecht (1960).
74. W. Hodgkiss, J. L. Shimwell and J. G. Carr: *Antonie v. Leeuwenhoek*, **28**, 357 (1962).
75. C. Rainbow and G. W. Mitson: *J. Gen. Microbiol.*, **9**, 371 (1953).
76. G. D. Brown and C. Rainbow: *J. Gen. Microbiol.*, **15**, 61 (1956).
77. A. H. Stouthamer: *Antonie v. Leeuwenhoek*, **25**, 241 (1959).
78. J. L. Shimwell and J. G. Carr: *Antonie v. Leeuwenhoek*, **26**, 430 (1960).
79. J. J. Joubert, W. Bayens and J. De Ley: *Antonie v. Leeuwenhoek*, **27**, 151 (1961).
80. J. G. Carr and J. L. Shimwell: *Antonie v. Leeuwenhoek*, **27**, 386 (1961).
81. J. De Ley: *J. Gen. Microbiol.*, **24**, 31 (1961).
82. K. E. Cooksey and C. Rainbow: *J. Gen. Microbiol.*, **27**, 135 (1962).
83. J. W. Riddle, P. W. Kabler, B. A. Kenner, R. H. Bordner, S. W. Rockwood and H. J. R. Stevenson: *J. Bacteriol.*, **72**, 593 (1956).
84. A. W. Scopes: *J. Gen. Microbiol.*, **28**, 69 (1962).
85. J. L. Shimwell: *Antonie v. Leeuwenhoek*, **25**, 49 (1959).
86. J. L. Shimwell and A. F. McIntosch: *Antonie v. Leeuwerhoek*, **28**, 49 (1962).
87. J. De Ley and J. Schell: *J. Gen. Microbiol.*, **33**, 243 (1963).
88. J. De Ley and S. Friedman: *J. Bacteriol.*, **88**, 937 (1964).
89. J. De Ley, I. W. Park, R. Tijtgat and J. Van Ermengen: *J. Gen. Microbiol.*, **42**, 43 (1966).
90. P. J. B. Williams and C. Rainbow: *J. Gen. Microbiol.*, **35**, 237 (1964).
91. T. Leisinger: *Cent. Bakt.*, 2 Abt., **119**, 329 (1965).
92. M. Ameyama, H. Fujisawa and K. Kondo: *J. Agr. Chem. Soc. Japan*, **39**, 427 (1965).
93. T. K. Walker and D. Kulka: *Wallerstein Lab. Commun.*, **12**, 12 (1949).
94. M. Schramm and S. Hestrin: *J. Gen. Microbiol.*, **11**, 123 (1954).
95. A. E. Creedy, P. Jowett and T. K. Walker: *Chem. & Ind.*, 1297 (1954).
96. D. Kulka and T. K. Walker: *Arch. Biochem. Biophys.*, **50**, 177 (1954).
97. R. Steel and T. K. Walker: *J. Gen. Microbiol.*, **17**, 12 (1957).
98. R. Steel and T. K. Walker: *J. Gen. Microbiol.*, **17**, 445 (1957).
99. J. L. Shimwell and J. G. Carr: *J. Inst. Brew.*, **64**, 477 (1958).
100. Z. Gromet, M. Schramm and S. Hestrin: *Biochem. J.*, **67**, 688 (1957).
101. J. L. Shimwell: *J. Inst. Brew.*, **63**, 45 (1957).
102. J. De Ley: *Antonie v. Leeuwenhoek*, **24**, 281 (1958).
103. J. Frateur and P. Simonart: IX Congress Intern. Industr. Agr., Roma (1952).
104. J. De Ley and A. H. Stouthamer: *Biochim. Biophys. Acta*, **34**, 171 (1959).
105. J. L. Shimwell: *J. Inst. Brew.*, **62**, 339 (1956).
106. J. G. Carr: *Nature*, **182**, 265 (1958).
107. J. G. Carr: *Antonie v. Leeuwenhoek*, **24**, 157 (1958).
108. M. W. Beijerinck: *Cent. Bakt.*, 2 Abt., **29**, 169 (1911).

109.   J. L. Shimwell and J. G. Carr: *Antonie v. Leeuwenhoek*, **26**, 169 (1960).
110.   J. Schell and J. De Ley: *Antonie v. Leeuwenhoek*, **28**, 445 (1962).
111.   J. L. Shimwell and J. G. Carr: *Nature*, **201**, 1051 (1964).
112.   T. Asai: *J. Agr. Chem. Soc. Japan*, **11**, 512 (1935).
113.   J. L. Shimwell, J. G. Carr and M. E. Rhodes: *J. Gen. Microbiol.*, **23**, 283 (1960)
114.   K. Komagata :   *J. Gen. Appl. Microbiol.*, **7**, 282 (1961).
115.   H. Iizuka and K. Komagata: *J. Gen. Appl. Microbiol.*, **9**, 73 (1963).
116.   T. Asai: *J. Agr. Chem. Soc. Japan*, **11**, 610 (1935).
117.   J. G. Carr and J. L. Shimwell: *Nature*, **186**, 331 (1960).
118.   M. R. Kimmit and P. J. L. Williams: *J. Gen. Microbiol.*, **31**, 447 (1963).
119.   E. Leifson and R. Hugh: *J. Bacteriol.*, **65**, 263 (1953).
120.   M. Ameyama and K. Kondo: *Agr. Biol. Chem.*, **31**, 724 (1967).
121.   J. A. Fewster :   *Biochem. J.*, **69**, 582 (1958).
122.   Y. Ikeda: *J. Agr. Chem. Soc. Japan*, **24**, 151 (1950).
123.   R. H. Vaughn: *Wallerstein Lab. Commun* , **5**, 5 (1942).
124.   O. K. Sebek and C. I. Randles: *J. Bacteriol.*, **63**, 693 (1952).
125.   K. A. Bisset: Bacteria, The Central Press Ltd., Aberdeen (1952), p. 15.
126.   J. De Ley and V. Vervloet: *Biochim. Biophys. Acta.*, **50**, 1 (1961).
127.   J. M. Wiame, R. Harpigny and R. G. Dothey: *J. Gen. Microbiol.*, **20**, 165 (1959).
128.   J. De Ley: *J. Appl. Bacteriol.*, **23**, 400 (1960).
129.   A. Janke: *Arch. Mikrobiol.*, **41**, 79 (1962).

Part II

BIOCHEMICAL ACTIVITIES OF ACETIC ACID BACTERIA

Chapter 1

NUTRITIONAL REQUIREMENTS

Assimilation of Ammonium Salts as Nitrogen Sources

Hoyer[1] reported on the ability of certain acetic acid bacteria to utilize ammonium salts as a sole source of nitrogen when either ethanol or acetic acid was used as the carbon source. Hoyer's solution was proposed as a synthetic medium for such organisms. It consists of ammonium phosphate 0.1 g, primary potassium phosphate 0.1 g, magnesium phosphate 0.1 g, sodium acetate 0.1 g, ethanol 3 ml and distilled water 100 ml.

Henneberg[2] and Beijerinck[3] found that *A. rancens* and *A. xylinum* were able to utilize nitrate as a nitrogen source when glucose was provided as a carbon source.

Janke,[4] in his classification system for acetic acid bacteria, called those bacteria capable of utilizing ammonium salts as a sole source of nitrogen " haplotrophic acetic acid bacteria," and differentiated these from the " symplotrophic acetic acid bacteria " with more complex organic nitrogen requirements.

The ability to grow in Hoyer's solution was adopted by Vaughn as a key in the classification of acetic acid bacteria. In *Bergey's Manual* (6th ed.), *A. aceti* was differentiated from *A. xylinum* and *A. rancens* on this basis.

Frateur[5] modified Hoyer's solution to the following composition: $(NH_4)_2SO_4$ 1 g; KH_2PO_4 0.9 g; K_2HPO_4 0.1 g; $MgSO_4 \cdot 7H_2O$ 0.25 g; 95% ethanol 30 ml and 0.5 ml of an aqueous solution (1% w/v) of $FeCl_3 \cdot 6H_2O$, in distilled water to make 1 liter. When examined in this modified Hoyer's solution *A. lovaniense* and *A. peroxydans* as well as *A. aceti* grew well.

Leifson[6] reported that *A. aceti* and four strains of *A. xylinum* having the characteristic leathery pellicle grew well in Hoyer's solution, and accordingly claimed that the differentiation of *A. aceti* from other species of acetic acid bacteria by its ability to grow in Hoyer's solution is meaningless.

Hall *et al.*[7] examined the ability of 46 strains of acetic acid bacteria to grow in the modified Hoyer's solution, and found that *A. peroxydans* NCIB 8138 and two strains of *A. aceti* (Pasteur) Beijerinck grew luxuriantly, and that two strains of *A. turbidans* also grew.

Shimwell,[8] however, suggested that inability to grow in the modified Hoyer's medium did not necessarily indicate an inability to utilize ammonium salts under any circumstances. Since these acetic acid bacteria are capable of growing in a glucose medium with ammonium salts as a nitrogen source, Hoyer's solution might not be a test of utilization of an ammonium salt as the sole source of nitrogen, but rather of utilization of ethanol as the sole source of carbon. A positive result, as with *A. aceti*, indicated inorganic N assimilation in addition to ethanol assimilation, but a negative result indicated merely that the carbon source is the limiting factor, for if glucose is supplied instead of ethanol, the ammonium salt is readily utilized as the sole source of nitrogen by many strains of *Acetobacter*.

In Shimwell's experiments, in fact, *A. xylinum* did grow well in a modified Hoyer's glucose medium (containing 2% glucose instead of ethanol). Only three species of those which grew in the modified Hoyer's ethanol medium, *A. peroxydans* NCIB 8618, *A. aceti* NCIB 8621, and *A. lovaniense* NCIB 8620, did not grow in the Hoyer's glucose medium, and even these could grow in the latter in the presence of yeast extract. These species are therefore prototrophic with ethanol, and auxotrophic with glucose as the sole carbon source. Most other *Acetobacter* species have the reverse properties, being auxotrophic with ethanol and prototrophic with glucose. The utilization of ammonium salts in glucose medium was observed also by Hall *et al.*[9] in *A. xylinoides*, *A. acetigenus*, *A. acetosus*, *A. kützingianus*, *A. mobile*, and *A. turbidans*. It was found that, with the exception of *A. gluconicus*, even the strains which did not grow in a glucose-salts medium were able to grow in the complete medium of Dunn *et al.*[10] containing various vitamins, amino acids and purines.

Gray and Tatum[11] also reported that a strain of *A. melanogenus* utilized ammonium nitrogen in the presence of thiamine, pantothenic acid, nicotinic acid, and *p*-aminobenzoic acid. In this respect, Rao and Stokes[12] made an interesting observation, namely that the ability to use ammonium nitrogen is more widespread among the acetic acid bacteria than had previously been suspected. Many strains of *A. suboxydans* and *A. melanogenus* were found to have this ability. However, most strains of *A. suboxydans* required the addition of pantothenic acid, nicotinic acid and *p*-aminobenzoic acid to the medium as growth factors, and most strains of *A. melanogenus* needed thiamine in addition (Table 1).

Table 1.   Defined growth medium for *Acetobacter*
(after Rao and Stokes[12]).

Ethanol, glucose, or other carbon source	2.0 g
$(NH_4)_2SO_4$	0.1 g
or	
Casein hydrolysate (vitamin-free)	0.5 g
(plus 10 mg of L-cysteine and 20 mg of DL-tryptophan)	
Salt solution A	0.5 ml
Salt solution B	0.5 ml
Vitamin mixture : 40 μg each of thiamine, riboflavin, nicotinic acid and pantothenic acid, 80 μg of pyridoxine, 10 μg of p-aminobenzoic acid, 1 μg of folic acid, 0.1 μg of vitamin B_{12}, 0.04 μg of biotin, and 0.5 mg of inositol	
Distilled water to	100 ml
Medium adjusted to pH 6.2–6.5.	

Salt solution A		Salt solution B	
KH_2PO_4	50 g	$MgSO_4 \cdot 7H_2O$	20 g
K_2HPO_4	50 g	NaCl	1 g
H_2O	500 ml	$FeSO_4 \cdot 7H_2O$	1 g
		$MnSO_5 \cdot 4H_2O$	1 g
		HCl, concentrated	1 ml
		H_2O	500 ml

Carbon sources suitable for the growth of these organisms were glucose, arabinose, mannitol, sorbitol, and glycerol. Ethanol, pyruvate, and lactate were not utilized under ordinary conditions, but were utilized in the presence of an unidentified factor or factors found in yeast extract and tryptone besides ammonium nitrogen and casein hydrolysate. Several strains of *A. gluconicus* and *A. rancens* apparently required a new and as yet unidentified factor or factors occurring in yeast autolysate.

Rao and Stokes[13] observed that those strains of *A. suboxydans* and *A. melanogenus* incapable of growing in a chemically defined medium with ethanol as a sole source of carbon, did grow with the addition of a small amount of yeast autolysate, peptone, or liver extract. They suggested that the growth-promoting activity of these biological materials was due to some reducing sugars in them and in fact, glucose, fructose, mannitol or glycerol were found capable of replacing yeast autolysate. These reducing sugars were believed to be necessary for initiating growth of the bacteria, which then would utilize ethanol as an additional source of carbon and energy, and oxidize it to acetic acid. Acetic acid bacteria were grown in an ethanol-glucose medium in the presence of $CH_3\,^{14}CH_2OH$, and ^{14}C from the ethanol was found to be incorporated into cell materials.

Recently Alian[14] reported that a strain identified as *A. aceti* was able to grow in a synthetic medium composed of 1 g/*l* each of $(NH_4)_2HPO_4$, KH_2PO_4, and $MgSO_4$, 10 g/*l* of glucose, 0.2 g/*l* of succinic acid, and 0.5 mg/*l* of pantothenic acid as efficiently as 3° Blg. wort in cell harvest, acid production, and growth rate coefficient.

Shchelkunova[15] reported that both *A. suboxydans* and *A. melanogenus* grew poorly with ethanol as a carbon source, although they possess ethanol oxidizing systems. The addition of a considerable amount of vitamin B complex (acid hydrolysate of baker's yeast), however, promoted the growth. This was accounted for by the buffering activity of the medium, because a small amount of vitamin B complex did not initiate growth in an ethanol-$(NH_4)_2SO_4$ medium, whereas the addition of phosphate- and acetate-buffer solution promoted the growth markedly.

Utilization of Carbon Sources in Relation to Nitrogen Sources

As discussed in the previous section, the assimilation of ammonium salts by acetic acid bacteria in a synthetic medium depends on the kind of carbon source used and, in some species, on the presence of amino acids, vitamins, purine bases, *etc.* as growth factors. Thus, the utilization of carbon sources must be discussed in relation to nitrogen sources or growth factors.

Rainbow and Mitson[16] used the following basic media for inoculation: (a) Glucose 2.0 g, syrupy lactic acid 1.0 g, casein acid hydrolysate (vitamin-free) 0.6 g, $(NH_4)_2SO_4$ 0.1 g, KH_2PO_4 0.3 g, $MgSO_4 \cdot 7H_2O$ 0.2 g, D-biotin 0.08 µg, calcium D-pantothenate 100 µg, thiamine 100 µg, pyridoxine-HCl 100 µg, nicotinic acid 100 µg, riboflavin 10 µg, *p*-aminobenzoic acid 10 µg, and the trace element solution of Emery *et al.*[17] 0.05 ml in 100 ml final volume. Adjustment to pH 5.8 was made with a KOH solution. (b) As above, but lactic acid omitted and 1.0 mg adenine/100 ml added. *A. acidum-mucosum* NCTC 6429, *A. mobile* NCTC 6428 and *A. suboxydans* NCTC 7069 were inoculated into medium (a) and *A. capsulatus* NCTC 4943, *A. gluconicus* NCTC 4739, *A. turbidans* NCTC 6249 and *A. viscosus* NCTC 6426 were inoculated into medium (b). The cells were harvested after 40–96 hr at 28°C, and were used to test carbon source requirement. For the experiments, the concentration of casein acid hydrolysate was reduced to one half that in the basal medium, and glucose and lactic acid were omitted. D-Glucose, D-fructose, D-galactose, D-mannose, D-arabinose, D-ribose, D-xylose, lactose, maltose, sucrose, D-mannitol,

D-sorbitol, erythritol, glycerol, acetate, citrate, fumarate, α-ketoglutarate, lactate, L-malate, or succinate was added (10 mg/ml) instead.

The results showed lactate to be the best source of carbon for the growth of *A. acidum-mucosum*, *A. mobile*, and *A. suboxydans* (*A. mobile* group). Sugars and sugar alcohols promoted relatively poor growth or no growth. On acids other than lactic acid *A. acidum-mucosum* grew only after a prolonged incubation, and *A. suboxydans* grew to a smaller extent on acetic acid. *A. mobile* also grew on succinate and malate to a smaller extent. Glycerol promoted good growth after an initial lag period in *A. mobile* and *A. suboxydans*.

A. capsulatus, *A. gluconicus*, *A. turbidans* and *A. viscosus* (*A. viscosus* group) failed to grow in the organic acids, but grew well on glucose, mannitol and (except for *A. gluconicus*) sorbitol, erythritol or glycerol. Fructose was a less effective substrate than glucose. *A. capsulatus* was the only strain that grew well on maltose.

For these seven organisms casein acid hydrolysate was a suitable source of nitrogen. In the absence of lactate the *A. mobile* group did not utilize ammonium salts as a sole source of nitrogen, but in its presence it did. The *A. viscosus* group, on the other hand, did not assimilate ammonia with or without lactate.

Brown and Rainbow[18] determined the nutritional patterns of 28 strains of acetic acid bacteria on lactate and glucose. Two groups, one with a predominant lactate metabolism (lactaphilic group) and the other with a predominant glucose metabolism (glycophilic group), were distinguished. Lactaphilic strains oxidized lactate and acetate and formed new ninhydrin-reacting substances when incubated with proline, glutamate, and aspartate. Glycophilic strains required nicotinate, pantothenate, and occasionally *p*-aminobenzoate for growth and did not form any new ninhydrin-reacting substances when incubated with proline, glutamate or aspartate. For the characteristics and classification of each group see Chapter I.

Fewster[19] examined various carbohydrates for their effectiveness as energy sources for growth in *A. suboxydans* ATTC 621, using a complex medium (containing peptone and yeast extract) and a semi-defined medium (containing casein acid hydrolysate, amino acids, minerals and growth factors). Maximum growth was obtained with D-mannitol and D-sorbitol, and good growth with glycerol and fructose and with glucose in the presence of $CaCO_3$. Ca D-gluconate, L-sorbose, D-mannose, D-galactose, D-xylose, L-arabinose, and sucrose gave only negligible growth.

The poor growth on D-gluconate supports the opinion of Stubbs *et al.*[20] that this organism's growth energy, on glucose, is mainly derived from the oxidation of glucose to gluconate, and the further oxidation of gluconate to a mixture of ketoacids is carried out only by non-proliferating cells.

All the carbohydrates which provided good energy sources for growth were extensively oxidized by washed cell suspensions of the organism. The converse was not true, however; gluconate and sorbose, though they were poor promoters of growth, were apparently extensively oxidized as intermediates in the oxidation of glucose and sorbitol, respectively.

Loitsyanskaya[21] reported that *A. schützenbachii, A. curvum* and *A. aceti*, which belong to a haplotrophic group active in the quick vinegar process, showed a much poorer growth with glucose as a carbon source than with ethanol, acetate or lactate. This was found to be true even with the addition of various nitrogen sources or growth factors.

Interestingly, Rasumofskaja and Bjelousowa[22] found CO_2 assimilation in *A. schützenbachii*. When the organism, freshly isolated from quick vinegar, was grown in a synthetic medium with ethanol as a carbon source, the removal of CO_2 resulted in delayed growth or even failure of growth. After preservation in a laboratory for a long period, however, the organism grew on a malt extract agar without ethanol in the absence of CO_2. Gsur[23] reported the incorporation of radioactivity into bacterial cell materials when a certain *Acetobacter* was grown in the presence of $^{14}CO_2$ in submerged culture.

According to King and Cheldelin,[24] intact cells of *A. suboxydans* ATCC 621 can oxidize sorbose, lactic and pyruvic acids, dihydroxyacetone and acetaldehyde, but cannot use these compounds as carbon sources for growth. Neither α-glycerol phosphate, glucose-6-phosphate, β-glycerol phosphate, fructose-1, 6-diphosphate, citraconic acid, mesaconic acid, itaconic acid, α-ketoglutaric acid, malic acid, fumaric acid, succinic acid, nor malonic acid was utilized as a source of carbon and energy.

Hall *et al.*[7] studied the assimilation of glucose and ethanol in acetic acid bacteria, using a modified Hoyer's medium which contained either ammonium sulfate or ammonium phosphate as a nitrogen source. Twenty-three strains representing 11 species utilized glucose, 9 strains representing 5 species utilized ethanol and 23 strains representing 11 species utilized neither glucose nor ethanol as a sole source of carbon. Three strains grew better on ethanol than on glucose, while the growth of three others on glucose was inhibited by the addition of ethanol.

Recently Asai *et al.*[25] studied the assimilation of glucose and ethanol in 25 strains of *Acetobacter* representing 11 species, and 29 strains of *Gluconobacter* representing 13 species. These were grown with shaking on a

Table 2.   Chemical composition of Foda and Vaughn's synthetic
medium modified (after Asai *et al.*).

Component	Amount in 100 ml	
Thiamine-HCl	100 μg	Salt solution A:
Nicotinic acid	100	K_2HPO_4 50 g
Ca DL-pantothenate	100	KH_2PO_4 50 g
p-Aminobenzoic acid	50	Redistil. water to 500 ml
L-Valine	10 mg	
L-Isoleucine	10	Salt solution B:
L-Alanine	10	$MgSO_4 \cdot 7H_2O$ 20 g
L-Proline	10	NaCl 1 g
L-Histidine	10	$FeSO_4 \cdot 7H_2O$ 1 g
L-Cystine	10	$MnSO_4 \cdot 4H_2O$ 1 g
L-Glutamic acid	10	Redistil. water to 500 ml
$(NH_4)_2SO_4$	1.0 g	
Salt solution A	0.5 ml	
Salt solution B	0.5 ml	
Glucose or ethanol (or 0.5 g each if a mixture of both is used)	1.0 g	

N.B.:   1) The water should be redistilled three times with quartz equipment.
2) pH should be adjusted to 6.0.
3) For the control experiment, 0.3 g of yeast extract should be added.
4) Foda and Vaughn's defined medium for growth of acetic acid bacteria is shown in Table 4.

modified Foda and Vaughn's synthetic medium[26] (Table 2) containing vitamins, amino acids, and ammonium sulfate.   As carbon sources ethanol, glucose, and ethanol plus glucose were used.   The strains which assimilated ethanol (Brown and Rainbow's lactaphilic group) were found to belong to the genus *Acetobacter*.   Only one *Acetobacter* strain (*A. capsulatus*) utilized glucose as a sole source of carbon.   Those strains which assimilated glucose (glycophilic group) were found solely in the *Gluconobacter*.   No *Gluconobacter* strains examined could utilize ethanol as a sole source of carbon.   It was concluded that although the selective assimilation of ethanol or glucose is a characteristic proper to *Acetobacter* or *Gluconobacter*, this property should not be considered as a criterion for distinguishing the two genera since there are *Acetobacter* strains which do not assimilate ethanol, and *Gluconobacter* strains which do not assimilate glucose.   Many of these strains could grow with the addition of both glucose and ethanol, but several strains did not grow in the synthetic medium even in the presence of the two carbon sources. *G. cerinus* 24 did grow on glucose when adenine was added to the medium.

Acetic acid bacteria which are capable of growing with ethanol as a source of carbon and energy can be shown to oxidize ethanol when measured manometrically or by other means, but the converse is not always true. The same situation can be seen with glucose.

Concerning the role of acetate on the growth of *A. gluconicus* 2 G, Goldman and Litsky[27] reported that sodium acetate at a concentration of 1.0% replaced yeast autolysate for the promotion of growth in a defined medium. They observed that both the acetate level and the reducing capacity of the medium decreased during cultivation. They then grew the organism in a casamino acids (vitamin-free)-mineral salts-medium containing 1% sodium acetate (methyl labeled) and 2% glucose for 120 hr at 28°C. The radioactivity distribution after cultivation was determined as follows: metabolic CO_2 56%, cells 11%, and medium 33%. Glucose was also utilized during cultivation. Accordingly they suggested that acetate acts, along with glucose, as a substrate for the growth of this organism.

Growth Factor Requirement

Vitamin requirement. In 1942 and 1943, Peterson's group carried out a series of experiments on the specific essential nutrients necessary for the growth of acetic acid bacteria, especially of *A. suboxydans*. Lampen *et al.*[28] and Underkofler *et al.*[29] found that *A. suboxydans* ATCC 621, when cultivated in a medium containing casein hydrolysate as a nitrogen source and glycerol as a carbon source, required pantothenic acid, *p*-aminobenzoic acid and nicotinic acid. Owing to the studies of Landy and Dicken[30] and Landy and Streightoff,[31] *A. suboxydans* ATCC 621 has become a test organism for the microbiological assay of *p*-aminobenzoic acid. Marshall and Postage[32] reported on an *A. suboxydans* strain which does not require *p*-aminobenzoate and is resistant against inhibition by sulfonamides.

Gray and Tatum[11] observed that the above three vitamins are required also by a strain of *A. melanogenus*. In the same paper they noted the requirement of proline by some of the X-ray induced mutants of *A. melanogenus* and of serine or glycine and adenosine by other mutants.

Foda and Vaughn[26] examined six strains of *A. melanogenus* and one strain each of *A. oxydans* and *A. rancens* for their vitamin requirements when grown in Foda and Vaughn's synthetic medium. All cultures required pantothenic acid, p-aminobenzoic acid, and nicotinic acid. *A. melanogenus* and *A. rancens* required thiamine in addition. According to Hall et al.,[33] *A. suboxydans* ATCC 621, when cultivated on Dunn's synthetic medium,[10] required nicotinic acid at pH 5.0 and 6.0 with glycerol as a carbon source. With glucose as a carbon source, however, nicotinic acid was not essential at pH 6.0 and was essential at pH 5.0.

For *A. xylinum* Litsky et al.[34] found that p-aminobenzoic acid was the only vitamin required in a synthetic medium containing casein hydrolysate as an amino acid source. *A. mobile* also required p-aminobenzoic acid (Rainbow and Mitson[16]).

Koft and Morrison[35] claimed that *A. suboxydans* formed an unknown metabolite from p-aminobenzoate which stimulated the growth of PABA-requiring lactic acid bacteria. The substance reduced the lag period in the growth of lactic acid bacteria by 30% and increased the cell yield by 50%; it was neither a member of the folic acid group nor a PABA-like compound.

In the experiments of Rainbow and Mitson[16] and of Brown and Rainbow[18] on the growth of glycophilic acetic acid bacteria, i.e. *A. capsulatus*, *A. turbidans*, *A. viscosus*, and *A. gluconicus*, nicotinic acid, pantothenic acid, and to a lesser extent p-aminobenzoic acid showed a growth-promoting effect in a glucose-casein hydrolysate-ammonium salt medium. The optimum concentration of p-aminobenzoic acid was lowered by the addition of purine bases, especially of hypoxanthine. Lactaphilic strains, on the contrary, did not require exogenous growth factors.

Biotin was reported by Litsky et al.[34] to be essential for the growth of *A. suboxydans*, but Karabinoos and Dicken[36] attributed the phenomenon to the possible contamination of biotin with nicotinic acid. According to Hall et al.,[9] however, biotin showed a good effect on the growth of *A. pasteurianus*, *A. ascendens*, and *A. acidum-mucosum*, and Baetsle[37] also reported on the effect of biotin on acetic acid bacteria.

Vitamin requirements in a synthetic medium used by Underkofler et al.[29] were investigated from the taxonomic viewpoint and led to the following findings by Ameyama and Kondo.[38] (The composition of the minimal medium is shown in Table 3.) *Gluconobacter* generally required pantothenic acid, and some strains required thiamine, nicotinic acid and p-aminobenzoic acid in addition, although a difference of requirements between strains was seen even in the same species. Riboflavin, pyridoxine, vitamin B_{12}, folic acid, biotin, and inositol were unnecessary for growth.

Table 3. Composition of minimal medium for *Acetobacter* growth
(after Underkofler *et al.*).

Glucose	0.5%
Amino acid	total 400 mg%
L-form	20 mg% each:

glycine, valine, leucine, isoleucine, serine, threonine, cystine, cysteine, aspartic acid, glutamic acid, arginine, lysine, histidine, phenylalanine, tryptophan, proline

DL-form	40 mg% each:

alanine, methionine

Bases	1 mg% each:

adenine, guanine, uracil, xanthine

Minerals

KH_2PO_4 50 mg%, K_2HPO_4 50 mg%, $MgSO_4 \cdot 7H_2O$ 20 mg%, NaCl 1 mg%, $FeSO_4 \cdot 7H_2O$ 1 mg%, $MnSO_4 \cdot 4H_2O$ 1 mg%

pH	6.0

Neither *Acetobacter* nor " intermediate " strains required vitamins for growth.

Later experiments[39] also gave the same results when ethanol, glycerol, or glucose was used as the carbon source. Jlli *et al.*[40] reported that four strains of *Acetomonas* (*Gluconobacter*) required pantothenic acid when glucose was the carbon source, and two strains required nicotinic acid in addition. Only one of ten strains of *Acetobacter* tested required *p*-aminobenzoic acid when glucose was the carbon source.

Goldman *et al.*[41] found an unknown substance or subtances in yeast autolysate essential for the growth of *A. gluconicus* 2G. This compound was stable in acid or alkali, more soluble in ethanol or acetone than in chloroform, extractable with ethyl ether at pH 2.0 but not at pH 10.0, and absorbable on anionic resins. Thus it was believed to be an acidic substance, but was not definitely identified. It is apparently different from the reducing sugar found by Rao and Stokes.[13]

Nicotinic acid, required for the growth of *A. suboxydans*, is necessary, according to Sarett and Cheldelin,[42] for the synthesis of pyridinenucleotide enzymes and alcohol dehydrogenase.

p-Aminobenzoic acid as a constituent of folic acid is essential for the synthesis of purine and pyrimidine bases, and hence of nucleotides, and is also concerned with the synthesis of dehydrogenases, an important group of enzymes for acetic acid bacteria.

Pantothenate is required for the growth of *A. ascendans*, *A. capsulatus*,

A. pasteurianus, A. turbidans (Hall *et al.*[9]) and *A. gluconicus* (Rainbow and Mitson[16]). It is an important constituent of the CoA molecule. The pantothenic acid requirement of acetic acid bacteria was studied with *A. suboxydans* ATCC 621 in relation to studies on reaction sequence of CoA synthesis.

King *et al.*[43,44] isolated an active acid-labile conjugate of pantothenate (PAC) from heart muscle. This compound showed an activity twice that of pantothenate for the growth of *A. suboxydans*, but unlike CoA, did not catalyze acetylation reactions. Novelli *et al.*[45] found that CoA itself was more active than pantothenate for the growth of *A. suboxydans* and that the cleavage products were also active. They believed that the PCA reported by King *et al.* was probably one of these cleavage products. Later Novelli[46] reported the presence, in the degradation products of CoA by potato pyrophosphatase, of a substance with an activity ten-fold that of pantothenate. He also isolated from the acid hydrolysate of CoA an effective substance which contained phosphate but not adenine or ribose, and called the substance " Acetobacter stimulatory factor (ASF)." The artificially synthesized phosphate of pantothenate or its cleavage products, however, did not support the growth of *A. suboxydans*, according to Baddiley and Thain[47] and King *et al.*[48,49,50]

Baddiley *et al.*[51] reported for the first time on an effective phosphate of pantothenate, pantetheine-4'-phosphate. According to Baddiley and Mathias[52] the synthetic pantothenyl-4'-phosphate had an activity 40% that of a nonphosphorylated compound.

Brown and Snell[53] found a substance of defined compotion, pantothenyl cysteine, which had a strong action on *A. suboxydans* ATCC 621. This substance was believed to be an intermediate of pantetheine synthesis from pantothenate. They[54] also found that the growth supporting activity of pantothenyl cystine and pantethine with *A. suboxydans* ATCC 621 in a pantothenic acid-free medium greatly increased with autoclaving of the medium or with the presence of reducing agents. Maximum activity was obtained with mercuric mercaptides of pantothenyl cysteine and pantetheine. It was concluded that –SH and not –S–S– compounds serve as growth factors. The activities of 4'-phosphopantetheine and CoA also increased with the addition of reducing agents.

Comparative growth-supporting activities of these pantoic acid derivatives were given as follows: CoA 10.8, 4'-phosphopantetheine 11.8, pantothenylcysteine Hg mercaptide 7.6, pantetheine Hg mercaptide 7.2, D-pantothenic acid 1.0, D-pantoic acid 1.0, D-pantoyl lactone 0. (Fig. 1)

The range of activity was in accord with the view that biosynthesis

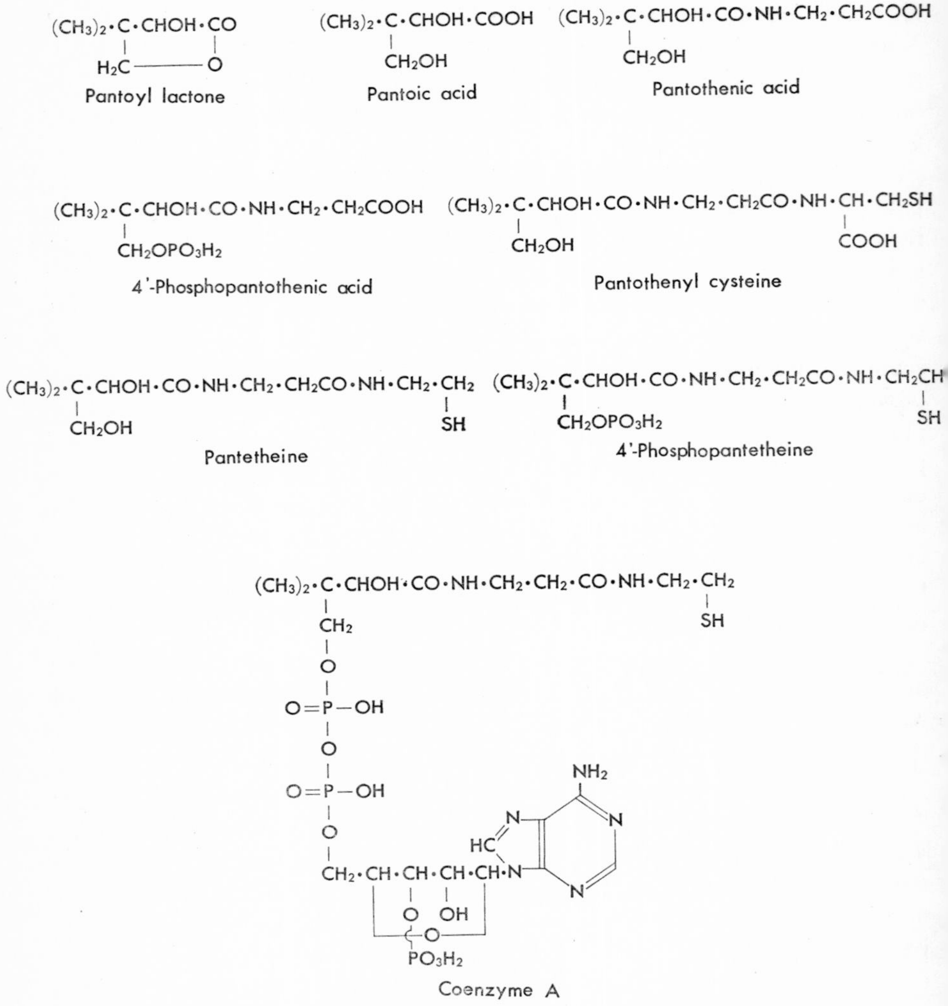

$(CH_3)_2 \cdot C \cdot CHOH \cdot CO$
$H_2C \text{———} O$

Pantoyl lactone

$(CH_3)_2 \cdot C \cdot CHOH \cdot COOH$
CH_2OH

Pantoic acid

$(CH_3)_2 \cdot C \cdot CHOH \cdot CO \cdot NH \cdot CH_2 \cdot CH_2COOH$
CH_2OH

Pantothenic acid

$(CH_3)_2 \cdot C \cdot CHOH \cdot CO \cdot NH \cdot CH_2 \cdot CH_2COOH$
$CH_2OPO_3H_2$

4'-Phosphopantothenic acid

$(CH_3)_2 \cdot C \cdot CHOH \cdot CO \cdot NH \cdot CH_2 \cdot CH_2CO \cdot NH \cdot CH \cdot CH_2SH$
$CH_2OH \qquad\qquad\qquad\qquad\qquad COOH$

Pantothenyl cysteine

$(CH_3)_2 \cdot C \cdot CHOH \cdot CO \cdot NH \cdot CH_2 \cdot CH_2CO \cdot NH \cdot CH_2 \cdot CH_2$
$CH_2OH \qquad\qquad\qquad\qquad\qquad\qquad\qquad SH$

Pantetheine

$(CH_3)_2 \cdot C \cdot CHOH \cdot CO \cdot NH \cdot CH_2 \cdot CH_2CO \cdot NH \cdot CH_2CH$
$CH_2OPO_3H_2 \qquad\qquad\qquad\qquad\qquad\qquad SH$

4'-Phosphopantetheine

$(CH_3)_2 \cdot C \cdot CHOH \cdot CO \cdot NH \cdot CH_2 \cdot CH_2 \cdot CO \cdot NH \cdot CH_2 \cdot CH_2$
$CH_2 \qquad\qquad\qquad\qquad\qquad\qquad\qquad\qquad SH$
O
$O = P - OH$
O
$O = P - OH$
O
$CH_2 \cdot CH \cdot CH \cdot CH \cdot CH \cdot N$
$O \quad OH$
O
PO_3H_2

Coenzyme A

Fig. 1. Pantoic acid and its derivatives

of CoA may proceed successively through pantoate, pantothenate, pantothenyl cysteine, pantetheine, and 4'-phosphopantetheine. It was thus concluded that the stimulatory factor for the growth of *A. suboxydans* in natural materials (pantothenic acid conjugates) is a mixture of such compounds.

King and Cheldelin[55] also reported the "stimulatory factor" to be multiple in nature.

Brown *et al.*,[56] from further studies, concluded that the activity of the compounds was attributable to 1) easy absorption, 2) the presence of –SH groups which permit efficient absorption despite the ionized carboxyl group, and 3) the ability of *A. suboxydans* to easily hydrolyze the conjugates of pantothenic acid. Sarett and Cheldelin[42] described the use of *A. suboxydans* in the bioassay of pantoyl lactone. Hall *et al.*[57] observed the growth-supporting activity of pantoyl lactone on other *Acetobacter* species in a pantothenic acid deficient medium. The activity was more apparent at an initial pH of 6.2 than at a lower pH.

Amino acid requirement. Stockes and Larsen[58] reported on the replacement of vitamin-free casein hydrolysate by a mixture of six amino acids in the growth of a strain of *A. suboxydans*. Valine, isoleucine and alanine were the essential amino acids, but with these alone growth did not take place. Addition of histidine gave a slight growth and addition of cystine or methionine supported a fairly good growth. With further addition of proline, growth increased to the level of that on casein hydrolysate. Development, however, was not so rapid and somewhat less extensive than with yeast extracts. It was also observed that ammonium sulfate was not adequate for growth, but that at suboptimal concentrations of the required amino acids, it stimulated growth.

In experiments by Foda and Vaughn[26] *A. oxydans* was shown to require the above six amino acids, *viz.* valine, isoleucine, alanine, cystine, histidine, and proline. Thus *A. suboxydans* resembles *A. oxydans*. *A. rancens* gave the best growth after further addition of aspartic or glutamic acid. Interestingly, all six strains of *A. melanogenus* examined did not require any amino acids. The cultures of tested strains were able to utilize ammonium salts as a source of nitrogen when supplied in the presence of other required compounds. Gray and Tatum[59] obtained X-ray induced mutants of *A. melanogenus* which lost the ability to synthesize the essential amino acids.

In contrast to the results of Stokes and Larsen,[58] Hall *et al.*[33] found that the amino acid requirement of *A. suboxydans* ATCC 621 is strongly dependent on the pH of the medium. In a glucose medium requirements are most complex at the lower pH values. In a pH range of 5.0–5.3 the alanine requirement changed from essential to merely stimulatory, and in a glycerol medium the valine requirement changed from non-essential to essential. Valine was not required in a glucose medium, but in a glycerol medium it was required at a pH above 5.3, indicating a function of glucose in valine synthesis not possessed by glycerol.

Tepper and Litsky[60] reported that two strains of *A. xylinum* required a combination of isoleucine, valine and alanine for growth.

According to experiments by Kerwar *et al.*,[61] extracts of *A. suboxydans* ATCC 621 can synthesize valine and isoleucine *via* acetolactate and acetohydroxybutyrate, respectively. Excess valine in the synthetic medium inhibited growth. The inhibition was reversed by the addition of isoleucine. Threonine deaminase was inhibited by valine and isoleucine but not by leucine. Repression of the deaminase by isoleucine but not by valine was indicated. Thus it was suggested that valine prevents growth

Table 4. Defined media for the growth of *A. melanogenus*, *A. oxydans* and *A. rancens* (after Foda and Vaughn).

Component	Amount in 100 ml		
	A. melanogenus	*A. oxydans*	*A. rancens*
KH$_2$PO$_4$	0.05 g	0.05 g	0.05 g
K$_2$HPO$_4 \cdot$3H$_2$O	0.05 g	0.05 g	0.05 g
MgSO$_4 \cdot$7H$_2$O	0.02 g	0.02 g	0.02 g
FeSO$_4 \cdot$7H$_2$O	0.001 g	0.001 g	0.001 g
MnSO$_4 \cdot$4H$_2$O	0.001 g	0.001 g	0.001 g
NaCl	0.001 g	0.001 g	0.001 g
(NH$_4$)$_2$SO$_4$	1.0 g	1.0 g	1.0 g
Pantothenic acid	100 μg	100 μg	100 μg
p-Aminobenzoic acid	20 μg	20 μg	20 μg
Nicotinic acid	20 μg	20 μg	20 μg
Thiamine-HCl	100 μg	—	100 μg
L-Valine*	—	20 mg	20 mg
L-Isoleucine	—	20 mg	20 mg
L-Alanine	—	20 mg	20 mg
L-Proline	—	20 mg	20 mg
L-Histidine	—	20 mg	20 mg
L-Cystine	—	20 mg	20 mg
L-Aspartic or glutamic acid	—	—	20 mg
Glucose	1.0 g	1.0 g	1.0 g
Redistilled water	to 100 ml	to 100 ml	to 100 ml
Final pH**	6.0±0.1	6.0±0.1	6.0±0.1

* With the DL-forms of the amino acids, 40 mg was used.

** Adjusted with NaOH so that the pH after sterilization of media at 15 lbs steam pressure for 15 min corresponded to this value.

through a false feedback inhibition of threonine deaminase, thereby limiting isoleucine biosynthesis.

According to Jlli et al.,[40] two strains of Acetobacter required alanine and one strain of Acetomonas (Gluconobacter) required cystine when glucose was the carbon source.

With ethanol as the carbon source, four of seven Acetobacter strains required alanine.

The amino acid requirement of Gluconobacter was investigated by Yamada et al.[62] using glucose, fructose, sorbitol or ethanol as an independent carbon source. All strains tested (40 in all) required glutamic acid essentially or accessorily; the strains were divided into the following three groups:

1) Strains which essentially required glutamic acid in a glucose medium: brown pigment-producing strains, viz. G. melanogenus, G. rubiginosus and A. aurantius (" intermediate " strain IFO 3245).

2) Strains which accessorily required glutamic acid in a glucose medium: G. suboxydans, G. roseus and G. cerinus.

3) Strains which essentially required glutamic acid in an ethanol

Table 5.   Basal growth medium for A. suboxydans (after Schamberger, thesis, Oregon State Univ., Corvallis, 1960).

Component	Amount
L-Histidine	400 mg
L-Glutamic acid	400 mg
L-Proline	400 mg
Ammonium sulfate	1 g
Ca pantothenate	20 mg
p-Aminobenzoic acid	20 mg
Nicotinic acid	20 mg
Glycerol	50 g
KH_2PO_4	500 mg
K_2HPO_4	500 mg
NaCl	150 mg
$MnSO_4$	10 mg
$FeSO_4 \cdot 7H_2O$	10 mg
Uracil	15 mg
Guanine sulfate	15 mg

All quantities per liter ; final pH 6.0.

medium: brown pigment-producing and acetate-oxidizing strains, *viz*. *G. liquefaciens* and two other strains ("intermediate" strains IAM 1835 and 1836).

The third group generally exhibited strong growth when fructose was given as a carbon source. Growth in glucose or in sorbitol was poor.

Recently Asai *et al.*[63] found that a strain of *Gluconobacter*, *G. cerinus* 24, which was unable to grow in a modified Foda and Vaughn's synthetic

Table 6. Some media for maintenance of *Acetobacter* cultures. Formulas employed by the American Type Culture Collection (Manual of Microbiological Methods, McGraw-Hill Co., New York [1957], p. 108).

---

1. *Acetobacter Agar (Glucose)*

Autolyzed yeast	10 g
CaCO$_3$	10 g
Agar	15 g
Distilled water	1,000 ml
Heat to 100°C, and add	
Glucose	3 g

In tubing, the CaCO$_3$ should be distributed evenly between tubes. After autoclaving, the tubes should be shaken, then cooled quickly and slanted so as to keep the CaCO$_3$ in suspension.

2. *Acetobacter Agar with Liver Extract*

Liver extract	100 ml
Tryptone	5 g
Agar	20 g
Distilled water	900 ml
Heat to 100°C, and add	
Glucose	20 g
CaCO$_3$	10 g

Observe precautions as in formula 1 to keep CaCO$_3$ evenly suspended.

3. *Acetobacter Agar (Mannitol)*

Yeast extract	5 g
Peptone	3 g
Agar	15 g
Distilled water	1,000 ml
Heat to 100°C, adjust pH to 7.4 and add	
Mannitol	25 g

medium with ethanol, glucose or a mixture of both as the carbon source, did grow in the presence of adenine. Adenosine was less effective as the growth supporting factor.

Brown and Fabian[64] reported on the growth promoting effect of mustard oil for *Acetobacter* and Fulmer *et al.*[65] reported on the substitution of alfalfa extracts for yeast autolysate in maintaining growth and chemical activities in *A. suboxydans*.

Some media for the growth or preservation of acetic acid bacteria, excluding those previously mentioned in the text, are listed in Tables 4, 5 and 6.

Chapter 2

PRODUCTION OF ANTIBACTERIAL
SUBSTANCES

Steel and Walker[66] reported on the spontaneous occurrence of cellu-loseless mutants of the cellulose-producing *A. acetigenus* NCIB 8132, 5346, and one strain of *A. xylinum* var. *africanum* (NCIB 7029). They[67] then carried out comparative studies between NCIB 8132, NCIB 5346 and NCIB 7029 strains and the 26 celluloseless mutants obtained from these three parent strains. Although the parent strains oxidized ethanol to acetic acid, the mutants did not. The optimal pH for growth was also shifted toward the alkaline side.

These workers[68, 69] found that the parent strains produced, in culture filtrates, a substance which inhibited the growth of all the mutants and of *Proteus vulgaris*. The yield of antibiotic produced by NCIB 8132 was not influenced by the nitrogen sources, but was greatly influenced by the carbon sources. The medium which promoted the highest yield was a glucose-yeast extract-peptone medium. The filtrate of the medium after incubation at 30°C for 10 days inhibited the growth of *Proteus vulgaris* by 50% with a 1 : 44 dilution.

The antibiotic is very alkali-labile and is destroyed if kept at pH 10, 18°C for 2 min. The antibiotic activity found in a culture filtrate of NCIB 8132 was not affected by $CaCO_3$, charcoal, cation- or anion-exchange resin treatments, but it decreased when the filtrate was extracted with *n*-butanol, chloroform, or methyl ethyl ketone at pH 3. The activity found in a culture filtrate of *A. xylinoides* NCIB 4940 reacted quite differently from that of NCIB 8132. The antibiotics produced by these two strains are thus different from each other. Antibiotic production did not occur when $CaCO_3$ was added to the culture medium.

Lethal action by an *Acetobacter* on yeasts has been reported by Gilliland and Lacey.[70] The isolated *Acetobacter* sp., a close relative of *A. mesoxydans*, prevented the growth of yeasts and caused them to die when both organisms were present in the ratio of 1 : 1. The lethal action was marked in beer, less marked in synthetic medium and absent in pH 5 phosphate buffer solution. The *Acetobacter* also showed a lethal action on strains of the genera *Pichia, Schizosaccharomyces, Zygosaccharomyces, Torula, Candida,* and *Brettanomyces* and on several species of the genus *Saccharomyces.* Lethal action was only observed under anaerobic conditions. Treatment with heat, chloroform, or ultraviolet rays or disintegration of the cells failed to kill the test organisms.

Addendum.

A new bacteriophage specific for *Acetobacter* species in rotting apples has been recently isolated and purified by Bradley.[71] It was found to resemble coliphage T3, but its morphology was of particular interest because of the prominence of the head capsomeres and the three-pronged tail.

Chapter 3

OXIDATION OF ETHANOL TO ACETIC ACID

Introduction

The formation of acetic acid from ethanol by acetic acid bacteria was first noted by Pasteur;[72] the reaction has since been termed " acetic acid fermentation," and is considered characteristic of these bacteria. The fermentation requires the presence of oxygen.

The highest ethanol concentration which allows the formation of acetic acid varies with the species and ranges, according to Henneberg,[73] between 5 and 11 per cent by volume. The maximum acetic acid production in various species is as follows:

A. oxydans	2.0%	*A. aceti*	6.6%
A. acetigenus	2.7%	*A. kützingianus*	6.6%
A. pasteurianus	6.2%	*A. schützenbachii*	10.9%
A. acetosus	6.6%	*A. xylinum*	4.5%
A. ascendens	9.0%		

Optimal temperature for acetic acid production lies between 20°C and 30°C and the reaction requires an extremely long time at a temperature either below 10°C or above 46°C. The optimal pH is 5~6 (Bertho[74]) but there is no significant difference in the acetic acid production between pH 3.6 and 7.6. Living cells of acetic acid bacteria can oxidize ethanol using methylene blue or quinone as a hydrogen acceptor instead of oxygen (Wieland[75]). When methylene blue is used, however, the oxidation rate is much slower than when oxygen or quinone is used. The oxidation of ethanol with oxygen as a hydrogen acceptor is inhibited initially by KCN (Wieland and Bertho[76]), but oxidation is restored once harmless aldehyde-cyanhydrin is formed. According to Tamiya and Tanaka[77] CO inhibits the oxidation of ethanol with oxygen, but not with quinone. This inhibition is reversed by light. The reduction of 2,6-dichlorophenol-indophenol by acetic acid bacteria is also inhibited by CO. The inhibition

124

is stronger in the dark than in the light. Toluene inhibits oxidation with oxygen, but not with quinone or methylene blue. From this evidence it was concluded that, as in the respiration of plants and animals, cytochromes play an important role as oxygen-regulators in the oxidation of ethanol by acetic acid bacteria in the presence of oxygen.

Tanaka[78] reported that KCN does not inhibit ethanol oxidation when either quinone or methylene blue is used as a hydrogen accetpor. Moreover, acetone-treated cells can oxidize ethanol with quinone or methylene blue, but not with oxygen. The same is true with KCN-treated cells. Thus the acetone treatment denatures the cytochrome system of the bacteria completely, leaving the dehydrogenase system intact. The acetone-treated cells can still carry out the dismutation of acetaldehyde. In this reaction acetaldehyde itself acts as a hydrogen acceptor and the dismutation reaction is nothing but a dehydrogenation reaction with acetaldehyde.

Bertho[79] studied the dehydrogenase activity of acetic acid bacteria with various substrates; the results are shown in Table 7.

Table 7. Dehydrogenase activities of *Acetobacter* for various alcohols and aldehydes (after Bertho).

Substrate	A. orleanense	A. pasteurianus	A. ascendens	A. aceti
Methyl alcohol	+	+		+
Formaldehyde	+	+		+
Ethyl alcohol	+++	+++	+++	+++
Acetaldehyde	++	+++	++	+++
Propyl alcohol	+++	+++	+++	+++
Propionaldehyde	++	+++		
Isobutyl alcohol	++	++		+++
Isobutyraldehyde	+	+++		+++
Isoamyl alcohol	+	++	+	++
Isovaleraldehyde		+		+
Benzyl alcohol	−	−		
Phenylethyl alcohol	+			
Saligenin	−			
Salicylaldehyde	−			
Isopropyl alcohol	++	++	++	++
Acetone	−	−		
Acetic acid	−	−		
Succinic acid	−			
Glucose		+++		+++

Table 8.   Oxidation of alcohols and aldehydes by *Acetobacter* (after Tanaka).

Substrate	*A. peroxydans*	*A. rancens*	*A. aceti*
Methyl alcohol	25	8	7
Formaldehyde	140	130	20
Ethyl alcohol	1000	1000	1000
Acetaldehyde	473	536	405
n-Propyl alcohol	700	766	960
n-Propionaldehyde	497	842	700
Isopropyl alcohol	48	9	8
n-Butyl alcohol	750	740	927
Isobutyraldehyde	231	272	170
Isobutyl alcohol	140	61	60
sec-Butyl alcohol	84	8	7
tert-Butyl alcohol	0	0	0
n-Amyl alcohol	600	640	850
Isoamyl alcohol	150	63	70
Methyl butyl alcohol	22	4	2
Dimethyl propyl alcohol	0	0	0
n-Hexyl alcohol	350	256	420
Methyl heptyl alcohol	12	3	2
n-Cetyl alcohol	2	0.2	0.3
Allyl alcohol	740	720	600
Crotonaldehyde	263	110	50
Ethylene glycol	200	35	40
Glycerol	15	17	10
Glyceraldehyde	30	36	—
Citronellol	31	13	22
Citronellal	19	15	2
Furfuryl alcohol	250	128	130
Furfural	134	235	200
Benzyl alcohol	35	2	3
Benzaldehyde	—	17	8
Phenylethyl alcohol	39	6	8
Phenylacetaldehyde	—	—	—
Phenylpropyl alcohol	45	37	17
Saligenin	4	3	1
Salicylaldehyde	—	1	4
Cinnamic alcohol	25	22	30
Cinnamaldehyde	17	15	5
Vanillin	8	14	6

Figures show relative activity compared to that for ethyl alcohol=1000. pH 6.0.

Tanaka[80] measured the oxygen uptake of resting cells of various acetic acid bacteria with a great variety of alcohols and aldehydes as substrates (Table 8).

As these studies indicate, acetic acid bacteria oxidize ethyl alcohol most rapidly, and n-propyl, n-butyl, n-amyl and n-hexyl alcohols with rates decreasing in that order.   The oxidizability becomes generally weaker from primary to secondary and tertiary alcohols, and also as the carbon-chains of the alcohols become longer.   The oxidizability of methyl alcohol is extremely weak.

Tanaka[81] further studied the effects of various metallic salts on the oxidation of ethyl alcohol and acetaldehyde, measuring the uptake of oxygen. $CuSO_4$, $AgNO_3$, $HgCl_2$ and $AuCl_3$ strongly inhibited the oxidation, but arsenites had very little effect.

Manometric studies (King and Cheldelin[82]) on the oxidation of ethanol by resting cells of $A.$ $suboxydans$, which belongs to the group incapable of oxidizing acetic acid, showed a quantitative oxidation to acetate without CO_2 production in all three strains tested (Table 9).

The same results were obtained by De Ley and Stouthamer.[83]

Mechanism of Acetic Acid Fermentation

The oxidation of ethanol by $Acetobacter$ is carried out in two steps. The first step is the oxidation of ethanol to acetaldehyde, as Henneberg[84] noted in $A.$ $industrius$.   Neuberg and Nord[85] verified the reaction by adding calcium sulfite to the fermentation medium and trapping the acetaldehyde.

Table 9.   Oxidation of ethanol by resting $A.$ $suboxydans$ cells
(after King and Cheldelin).

Substrate	Amount added in μM	Organism strain no.	Duration of experiment in min	O_2 uptake in microatoms	CO_2 produced in μM	Acetic acid formed in μM
Ethanol	50	621U	60	99	1	52
„	100	621U	120	182	8	106
„	50	621	100	98	0	49
„	100	621	120	193		99
„	50	9322	100	94	2	48

System : 0.05 M phosphate, 0.01 M $MgCl_2$, 10 mg (dry weight) of resting cells washed in a cold room. Total volume : 2.8 ml. pH : 6.0.

The second step is the formation of acetic acid from acetaldehyde. Neuberg and Windisch[86] verified the anaerobic formation of equimolar amounts of ethanol and acetic acid from acetaldehyde in three species of bacteria, *A. ascendens, A. pasteurianus* and *A. xylinum.* This is a dismutation reaction in which one mole of acetaldehyde is reduced to ethanol while another mole is oxidized to acetic acid (Neuberg[87]).

$$
\begin{array}{c}
CH_3-CHO \\
CH_3-CHO
\end{array}
\quad + \quad
\begin{array}{c}
H_2 \\
O
\end{array}
\quad = \quad
\begin{array}{c}
CH_3-CH_2OH \\
CH_3-COOH
\end{array}
$$

This reaction proceeds even under aerobic conditions (Neuberg and Morinari[88]) but only up to 50%, the remaining half of the acetaldehyde being oxidized directly to acetic acid. Therefore under these conditions 75% of the acetaldehyde is converted to acetic acid and 25% to ethanol.

With good aeration the oxidation and dismutation proceed side by side converting all the acetaldehyde to acetic acid.

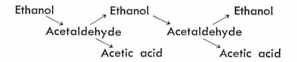

Later, Wieland and Bertho[76] reported an extremely low activity of dismutation in *A. orleanense, A. pasteurianus* and *A. ascendens.* In a cell suspension of *A. orleanense* the acetic acid production by dismutation was only 77% after six days, while under aerobic conditions there was a 100% formation of acetic acid after 340 minutes. They then proposed a pathway by which acetaldehyde was oxidized rather than dismutated to acetic acid. Since Wieland[89] had already demonstrated the formation of acetic acid in the absence of oxygen with the addition of such hydrogen acceptors as quinone and methylene blue, and with very little inhibition by cyanide and carbon monoxide, it was concluded that acetic acid was formed by the action of dehydrogenases. The following equations for the oxidation of acetaldehyde to acetic acid were proposed, assuming the formation of acetaldehyde hydrate as an intermediate product.

According to Simon[90] the ratio of dismutation to direct oxidation is dependent on the environmental pH, with dismutation increasing in basic conditions and decreasing in acidic conditons. Neuberg *et al.* had carried out their experiments at pH 8.1 in the presence of $CaCO_3$, while Wieland had done his experiments at pH 5.5.

$$\underset{\text{Ethanol}}{CH_3-\overset{\displaystyle OH}{\underset{\displaystyle H}{\overset{|}{\underset{|}{C}}}}-H} \quad +O \longrightarrow CH_3-\overset{\displaystyle O}{\overset{\|}{C}}-H \quad +H_2O$$

$$\underset{\text{Acetaldehyde}}{CH_3-\overset{\displaystyle O}{\overset{\|}{C}}-H} +H_2O \; \underset{\longleftarrow}{\longrightarrow} \; CH_3-\overset{\displaystyle OH}{\underset{\displaystyle H}{\overset{|}{\underset{|}{C}}}}-OH \quad +O \longrightarrow$$

$$\underset{\text{Acetic acid}}{CH_3-\overset{\displaystyle OH}{\overset{|}{C}}=O} \quad +H_2O$$

Bertho and Basu,[91] however, stated that the dismutation reaction was independent of pH.

Oxido-reductive reactions in acetic acid bacteria have been reported, in addition to acetaldehyde, on n-butyraldehyde and isovaleraldehyde,[86] and on benzaldehyde, anisaldehyde, cinnamaldehyde, propionaldehyde, citronellal and furfural.[88]

Acetic acid bacteria in their normal growth do not split glucose into 3-carbon compounds as in alcoholic fermentation. Simon,[90] however, reported the anaerobic production of roughly equimolar ethanol and CO_2 from glucose by heavy cell suspensions of *A. pasteurianus* or *A. suboxydans*. The presence of pyruvate decarboxylase, ketoaldehyde mutase (glyoxalase) and the glycolytic enzyme was demonstrated in dried cell preparations of these bacteria. The inability of acetic acid bacteria to carry out a normal anaerobic fermentation is probably, therefore, due to their inability to grow anaerobically.

The strict aerobic character of acetic acid bacteria can also be modified with regard to growth. It was reported by Cozic[92] that *A. xylinum* was able to grow anaerobically in ethanol in the presence of reducible dyes.

Alcohol and Acetaldehyde Dehydrogenases

The oxidation of ethanol to acetic acid proceeds first through alcohol dehydrogenase and second through acetaldehyde dehydrogenase.

Early observations on alcohol dehydrogenase were made by Lutwak-Mann,[93] who reported that the enzyme of *A. suboxydans* was NAD-linked.

King and Cheldelin[94] studied the influence of phosphate and dinitrophenol on ethanol oxidation in the same organism and concluded that the oxidation of ethanol to acetic acid did not depend upon coenzyme-linked phosphorylation.

Later the same workers[95] purified the alcohol dehydrogenase from cell-free extracts of *A. suboxydans* up to 14,000 Racker units/mg. The purified enzyme was NAD-linked and was inactive towards acetaldehyde. They[96] also attempted to purify the acetaldehyde dehydrogenase of the same organism, and obtained from the soluble fraction of cell-free extracts a dehydrogenase preparation purified over 100-fold. This preparation had no alcohol dehydrogenase activity, but was active towards propyonaldehyde and *n*-butyraldehyde. The acetaldehyde dehydrogenase required NADP as a coenzyme. NAD had only one fourth of the coenzymic activity of NADP with the most purified preparation. It was thought impossible to state with certainty whether the two dehydrogenases were present. The optimal pH of this enzyme was 8.7 for both NADP and NAD. The Michaelis constant, substrate specificity, and other properties are shown in Table 10.

Rao and Gunsalus[97] reported that both *A. suboxydans* and *A. aceti*

Table 10.   Properties of acetaldehyde dehydrogenase
(*A. suboxydans*, after King and Cheldelin).

Effect of KCl and MgCl$_2$	No stimulatory effect at 4×10^{-2} M (either alone or in combination).
Effect of phosphate and CoA	No effect.
Effect of cystein, glutathione or EDTA	Stimulatory effect.
Km	9.8×10^{-5} moles/litre for acetaldehyde (NADP). 12.8×10^{-5} moles/litre for acetaldehyde (NAD).
Substrate specificity	Catalyzed the oxidation of propionaldehyde and *n*-butyraldehyde. DL-glyceraldehyde is completely inert.
Inhibition	Ag$^+$, Hg^{++} and Cu^{++} are toxic. Iodosobenzoate is not as potent as the heavy metals mentioned.
Stability (in crude extracts)	Not impaired by heating at 56°C for 10 minutes, 30°C for 4 hours, by standing at room temperature for 14 hours, or by lyophilization. More than 80% of original activity was retained for 20 months at -10°C.

possess an NADP-linked acetaldehyde dehydrogenase, and that *A. aceti* possesses an NAD-linked enzyme in addition.

Wieland and Pistor[98] studied *A. peroxydans*, a rather peculiar member of the acetic acid bacteria (because of its lack of catalase, failure to produce acid from glucose, *etc.*), and reported that a cell suspension of the organism oxidized ethanol with molecular oxygen and hydrogen peroxide as hydrogen acceptors, and that acetaldehyde was the primary product of the oxidation. But carbohydrates and related compounds could not be oxidized.

According to Atkinson,[99] the alcohol dehydrogenase of this organism is NAD-linked. The resting cells oxidized ethanol with a pH optimum of 5.5, but the oxidation stopped at the level of acetic acid. In cell-free extracts also, the oxidation took up one mole of oxygen per mole of ethanol and stopped at acetic acid. The growing cells could oxidize ethanol to CO_2 and H_2O. The failure of the resting cells and the cell-free extracts to do so was probably due to the inactivation of an enzymatic system involved in the oxidation of intermediates participating in the acetic acid oxidation. Pyruvic acid and members of the TCA cycle were completely oxidized by the organism and the presence of aconitase and isocitrate dehydrogenase was confirmed (Atkinson[100]).

Tanenbaum[101] made detailed studies on the oxidative enzymes of *A. peroxydans*. In his experiments freshly washed cells, cells aged in the cold, and dried cells of *A. peroxydans* NCIB 8618 all oxidized ethanol to CO_2 and H_2O. Cyanide and azide at $10^{-3}M$ inhibited oxidation. The cell-free extracts also usually oxidized ethanol and acetaldehyde. Treatment of the cell-free extracts with up to 20% by volume of Dowex-I for removal of CoA did not diminish their ability to oxidize either substrate. CoA is probably not involved in the oxidation of acetaldehyde to acetate in this species. The oxidation of ethanol to acetaldehyde by washed cells was unaffected by the addition of $10^{-3}M$ arsenate, whereas acetaldehyde oxidation was completely arrested. Oxidation of ethanol and acetaldehyde by a crude cell-free preparation was NADP-linked. NAD could not substitute for NADP. The presence of reduced NADP dehydrogenase, reduced NADP cytochrome *c* reductase, diaphorase and the H_2O_2 peroxidations of reduced NADP and of reduced cytochrome has been demonstrated in cell-free extracts. The entire electron transport chain involved in ethanol oxidation was antimycin A insensitive.

Tanenbaum[102] confirmed the reduction of H_2O_2, dyes and molecular oxygen in the presence of molecular hydrogen by intact cells and cell-free extracts of a strain of *A. peroxydans*. With the cell-free preparation, reduction of cytochrome *c* with hydrogen was also observed. But reduction of pyridine nucleotides following hydrogen activation was not confirmed. Tamiya et al.[103] reported the reduction of methylene blue by cell-free extracts of A. peroxydans, but there was no evolution of hydrogen from reduced methyl viologen in their manometric experiments.

Prieur[104] claimed that there were two systems involved in ethanol oxidation in A. xylinum, one showing maximal activity at pH 5.7 and not requiring NAD, and the other showing maximal activity at pH 8.1 and requiring NAD. Through ammonium sulfate fractionation of cell-free extracts, the pH 5.7 system was precipitated between 0–30% saturation and the pH 8.1 system at over 50% saturation.

Recently, Nakayama[105] studied the enzymatic oxidation of ethanol biochemically, using an Acetobacter species used for the production of Japanese vinegar. (The organism has properties very similar to A. suboxydans(?), and can oxidize acetic acid, though much more slowly than ethanol.)   He found that the ethanol- and acetaldehyde-oxidizing activities were very tightly bound to the particulate cell fragments.   The oxidation of ethanol was inhibited by cyanide and carbon monoxide and hence a cytochrome oxidase was considered to be working as the terminal oxidase. But since the inhibition by cyanide was not reversed by the addition of methylene blue it was believed that there is an intimate relationship between the ethanol-oxidizing enzymes and the terminal electron-transferring system, and that the electrons liberated from ethanol were not transferred to methylene blue.   As long as ethanol was present the bacterial cells did not metabolize acetic acid.   Cytochromes in the cells were rapidly reduced by the addition of ethanol or acetaldehyde.

With respect to the mechanism of the accumulation of large quantities of acetic acid as a metabolic intermediate of ethanol, Nakayama sugested either that " bound coenzymes " in the cells were all used for the oxidation of ethanol to acetate, thus making the enzymes of acetate oxidation nonfunctionable, or that the oxidation of ethanol to acetate and the further oxidation of acetate were regulated in an unknown way by particle-bound enzymes, coenzymes and the cytochrome system.

Nakayama[106] studied the aldehyde dehydrogenase of this strain further and succeeded in extracting the enzyme in a water soluble form from cell-free preparations as well as the cell acetone powder.   The cytochrome-oxidizing system was not present in this water-soluble fraction, but was always found in the cellular fragments together with most of the haemoproteins.   The TCA cycle enzymes examined were all detected also in the cellular fragments rather than in the water-soluble fraction.   Through rivanol treatment of the extract and column chromatography on Amberlite CG-50 Nakayama succeeded in increasing the specific activity of aldehyde dehydrogenase 200-fold. This enzyme was NADP-linked and NAD did not replace NADP. CoA was not required for the activity. The optimum pH lay at 8.5–9.5 (see Table 11).

Table 11.   Properties of acetaldehyde dehydrogenase
(*Acetobacter* sp., after Nakayama).

Effect of metals	Co^{++}, Mg^{++}, Mn^{++} and phosphate ion reactivated the enzyme previously inactivated by dialysis, but Zn^{++} and Fe^{++} could not.
Substrate specificity	Catalyzed the oxidation of acet-, propion-, *n*-butyr-, *n*-valer-, *n*-capron-, *n*-enanth- and *n*-capryl-aldehydes but not glyceraldehyde.
Km	2.44×10^{-4} moles/liter for acetaldehyde. 0.94×10^{-4} moles/liter for propionaldehyde. 4.65×10^{-4} moles/liter for *n*-butyraldehyde.
Km for NADP	3.33×10^{-5} moles/liter in the presence of acetaldehyde. 5.10×10^{-5} moles/liter in the presence of propionaldehyde. 8.38×10^{-5} moles/liter in the presence of *n*-butyraldehyde.
Inhibition	Inhibited by the presence of EDTA, pyrophosphate, slightly inhibited by azide, while strongly inhibited by cyanide, semicarbazide, PCMB, monoiodoacetate, $NaHCO_3$, arsenate and HA. Not inhibited by *o*-phenanthroline, 8-oxyquinoline and dizizone.
Stability	Fairly stable at pH 4.0–8.5, but not at a higher pH range, very stable at pH 7.0 at temperatures lower than 40°C, whereas rapidly denatured at temperatures higher than 50°C. More than half the activity was lost by heating at 65°C for two minutes, or at 50°C for eight minutes.

EDTA, ethylenediamine tetraacetate; PCMB, *p*-chloromercuribenzoate; HA, hydroxylamine.

Kida and Asai,[107] using intact cells of *A. dioxyacetonicus* A 15, studied the effects of initial pH on the oxidation of ethanol. There was little difference in the activity between pH 3.0 and 8.0, and ethanol was completely oxidized to CO_2 and H_2O. With washed cells grown in a medium containing lactate, ethanol, yeast extract, and mineral salts, the alcohol dehydrogenase was most active at pH 7.0 and was considerably less active at a pH above 9. The activities of alcohol dehydrogenase and aldehyde dehydrogenase in dialyzed cell-free preparations are shown in Figs. 2 and 3. The alcohol dehydrogenase is NADP-linked and is stimulated by Mg^{++} and Mn^{++}, the latter being more effective. Aldehyde dehydroge-

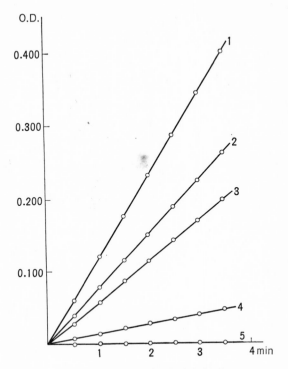

Fig. 2.   Activities of alcohol dehydrogenase
(*A. dioxyacetonicus* A 15, after Kida and Asai).

1.   Ethanol 10 μmoles, NADP 250 γ, MnCl$_2$·4H$_2$O 2μmoles, enzyme extracts
0.2 ml, total volume 3.0 ml.
2.   Ethanol 10 μmoles, NADP 250 γ, MgSO$_4$·7H$_2$O 2 μmoles, enzyme extracts
0.2 ml, total volume 3.0 ml.
3.   Ethanol 10 μmoles, NADP 250 γ, enzyme extracts 0.2 ml, total volume
3.0 ml.
4.   NADP 250 γ, enzyme extracts 0.2 ml, total volume 3.0 ml.
5.   Ethanol 10 μmoles, NAD 250 γ, enzyme extracts 0.2 ml, total volume
3.0 ml.
Enzyme extracts 0.2 ml=0.35 mgN.

nase is also NADP-linked and is stimulated by the addition of Mn^{++}.
NAD does not substitute for NADP in either enzyme activity.

Nakayama[108] reported on a coenzyme-independent aldehyde dehydrogenase in an *Acetobacter* species, different from the previously known
NADP-linked one.   The enzyme was purified 30-fold.   It reduced both
uni- and di-electron acceptors such as ferricyanide, 2,6-dichlorophenol-

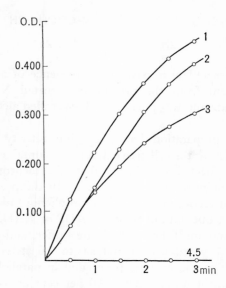

Fig. 3. Activities of acetaldehyde dehydrogenase
(*A. dioxyacetonicus* A 15, after Kida and Asai).

1. Acetaldehyde 10 μmoles, NADP 250 γ, MnCl₂·4H₂O 2 μmoles, enzyme extracts 0.05 ml, total volume 3.0 ml.
2. Acetaldehyde 10 μmoles, NADP 250 γ, enzyme extracts 0.05 ml, total volume 3.0 ml.
3. Acetaldehyde 10 μmoles, NADP 250 γ, MgSO₄·7H₂O 2 μmoles, enzyme extracts 0.05 ml, total volume 3.0 ml.
4. NADP 250 γ, enzyme extracts 0.05 ml, total volume 3.0 ml.
5. Acetaldehyde 10 μmoles, NAD 250 γ, enzyme extracts 0.05 ml, total volume 3.0 ml.

indophenol, thionine and methylene blue in the presence of acetaldehyde, but not NADP and NAD.

The substrate specificity was very broad and the enzyme oxidized acetaldehyde, propionaldehyde, *n*-butyraldehyde, and crotonaldehyde at almost the same rate. The Km for acetaldehyde was 8.7×10^{-5} moles/ liter and for ferricyanide 8.2×10^{-4} moles/liter. The pH optimum was 7.0 in citrate-phosphate buffer. The enzyme was fairly stable at pH 5.0 at a low temperature, but not above 40°C. Its activity was strongly inhibited by *p*-chloromercuribenzoate, but it was little affected by hydroxylamine. The enzyme reaction can be expressed by the equations:

$$R \cdot CHO + H_2O \longrightarrow R \cdot COOH + 2H$$

$$2H \longrightarrow 2H^+ + 2e^-$$

The same worker[109] also verified the presence of a new alcohol dehydrogenase different from the previously reported NAD- or NADP-linked one. The enzyme was purified and was called alcohol-cytochrome-553 reductase.

The purified preparation had a specific activity more than one hundred times as high as the cell-free extract. It contained a haemoprotein with an absorption spectrum similar to that of cytochrome c, its reduced α-absorption peak being located at 553 mμ. In the presence of ethanol, the enzyme reduced ferricyanide, 2,6-dichlorophenol-indophenol, thionine and methylene blue, but did not reduce NADP or NAD.

The enzyme was most active in citrate buffer, and the optimum pH was 3.8. It was fairly stable from pH 5.0 to 7.0 at temperatures below 45°C. Inactivation of the enzyme proceeded in parallel with the breakdown of the haemoprotein, and it lost 40 per cent of its activity after 24 hours' dialysis against deionized water at 3–4°C. Mg^{++} activated the enzyme slightly, while NH_4^+, Zn^{++}, Fe^{+++} and PO_4^{---} strongly inhibited its activity. The sulfhydryl reagents, including monoiodoacetate, p-chloromercuribenzoate and arsenite, did not inhibit the enzyme activity, while carbonyl and metal-chelating reagents, including semicarbazide and fluoride, more or less inhibited the activity. When ferricyanide was used as an electron acceptor, the enzyme showed a broad substrate specificity: the saturated and unsaturated straight chain monoalcohols, including ethyl alcohol, n-propyl alcohol, n-butyl alcohol, n-pentyl alcohol, n-hexyl alcohol, n-heptyl alcohol, n-octyl alcohol, n-nonyl alcohol, n-decyl alcohol, allyl alcohol, and propagyl alcohol were oxidized, but methyl alcohol and the iso-alcohols were not; 1,4-butyleneglycol, phenyl propyl alcohol, and cinnamyl alcohol were oxidized while the monosaccharides were not.

The Michaelis constant (Km) was 2.1×10^{-3} for ethanol in the presence of ferricyanide and 3.5×10^{-4} for ferricyanide with ethanol.

From the results of these experiments and from related reports, Nakayama devised the following scheme for ethanol oxidation in *Acetobacter* sp. (see also Fig. 4).

Ethanol is oxidized to acetaldehyde by E_1 and the resulting electrons are successively delivered to the heme iron of cytochrome 553 which is E_1 itself. The acetaldehyde thus formed is oxidized further by E_2 or by E_3. *Via* the former, the liberated electrons from acetaldehyde are transformed

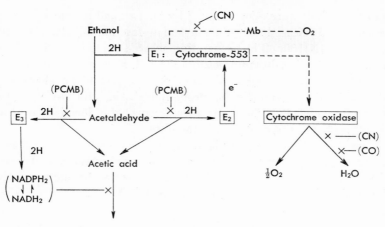

Fig. 4.   The scheme of ethanol oxidation by *Acetobacter* sp.
(after Nakayama).

E_1 : alcohol-cytochrome-553 reductase
E_2 : coenzyme-independent aldehyde dehydrogenase
E_3 : NADP-dependent aldehyde dehydrogenase
PCMB : *p*-chloromercuribenzoate

to the heme bound to E_1, cytochrome-553, and *via* the latter they reduce NADP. Reduced cytochrome-553 is then oxidized by a cytochrome oxidase.   Cyanide inhibits both the cytochrome oxidase and E_1, in agreement with the results of previous experiments,[105] in which methylene blue did not restore the ethanol oxidizing activity of either whole cells or a cell-free preparation after treatment with cyanide.   Probably *p*-chloromercuribenzoate inhibits the ethanol oxidation of whole cells and a cell-free preparation at E_2.   Thus, presumably, the cytochrome system operates in the ethanol and acetaldehyde oxidation by E_1 and E_2, whereas the $NADPH_2$ produced by E_3 inhibits further oxidation of acetic acid through the TCA cycle by upsetting the equilibrium ($NADPH_2 \rightleftharpoons NADH_2$), thus favoring back reaction of coenzyme-linked dehydrogenases in the TCA cycle. The acidic pH optima of E_1 and E_2 must also favor acetic acid accumulation by *Acetobacter* sp., since generally, enzymes involving the reaction of coenzymes exhibit pH optima mostly at neutral or higher pHs.

Dupuy and Maugenet[110] suggested the presence in *A. rancens* of two pathways of ethanol oxidation, one similar to that of yeast and liver, the other working only under acidic conditions without need for a soluble coenzyme.   The cells grown at pH 3.8 and 6.0 were able to oxidize ethanol.

The particular enzymes responsible for the oxidation of primary and secondary alcohols were demonstrated by De Ley and Kersters[111] in a purified preparation released from the particulate fraction of cells of *G. suboxydans* by Triton X 100. The preparation showed oxidative activity towards ethanol, *n*-propanol, *n*-butanol, *n*-amyl alcohol, *n*-hexanol, *n*-octyl alcohol, isobutanol, and allyl alcohol. Methanol and secondary alcohols were weakly oxidized. Acetaldehyde, propionaldehyde, and *n*-butyraldehyde were readily oxidized. These enzymes are probably localized on the cell envelope (probably the cytoplasmic membrane).

King *et al.*[112] found, in crude cell-free extracts of *A. suboxydans* separated from oxidative enzymes, pyruvate decarboxylase activity and an activity to form acetoin from acetaldehyde. The apoenzyme was activated by the addition of thiamine pyrophosphate and Mg^{++}.

Chapter 4

OXIDATION OF OTHER ALIPHATIC MONOALCOHOLS

Acetic acid bacteria oxidize other monohydric alcohols besides ethyl alcohol. Primary alcohols are oxidized to the corresponding acids *via* aldehydes, and secondary alcohols are oxidized to carbonyl compounds. There has been little work on the oxidation of tertiary alcohols.

Methyl alcohol. $H \cdot CH_2OH$ Brown[113] and Seifert[114] reported on the non-oxidizability of methyl alcohol by acetic acid bacteria. Visser't Hooft,[115] however, reported that *A. suboxydans* could oxidize methyl alcohol, though weakly, and form formic acid, and that *A. rancens* oxidized methyl alcohol to CO_2 and H_2O.   Müller[116] found that the resting cells of *A. pasteurianus* oxidized methyl alcohol and Krehan[117] made the same observation with growing cells.   Bertho[74] reported the oxidation of methyl alcohol, with methylene blue as a hydrogen acceptor, by resting cells of *A. ascendens*, *A. aceti*, *A. orleanense* and *A. pasteurianus*.   Chauvet[118] reported the oxidation of methyl alcohol by some strains of wine acetic acid bacteria. Tanaka[119] observed that resting cells of *A. peroxydans* took up one-fortieth the amount of oxygen with methyl alcohol that they did with ethyl alcohol.   He suggested that the failure of acid formation with methyl alcohol was due to the further oxidation of formic acid to CO_2 and H_2O.

n-Propyl alcohol. $CH_3 \cdot CH_2 \cdot CH_2OH$ *n*-Propyl alcohol is oxidized by most acetic acid bacteria.   Henneberg[120] reported its oxidation to propionic acid by almost all the acetic acid bacteria he examined: *A. ascendens*, *A. acetosus*, *A. oxydans*, *A. industrius*, *A. kützingianus*, *A. pasteurianus* and *A. acetigenus*.   Brown[121] observed the formation in the reaction of trace amounts of some nonvolatile acid in addition to propionic acid.   Visser't Hooft,[115] Krehan,[117] Mosel,[122] Hermann and Neuschul,[123] Asai,[124] and Kondo and Ameyama[125] also reported the oxidation of *n*-propyl alcohol by many acetic acid bacteria, as did Wieland and Pistor[98] and Tošić.[126] Polesofsky[127] reported the oxidation of a 4% propanol solution by shake cultures of acetic acid bacteria.

Isopropyl alcohol. $CH_3 \cdot CHOH \cdot CH_3$ Isopropyl alcohol was reported by Visser't Hooft[115] to be oxidized by the growing cells of acetic acid bacteria.   Bertho[128] and Müller[129] observed the oxidation of isopropyl alcohol to acetone by a bacterial cell suspension.   Tanaka[119] observed that the oxidation rate of isopropyl alcohol by the resting cells of *A. peroxydans* was one-twentieth that of ethanol.

Visser't Hooft[115] verified the formation of acetone from isopropyl alcohol in *A. xylinum*, *A. melanogenus* and *A. suboxydans*. Müller[129] reported the oxidation of 3% isopropyl alcohol to acetone by *A. pasteurianus*. Asai[130] found that nearly all the acetic acid bacteria isolated from fruits could oxidize isopropyl alcohol to acetone, but there was no formation of acid, which agrees with the results of Uemura and Kondo.[131]   Kondo and Ameyama[125] observed the oxidation of isopropyl alcohol by almost all the acetic acid bacteria examined.

According to Müller,[132] quinone can replace oxygen as a hydrogen acceptor in the oxidation of isopropyl alcohol by *A. pasteurianus*.

n-Butyl alcohol. $CH_3 \cdot CH_2 \cdot CH_2 \cdot CH_2OH$ Seifert,[114] Visser't Hooft,[115] Krehan,[117] Mosel,[122] and Wieland and Pistor[98] reported the oxidation of the alcohol to the corresponding acid.   According to Asai,[130] two of 38 strains of acetic acid bacteria isolated from fruits, *Gluconobacter roseus* and *G. liquefaciens*, could not form acid from *n*-butyl alcohol.

Cozic[133] observed that *A. xylinum* could not oxidize *n*-butanol.   According to Tanaka,[119] the oxidative activities of *A. peroxydans*, *A. rancens* and *A. aceti* toward *n*-propanol and *n*-butanol were nearly the same, and were 70–90% of the activity toward ethanol.

Isobutyl alcohol. $CH_3 \cdot CH(CH_3) \cdot CH_2OH$ Seifert[114] found that *A. pasteurianus* and *A. kützingianus* oxidized isobutyl alcohol in a beer wort medium, but not in a yeast extract medium.   Visser't Hooft,[115] Mosel,[122] Asai[134] and Tanaka[119] observed the oxidation of isobutyl alcohol to the corresponding acid.

sec-Butyl alcohol. $CH_3 \cdot CH_2 \cdot CHOH \cdot CH_3$ The oxidation of *sec*-butyl alcohol by resting cells of *A. peroxydans* was reported by Tanaka.[119]

tert-Butyl alcohol. $(CH_3)_3C \cdot OH$ Visser't Hooft[115] reported that *A. suboxydans* did not oxidize this alcohol, but that *A. rancens* did.

n-Amyl alcohol. $CH_3 \cdot CH_2 \cdot CH_2 \cdot CH_2 \cdot CH_2OH$ Tanaka[119] reported the oxidation of *n*-amyl alcohol by resting cells of *A. peroxydans*, *A. rancens*, and *A. aceti*.   According to Kondo and Ameyama[125] all 19 acetic acid bacteria examined, except for *A. suboxydans*, oxidized *n*-amyl alcohol. Asai and Shoda[135] reported rapid oxidation of the alcohol and the accumulation of acid by *A. ascendens* and *A. aceti*.

Isoamyl alcohol. $CH_3 \cdot CH(CH_3) \cdot CH_2 \cdot CH_2OH$ Seifert[114] reported the formation of acid from isoamyl alcohol by *A. kützingianus*.   According to Krehan[117] *A. acetigenoideus* oxidized isoamyl alcohol completely in 0.5% concentration.   Asai and Shoda[135] observed strong oxidative activity in *A. aceti* and *A. albuminosus*.   Polesofsky[127] reported that cultures of acetic acid bacteria were able to oxidize isoamyl alcohol to acid even in 1% concentration.   Tanaka[136] reported oxidation by resting cells of *A. peroxydans*, *A. rancens* and *A. aceti*.

n-Hexyl alcohol. $CH_3 \cdot (CH_2)_4 \cdot CH_2OH$ Tanaka reported the oxidation of *n*-hexyl alcohol by *A. rancens*, *A. aceti* and *A. peroxydans* at a rate one-third that of ethanol.   But methyl heptyl alcohol and *n*-cetyl alcohol were very little oxidized, if at all.

Unsaturated monoalcohols.   Tanaka[80] reported the rapid oxidation of allyl alcohol by three species of *Acetobacter*, and Nakayama[109] reported the oxidation of allyl alcohol and propagyl alcohol (added as an emulsion) by an alcohol dehydrogenase (alcohol-cytochrome-553 reductase) of *Acetobacter* sp.

Addendum.   Oxidation of Aromatic Alcohols

Tanaka[136] reported the oxidation of benzyl alcohol [$C_6H_5 \cdot CH_2OH$], β-phenyl ethyl alcohol [$C_6H_5 \cdot CH_2 \cdot CH_2OH$], and furfuryl alcohol [C_4H_3O (CH_2OH)] by acetic acid bacteria.   Polesofsky[127] found oxidation of β-phenyl ethyl alcohol and furfuryl alcohol to phenyl acetic acid and pyromucic acid, respectively.

Chapter 5

OXIDATION OF GLYCOLS AND ALIPHATIC POLYALCOHOLS

Glycols and aliphatic polyalcohols with more than two OH groups per molecule are also oxidized by acetic acid bacteria to corresponding acids or ketogenic compounds. The ketogenic reaction is shown by the equation:

$$-CHOH \cdot CH_2OH \longrightarrow -CO \cdot CH_2OH + 2H$$

This reaction was first noted by Bertrand[137] in *A. xylinum*, and has since been considered a characteristic of the sorbose bacteria. Bacteria with this characteristic were termed "ketogenic acetic acid bacteria" by Hermann and Neuschul.[123]

Oxidation of Glycols

1,2-Ethanediol (*Ethylene glycol*). $CH_2OH \cdot CH_2OH$ The oxidation of ethylene glycol to glycolic acid was reported by Brown[138] in *A. aceti*, by Seifert[114] in *A. pasteurianus* and *A. kützingianus*, and by Henneberg,[139] Visser't Hooft,[115] Krehan,[117] and Mosel[122] in the other acetic acid bacteria examined. Müller[129] reported oxidation by resting cells of acetic acid baceria. Tanaka[136] observed oxidation in *A. peroxydans*, *A. rancens* and *A. aceti* resting cells. The oxidation was inhibited by HCN. According to Visser't Hooft,[115] glycolaldehyde is formed as an intermediate, and *A. rancens* oxidizes this partly to oxalic acid. Banning[140] reported that many strains were able to form oxalic acid from ethylene glycol. According to Kasai *et al.*,[141] however, the cell-free extract of *A. dioxyacetonicus* strain A 15 oxidizes glyoxylic acid, but not glycolic acid, to oxalic acid.

Kaushal *et al.*[142] reported the oxidation of ethylene glycol to glycolaldehyde by *A. acetigenus*. Kersters and De Ley[143] observed the oxidation of ethylene glycol by resting cells of 15 strains, representing the entire taxonomic range, and found that none of the strains oxidized the substrate beyond glycolic acid.

Polesofsky[127] reported the oxidation of diethylene glycol [$O(CH_2 \cdot CH_2OH)_2$] by a submerged culture of acetic acid bacteria. The oxidation shown below took place, though slowly, in strains of *G. suboxydans*.

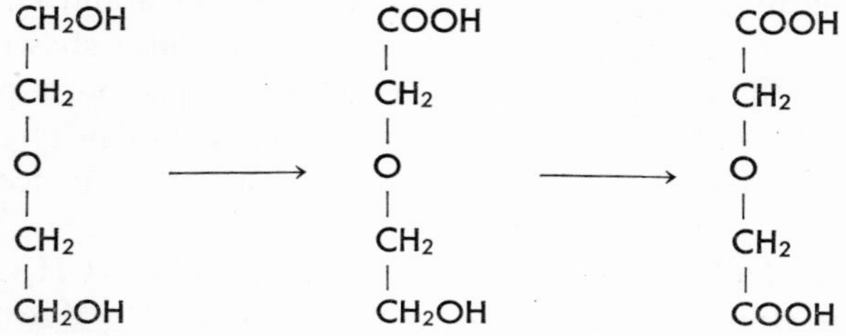

$$
\begin{array}{ccccc}
\text{CH}_2\text{OH} & & \text{COOH} & & \text{COOH} \\
| & & | & & | \\
\text{CH}_2 & & \text{CH}_2 & & \text{CH}_2 \\
| & & | & & | \\
\text{O} & \longrightarrow & \text{O} & \longrightarrow & \text{O} \\
| & & | & & | \\
\text{CH}_2 & & \text{CH}_2 & & \text{CH}_2 \\
| & & | & & | \\
\text{CH}_2\text{OH} & & \text{CH}_2\text{OH} & & \text{COOH}
\end{array}
$$

Diethylene glycol   Diglycolic acid

Diethylene glycol monomethyl ether was also oxidized by the same strain (Hromatka and Polesofsky[144]).

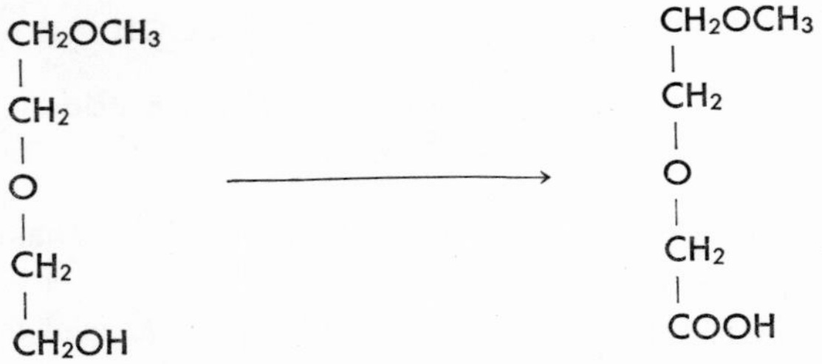

$$
\begin{array}{ccc}
\text{CH}_2\text{OCH}_3 & & \text{CH}_2\text{OCH}_3 \\
| & & | \\
\text{CH}_2 & & \text{CH}_2 \\
| & & | \\
\text{O} & \longrightarrow & \text{O} \\
| & & | \\
\text{CH}_2 & & \text{CH}_2 \\
| & & | \\
\text{CH}_2\text{OH} & & \text{COOH}
\end{array}
$$

2-(2-Methoxy)-ethoxy-acetic acid

According to Kersters and De Ley,[143] triethylene glycol $HOCH_2 \cdot CH_2 \cdot O \cdot CH_2 \cdot CH_2 \cdot O \cdot CH_2 \cdot CH_2OH$ was oxidized rapidly by resting cells of *G. suboxydans* with the uptake of 1 mole of O_2 per mole of substrate. The end product was not identified.

Thiodiethylene glycol $CH_2OH \cdot CH_2 \cdot S \cdot CH_2 \cdot CH_2OH$ was slowly oxidized by resting cells of *G. suboxydans* ATCC 621, according to Cummins.[145] The end product was not identified.

1,2-Propanediol (*Propylene glycol*). $CH_3 \cdot CHOH \cdot CH_2OH$ Kling[146] reported that sorbose bacteria and "mycoderme race d'Orléans" oxidized only the levorotatory propylene glycol of the racemic mixture to acetol ($CH_3 \cdot CO \cdot CH_2OH$).

Recently Copet *et al.*[147] reported oxidation of the D(−) form by *A. aceti* and *A. xylinum* and oxidation of the L(+) form by *G. suboxydans* as summarized below:

$$D(-)1,2\text{-Propanediol} \longrightarrow \text{Acetol}$$
$$L(+)1,2\text{-Propanediol} \longrightarrow \text{Acetol}$$

Van Risseghem[148] found that the $(-)$ isomer was oxidized prefentially by another strain of *G. suboxydans.*

According to Janke,[149] however, both *A. xylinum* and *G. suboxydans* oxidized both optical isomers to acetol.   This was also shown by Bultin and Wince[150] with *G. suboxydans* under intensive aeration, a pH between 4.5 and 6.5, and a suitable carbon source such as glucose or glycerol.   The quantitative oxidation to acetol was reported in a substrate concentration up to 15%.

1,3-Propanediol (*Trimethylene glycol*). $CH_2OH \cdot CH_2 \cdot CH_2OH$ The *G. suboxydans* strains of Hromatka and Polesofsky[127,144] oxidized this compound by the following reaction.

CH₂OH   COOH   COOH
| | |
CH₂   ———→   CH₂   ———→   CH₂
| | |
CH₂OH   CH₂OH   COOH

Hydracrylic acid   **Malonic acid**
(β-hydroxypropionic acid)

Kersters and De Ley[143] obtained hydracrylic acid by shaking the substrate with resting cells of *G. suboxydans.*

DL-1,3-Butanediol. $CH_2OH \cdot CH_2 \cdot CHOH \cdot CH_3$ *Gluconobacter* and strains of the *Mesoxydans* group of *Acetobacter* oxidized this compound to DL-β-hydroxybutyric acid (Kersters and De Ley,[143] Verloove[151]).

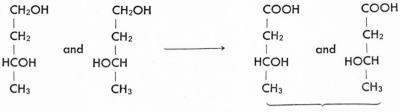

1,4-Butanediol (*Tetramethylene glycol*). $CH_2OH \cdot CH_2 \cdot CH_2 \cdot CH_2OH$ The formation of γ-hydroxybutyric acid and succinic acid was observed by Hromatka and Polesofsky[144] in a strain of *G. suboxydans.*   The same observation was made by Kersters and De Ley.[143]

CH₂OH   COOH   COOH

The reaction sequence:

$$CH_2OH\text{-}CH_2\text{-}CH_2\text{-}CH_2OH \longrightarrow COOH\text{-}CH_2\text{-}CH_2\text{-}CH_2OH \longrightarrow COOH\text{-}CH_2\text{-}CH_2\text{-}COOH$$

γ-Hydroxy butyric acid   Succinic acid

2,3-Butanediol (*Butylene glycol*). $CH_3 \cdot CHOH \cdot CHOH \cdot CH_3$ According to Bertrand's rule, only the *meso* form should be oxidized. However, experiments have revealed that both the *meso* and D(−) forms are readily oxidized, whereas the L(+) form is very slowly oxidized. Grivsky[152] obtained results supporting the formation of L(+)-acetoin from *meso*-2,3-butanediol with strains of *A. xylinum* and *A. aceti*. Fulmer *et al.*[153] observed the same results with *G. suboxydans*. Kersters and De Ley[143] found that resting cells of *Gluconobacter* and *G. liquefaciens* oxidized the substrate to acetoin. *Mesoxydans-* and *oxydans-* strains of *Acetobacter* oxidized it beyond the acetoin stage.

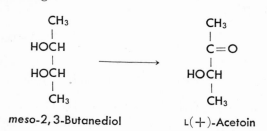

meso-2, 3-Butanediol   L(+)-Acetoin

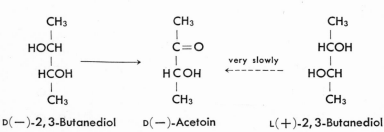

D(−)-2, 3-Butanediol   D(−)-Acetoin   L(+)-2, 3-Butanediol

D(−)-2,3-Butanediol was shown to be oxidized to D(−)-acetoin by Kling,[154] with strains of *A. xylinum* and *A. aceti*. He used the DL mixture for the experiment and the results showed that the D(−) form is oxidized preferentially over the L(+) form.

According to Kersters and De Ley[143] a strain of *G. suboxydans* oxidized both $D(-)$ and $L(+)$ forms of 2,3-butanediol to $D(-)$-acetoin.

1,5-Pentanediol. $CH_2OH \cdot (CH_2)_3 \cdot CH_2OH$ Kersters and De Ley[143] observed oxidation of 1,5-pentanediol by resting cells of *G. suboxydans*. The oxidation proceeded in two stages corresponding to the following reactions:

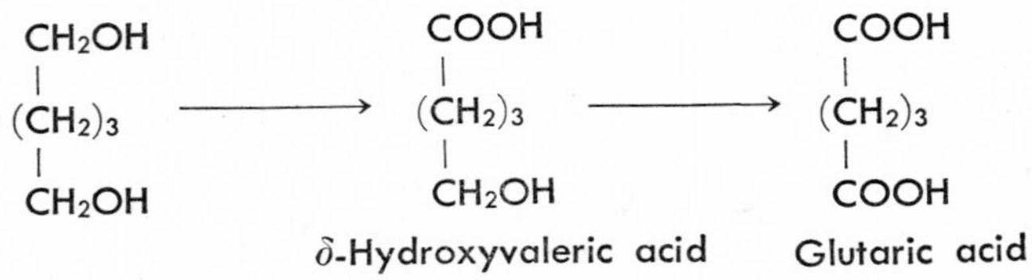

$$
\begin{array}{ccccc}
CH_2OH & & COOH & & COOH \\
| & & | & & | \\
(CH_2)_3 & \longrightarrow & (CH_2)_3 & \longrightarrow & (CH_2)_3 \\
| & & | & & | \\
CH_2OH & & CH_2OH & & COOH \\
 & & \eth\text{-Hydroxyvaleric acid} & & \text{Glutaric acid}
\end{array}
$$

1,6-Hexanediol. $CH_2OH \cdot (CH_2)_4 \cdot CH_2OH$ This compound, according to Kersters and De Ley,[143] was rapidly oxidized to adipic acid by resting cells of *G. suboxydans via* the intermediate formation of ε-hydroxycaproic acid. Polesofsky[127] reported the same oxidation.

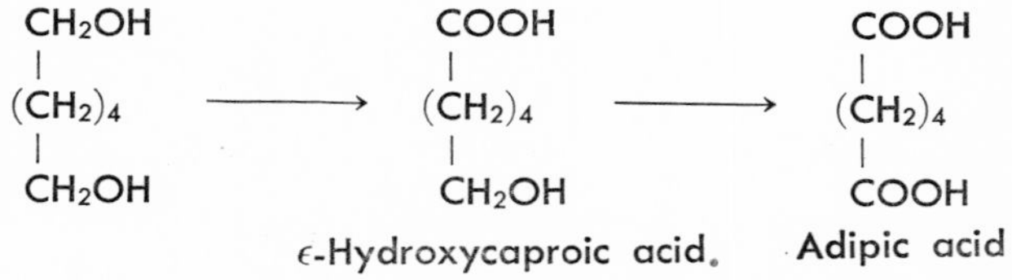

$$
\begin{array}{ccccc}
CH_2OH & & COOH & & COOH \\
| & & | & & | \\
(CH_2)_4 & \longrightarrow & (CH_2)_4 & \longrightarrow & (CH_2)_4 \\
| & & | & & | \\
CH_2OH & & CH_2OH & & COOH \\
 & & \varepsilon\text{-Hydroxycaproic acid.} & & \text{Adipic acid}
\end{array}
$$

3,4-Hexanediol. $CH_3 \cdot CH_2 \cdot CHOH \cdot CHOH \cdot CH_2 \cdot CH_3$ It was observed by Van Risseghem[155] that this compound was readily oxidized by *A. xylinum* and *A. aceti* to $L(+)$-ethylpropionyl carbinol and a small amount of dipropionyl.

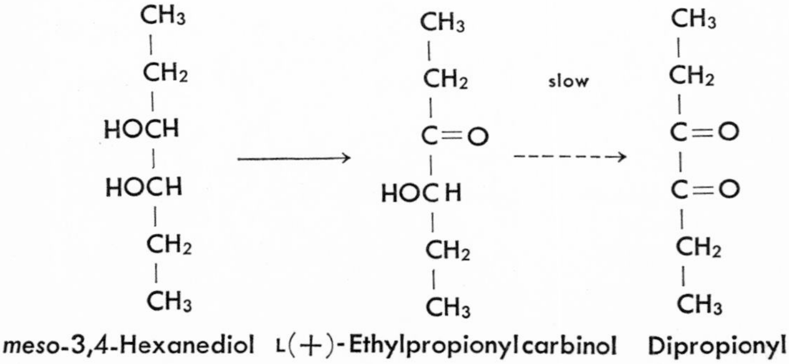

$$
\begin{array}{ccccc}
CH_3 & & CH_3 & & CH_3 \\
| & & | & & | \\
CH_2 & & CH_2 & \text{slow} & CH_2 \\
| & & | & & | \\
HOCH & \longrightarrow & C{=}O & \dashrightarrow & C{=}O \\
| & & | & & | \\
HOCH & & HOCH & & C{=}O \\
| & & | & & | \\
CH_2 & & CH_2 & & CH_2 \\
| & & | & & | \\
CH_3 & & CH_3 & & CH_3 \\
\textit{meso-}\text{3,4-Hexanediol} & & L(+)\text{-Ethylpropionylcarbinol} & & \text{Dipropionyl}
\end{array}
$$

Kersters and De Ley[143] also observed quantitative conversion of the *meso* compound into L(+) carbinol by resting cells of *G. suboxydans*. Contrary to Bertrand's rule, the D(−) form was shown to be attacked by *A. aceti* (and weakly by *A. xylinum*) and converted into D(−)-ethylpropionyl carbinol. A small amount of dipropionyl was also formed (Van Risseghem[155]). L(−)-3,4-hexanediol was not attacked.

2,5-Hexanediol. $CH_3 \cdot CHOH \cdot CH_2 \cdot CH_2 \cdot CHOH \cdot CH_3$ This compound was oxidized by resting cells of *G. suboxydans*, and probably converted into 2,5-diketohexane ($CH_3 \cdot CO \cdot CH_2 \cdot CH_2 \cdot CO \cdot CH_3$) (Kersters and De Ley[143]).

DL-1,2,6-Hexanetriol. $CH_2OH \cdot CHOH \cdot CH_2 \cdot CH_2 \cdot CH_2 \cdot CH_2OH$ This compound was oxidized by resting cells of *G. suboxydans*, possibly with the formation of 5,6-dihydroxycaproic acid, followed by a slower oxidation at the end of the molecule (Kersters and De Ley[143]).

1,7-Heptanediol. $CH_2OH \cdot (CH_2)_5 \cdot CH_2OH$ Resting cells of *G. suboxydans* oxidized this compound into pimelic acid, probably *via* 7-hydroxyheptylic acid (Kersters and De Ley[143]).

$$
\begin{array}{ccc}
CH_2OH & COOH & COOH \\
| & | & | \\
(CH_2)_5 & \longrightarrow \quad (CH_2)_5 & \longrightarrow \quad (CH_2)_5 \\
| & | & | \\
CH_2OH & CH_2OH & COOH \\
& \text{7-Hydroxyheptylic acid} & \text{Pimelic acid}
\end{array}
$$

Cummins[145] reported the oxidation of the following glycols by resting cells or cell-free extracts of *G. suboxydans*: DL-1,2-propanediol, 2-methyl-2-nitro-1,3-propanediol, 2-butene-1,4-diol, 1,2,4-butanediol, 1,3-pentanediol, 1,5-pentanediol, hexylene glycol, 1,2,6-hexanetriol, 2,5-hexanediol. 2,5-Dimethyl-hexyn-3-diol-2,5, pentaerythritol, dipropylene glycol, diethylene glycol, styrene glycol and cyclohexane-1,4-diol were not oxidized.

Enzymes relating to glycol oxidation.

Goldschmidt and Krampitz[156] reported that an NAD-linked dehydrogenase for ethylene glycol and another for 1,2-propanediol and 2,3-butanediol were present in *G. suboxydans*. Kersters and De Ley[143] carried out enzymatic studies on the oxidation of several glycols by *G. suboxydans* strain SU. The results indicated that all glycols which were oxidized by resting cells were also oxidized by the particulate fraction. The particulate fraction oxidized D(−)- and L(+)-1,2- propanediol and D(−)-2,3-butanediol to acetol and D(−)- and L(+)- acetoin, respectively.

The distribution of soluble alcohol dehydrogenase was also investigated and the results showed that a soluble NAD-linked primary alcohol dehydrogenase oxidized monohydric primary alcohols and ω-diols and a soluble NAD-linked secondary alcohol dehydrogenase oxidized monohydric secondary alcohols and the secondary alcohol function of the following glycols: *meso*-2,3-butanediol, DL-2,3-butanediol, DL-1,2-propanediol, L(+)-1,2- propanediol, and *meso*-3,4-hexanediol.

DL-1,3-Butanediol was oxidized slowly at the C-1 position by the primary alcohol dehydrogenase. *Meso*-2,3-butanediol and *meso*-3,4-hexanediol were oxidized to L(+)-acetoin and L(+)-ethylpropionyl carbinol respectively by the secondary alcohol dehydrogenase. Purification and chromatographic separation of both soluble dehydrogenases were also described. These three enzyme systems at least appear to be involved in the oxidation of aliphatic glycols: a soluble NAD-linked primary alcohol dehydrogenase, a soluble NAD-linked secondary alcohol dehydrogenase and at least two particulate oxidative systems (De Ley and Kersters[111]). Since neither of the purified soluble dehydrogenases oxidized polyols such as mannitol, sorbitol, glycerol, *etc.*, it was determined that glycol oxidation has no connection with Bertrand's rule.

Oxidation of Polyalcohols

Glycerol. Since Bertrand's report[157] of the oxidation of glycerol to dihydroxyacetone by a sorbose bacterium that he had isolated, there have been many similar reports of glycerol oxidation by acetic acid bacteria, especially by groups with strong ketogenic activities. To cite them all here would be superfluous. It may be noted, however, that among various fruit acetic acid bacteria examined, Takahashi and Asai[158] obtained L-glyceric acid as the oxidation product of glycerol by *G. cerinus*. This organism possesses both of the following pathways and gave as oxidation products, in addition to the compounds above, acetic acid, glycolic acid, and succinic acid.

The yields of these products from 210 g glycerol with a stationary culture were 82.6 g dihydroxyacetone, 20 g calcium acetate, 1.2 g succinic acid, 6.0 g calcium glycollate, and 32 g calcium L-glycerate. A substance was also obtained which gave a reddish-violet color reaction with $FeCl_3$, but it was not identified further. Later, Ikeda[159] reported the production by *G. roseus* from glycerol and from dihydroxyacetone of a substance which gave a purple color with $FeCl_3$ and reduced 2,6-dichlorophenolindophenol.

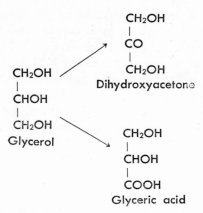

This substance behaved very much like Euler's reductone (the enol-form of glycerosone) in qualitative tests, but could not be obtained in a crystalline form. Ikeda found another unknown reducing substance in the oxidation products. The following schema was proposed for the initial steps of glycerol oxidation.

CH₂OH | CHOH | CH₂OH	−2H →	CH₂OH | CO | CH₂OH	⇄	CHOH || COH | CH₂OH	−2H →	CHO | CO | CH₂OH	⇄	CHO | COH || CHOH
Glycerol		Dihydroxyacetone		Enol-form		Glycerosone		Reductone

Kondo and Ameyama[160] reported the formation of hydroxypyruvic acid.

King and Cheldelin[161] made a series of studies on the oxidation of glycerol by *A. suboxydans* ATCC 621 using cell-free preparations. 2,4-Dinitrophenol at 10^{-4} M did not inhibit the oxidation of glycerol to dihydroxyacetone, but inhibited the further oxidation of dihydroxyacetone. Inorganic phosphate and NAD did not participate in the oxidation of glycerol to dihydroxyacetone. The TCA cycle and the C-4-dicarboxylic acid cycle were absent. The oxidation of dihydroxyacetone required CoA (Cheldelin *et al.*[162]). From these results they proposed the presence of two pathways in glycerol oxidation by this organism. One is independent of ATP and NAD and brings about the accumulation of dihydroxyacetone at pH 6.0. The other has an optimum around pH 8.5 and forms in the presence of ATP and Mg⁺⁺ glycerol-α-phosphate which is further oxidized to dihydroxyacetone phosphate by an NAD-linked dehydrogenase

(Hauge *et al.*[163]). The dihydroxyacetone phosphate thus formed is converted to hexose diphosphate by isomerase and aldolase and further through dephosphorylation and isomerization to glucose-6-phosphate, which is oxidized in the presence of NAD or NADP to 6-phosphogluconate and to ribose-5-phosphate, entering Horecker's pentose cycle (Hauge *et al.*[164]). This is shown in Figure 5. With regard to *G. suboxydans,* however, this cycle is overshadowed during growth by the non-phosphorylative oxidation, which accumulates dihydroxyacetone more rapidly than it can be metabolized.

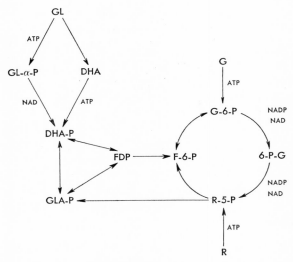

Fig. 5.   Oxidative pathway of glycerol *via* the
pentose cycle in *A. suboxydans*
(after Cheldelin).

GL, glycerol; GL-α-P, glycerol-α-phosphate; GLA, glyceraldehyde; GLA-P, glyceraldehyde-3-phosphate; DHA, dihydroxyacetone; DHA-P, dihydroxyacetone phosphate; FDP, fructose diphosphate; F-6-P, fructose-6-phosphate; G-6-P, glucose-6-phosphate; G, glucose; 6-P-G, 6-phosphogluconate; R-5-P, ribose-5-phosphate; R, ribose.

A list of enzyme systems and intermediate substances of metabolism whose presence in *G. suboxydans* cell-free extracts was either proved or suggested is given below.

    1.   DHA kinase; the formation of DHA-P from DHA in the presence of ATP.

2. Aldolase and triosephosphate isomerase; the formation of fructose from DHA in the presence of ATP and Mg^{++}.
3. The formation of F-6-P and G-6-P from FDP in the presence of Mg^{++} and the release of inorganic phosphate.
4. Phosphohexose isomerase.
5. The formation of 6-P-G and R-5-P accompanying the production of CO_2 from G-6-P.
6. The formation of sedoheptulose phosphate, GLA-P and F-6-P from R-5-P.
7. Glycerokinase; the formation of GL-α-P from glycerol in the presence of ATP.
8. GL-α-P dehydrogenase; the formation of DHA-P from GL-α-P in the presence of NAD and Mg^{++}.
9. The presence of ribokinase was indicated by the fact that ribose itself is not oxidized unless ATP is present.
10. Glucokinase; the formation of G-6-P from glucose. Later Klungsöyr et al.[165] verified phosphorylative oxidation by intact cells of this organism using $^{32}PO_4$.

Kaushal and Walker[142] detected glyceraldehyde as an intermediate during biosynthesis of cellulose from glycerol by *A. acetigenus*. The formation of a small amount of acetoin from glycerophosphate was shown by Federico and Gobis[166] with a strain of *Acetobacter*. Cozic[133] made studies on the oxidation of glycerol derivatives by intact cells of *A. xylinum*. The monoacetyl-compound was oxidized nearly as well as glycerol, but the diacetyl- and triacetyl-derivatives were oxidized very little.

Wine-red pigment formation by *A. acetigenus* in a lactate-buffered glycerol medium was reported by Ramamurti and Jackson.[167] The coloration of the medium appeared only in stationary cultures and not in shaken cultures. Although chemical characterization of the coloring matter was not attempted, it is presumably a manifestation of some metabolic by-product, which reacts with ferrous sulfate in the medium and consequently manifests a red coloration provided that the substance is an γ-pyrone compound.

Erithritol. The oxidation of *meso*-erythritol to L-erythrulose by acetic acid bacteria was shown by Bertrand,[168] Hermann and Neuschul[123] and Müller et al.[169] Whistler and Underkofler[170] also studied the oxidation of *meso*-erythritol to L-erythrulose by *A. suboxydans* and found that the concentration of substrate should not exceed over 4.5%. Under optimum conditions the yield of L-erythrulose was practically quantitative in seven days in stationary culture.

Visser't Hooft[115] reported the oxidation of erythrulose to CO_2 and H_2O by *A. rancens*.   Cozic[133] also reported on the oxidation of erythritol by *A. xylinum*.

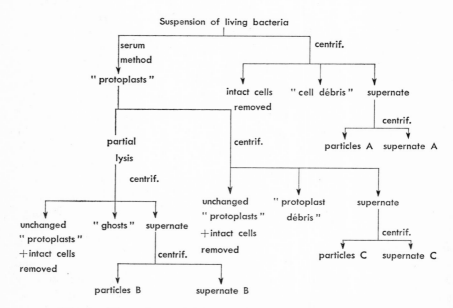

According to De Ley and Dochy[171] intact cells of *G. liquefaciens* oxidized *meso*-erythritol, but the protoplasts were inactive.   Both the intact cells and protoplasts oxidized sorbitol and mannitol, although rather slowly in the case of the protoplasts.

Particles A and cell débris oxidized *meso*-erythritol and glycerol to the corresponding keto-compounds with final uptake of 0.5 mole 0_2/mole

Fig. 6.   Flow-sheet of the preparation of " protoplasts " and subcellular fractions from *G. liquefaciens* (after De Ley and Dochy).

substrate. Ghosts and protoplast débris were scarcely able to carry out this oxidation. The difference is probably due to the temperature-lability of the two oxidases responsible for the substrates. Treatment of particles A at 37°C for 3 hours reduced the oxidation rate of erythritol by 75% and that of glycerol by 60%. It was therefore assumed that during the treatment of protoplasts at 37°C, both oxidase systems, very possibly located on the ghosts, were almost completely destroyed (see Fig. 6 for the preparation of " protoplasts " and various subcellular fractions).

Pentitols. The oxidation of L-arabitol to L-xylulose was noted by Bertrand[157] in *A. xylinum*. D-Arabitol is oxidized to D-xylulose by *A. suboxydans* (Hann *et al.*[172]). At alkaline pH (pH=8), the cell-free preparations of this organism oxidize, though slowly, the two optically active forms of arabitol (Arcus and Edson[173]). Visser't Hooft[115] reported the

L-Arabitol → L-Xylulose (−2H)
D-Arabitol → D-Xylulose (−2H)
Adonitol → L-Adonose (−2H)

oxidation of adonitol by *A. suboxydans*. Reichstein[174] reported the oxidation of adonitol to L-adonose by an unidentified acetic acid bacterium. Arcus and Edson[173] noted a rapid oxidation of xylitol by cell-free extracts of *A. suboxydans* under alkaline conditions.

Deoxy sugar alcohols. Votoček *et al.*[175] reported on the inability of

A. xylinum to oxidize rhodeitol and L-rhamnitol although their configurations follow Bertrand's rule. Anderson and Lardy,[176] however, observed the oxidation of D-rhamnitol to D-fructomethylose with 80% yield by *A. suboxydans.* Müller and Reichstein[177] observed a slight oxidation of L-gulomethylitol, and Hann *et al.*[172] found that *A. suboxydans* oxidized L-fucitol. They did not identify the oxidation product, but Richtmyer *et al.*[178] identified it as L-fuco-4-ketose.

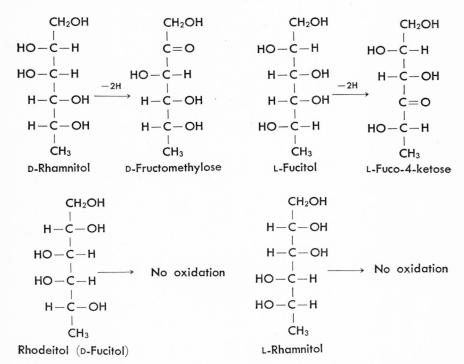

The following structural formulas are shown:

D-Rhamnitol $\xrightarrow{-2H}$ D-Fructomethylose

L-Fucitol $\xrightarrow{-2H}$ L-Fuco-4-ketose

Rhodeitol (D-Fucitol) $\longrightarrow$ No oxidation

L-Rhamnitol $\longrightarrow$ No oxidation

Bollenback and Underkofler[179] reported that D-lyxomethylitol, L-gulomethylitol, and D-rhamnitol were oxidized rapidly and almost quantitatively to the corresponding ketogenic compounds by growing cells of *A. suboxydans,* but L-lyxomethylitol, D-allomethylitol and L-fucitol were oxidized slowly and gave small but definite amounts of reducing compounds. In the oxidation of L-fucitol and L-lyxomethylitol the addition of a small amount of sorbitol to the media and the reinoculation with active cells of *A. suboxydans* after an initial growth period raised the oxidation activity considerably. L-Gulomethylitol, D-gulomethylitol, D-idomethylitol and L-rhamnitol were not oxidized by this organism.

$$
\begin{array}{ccccc}
\text{CH}_2\text{OH} & \text{CH}_2\text{OH} & \text{CH}_2\text{OH} & \text{CH}_2\text{OH} & \text{CH}_2\text{OH} \\
\text{HO}-\text{C}-\text{H} & \text{HO}-\text{C}-\text{H} & \text{H}-\text{C}-\text{OH} & \text{H}-\text{C}-\text{OH} & \text{HO}-\text{C}-\text{H} \\
\text{HO}-\text{C}-\text{H} & \text{HO}-\text{C}-\text{H} & \text{H}-\text{C}-\text{OH} & \text{H}-\text{C}-\text{OH} & \text{H}-\text{C}-\text{OH} \\
\text{H}-\text{C}-\text{OH} & \text{H}-\text{C}-\text{OH} & \text{HO}-\text{C}-\text{H} & \text{H}-\text{C}-\text{OH} & \text{H}-\text{C}-\text{OH} \\
\text{CH}_3 & \text{HO}-\text{C}-\text{H} & \text{CH}_3 & \text{H}-\text{C}-\text{OH} & \text{HO}-\text{C}-\text{H} \\
 & \text{CH}_3 & & \text{CH}_3 & \text{CH}_3 \\
\text{D-Lyxomethylitol} & \text{L-Gulomethylitol} & \text{L-Lyxomethylitol} & \text{D-Allomethylitol} & \text{L-Fucitol}
\end{array}
$$

attacked rapidly   attacked slowly

α-Rhamnohexitol (L-manno-L-*gala*-7-deoxy-heptitol) and β-rhamnohexitol (L-manno-L-*talo*-7-deoxy-heptitol) were not oxidized by *A. xylinum*, according to Votoček *et al.*,[175] and 2-deoxy-D-sorbitol was shown by Regna[180] to be oxidized to 5-deoxy-L-sorbose by *A. suboxydans*.

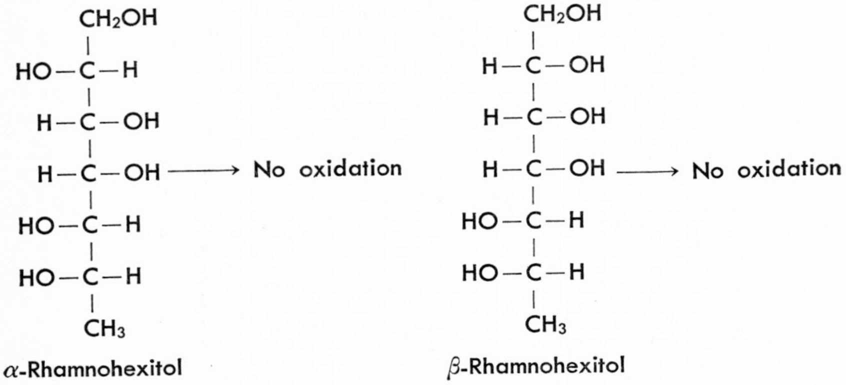

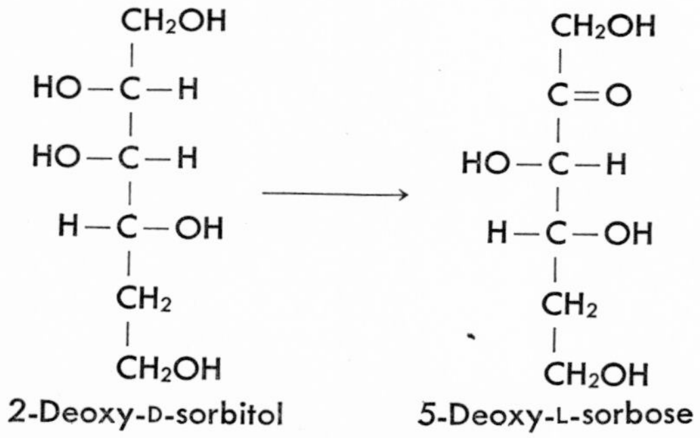

CH₂OH
|
HO—C—H
|
H—C—OH
|
H—C—OH ⟶ No oxidation
|
HO—C—H
|
HO—C—H
|
CH₃

α-Rhamnohexitol

CH₂OH
|
H—C—OH
|
H—C—OH
|
H—C—OH ⟶ No oxidation
|
HO—C—H
|
HO—C—H
|
CH₃

β-Rhamnohexitol

CH₂OH
|
HO—C—H
|
HO—C—H
|
H—C—OH
|
CH₂
|
CH₂OH

2-Deoxy-D-sorbitol

⟶

CH₂OH
|
C=O
|
HO—C—H
|
H—C—OH
|
CH₂
|
CH₂OH

5-Deoxy-L-sorbose

D-Sorbitol.   The oxidation of D-sorbitol to L-sorbose by a certain oxidative bacterium was discovered by Bertrand;[137] the bacterium was called " sorbose bacterium," and was later shown to be *A. xylinum*. This oxidation is called sorbose fermentation and is used for the commercial production of sorbose as a vitamin C intermediate. Later the sorbose fermentation was studied by many workers, and it was shown that *A. xylinoides*, *A. gluconicus*, *A. suboxydans*, *A. melanogenus* and *G. roseus* have a strong oxidative activity. According to Asai *et al.*,[181] *G. roseus* converts D-sorbitol almost quantitatively to L-sorbose.

Nehira[182] reported on the various conditions of sorbose fermentation in stationary culture by *G. liquefaciens*.

The formation of fructose and sorbose from D-sorbitol was reported by Cummins et al.[183] It was shown that cell-free extracts of A. suboxydans formed fructose from D-sorbitol in the presence of NAD and sorbose in the presence of NADP. The enzymatic characteristics of these reactions will be discussed later. Arcus and Edson[173] reported that cell-free extracts of A. suboxydans oxidized sorbitol to sorbose at pH 5.5 and to fructose at pH 8. Mikhlin and Golysheva[184] found that the oxidation of D-sorbitol by A. suboxydans and A. melanogenus was stimulated by catalase. They[185] also reported that the addition of an adequate amount of methylene blue stimulated the production of sorbose. Rasumovskaya and Zhdan-Pushkina[186] studied the influence of inoculating cells on sorbitol oxidation. Cells harvested during a logarithmic phase had a weaker oxidizing activity than those obtained from a stationary phase. Vigorous aeration produced active cells. Cells grown in a nitrogen-rich medium had weak oxidation activity.

Mitjuschowa[187] studied the oxidation of sorbitol by A. suboxydans cells grown in a yeast extract medium and a medium containing 0.03% NH_4NO_3 and 0.05% vitamin B complex (both containing 10% sorbitol). There were two simultaneous reactions: 1) oxidation of sorbitol to sorbose, and 2) formation of acetic acid from sorbitol up to a ratio of 1:100. According to Zhdan-Pushkina,[188] when the concentration of vitamin B complex was raised to 0.5–5%(?), growth of the organism was stimulated, reaching a cell number of 2,000 to 3,000 million/ml. Furthermore when the formed acid was neutralized, the cell number reached a peak of 5,000 million. The oxidation of sorbitol, however, dropped from 96.1% to 91.0%.

According to Görlich,[189] the addition of 0.01–0.03% sorbic acid to the fermentation medium of A. suboxydans in stationary or submerged culture was effective in preventing yeast infection without inhibiting the activity of sorbose fermentation.

Shchelkunova[190] investigated the effect of phosphate concentration on the multiplication of A. suboxydans and A. melanogenus and on their oxidation of sorbitol in a synthetic medium containing ammonium sulfate and the vitamin B complex. They observed that both organisms multiplied rapidly and oxidized sorbitol strongly when the amount of phosphate (KH_2PO_4) was relatively low (ca. 0.1%). An increase to 0.4–0.8% phosphate inhibited the multiplication and the oxidative activity of the organisms. The adverse effect of the increased amount of phosphate was attributed in part to the increase in buffer capacity of the medium.

grown in media with a poor vitamin supply showed delayed reproduction but oxidized sorbitol to sorbose more effectively than those grown in vitamin-enriched media.

In recent studies by Kulhanek and Sevcikova,[192] G. suboxydans and G. melanogenus produced a diketohexose having an R_f value of 0.29 (in paper chromatography with an H_2O-saturated phenol solution) from D-sorbitol. This substance appeared when 80% of the D-sorbitol had been converted to L-sorbose and D-fructose. Mannitol gave the same diketohexose.

The formation of 2-keto-L-gulonic acid and the other conversion products of D-sorbitol will be discussed later in the section on sorbose oxidation.

The industrial production of sorbose from sorbitol by submerged growth of A. suboxydans in a rotary drum fermentor has been successfully carried out by Wells et al.[193,194] and by Matsukura et al.[181] in the case of G. roseus. According to the latter authors, the yield of crystal sorbose based on sorbitol supplied was approximately 76% in 18 hours, when 10% solutions of sorbitol were fermented. Solutions containing 20% sorbitol required 28 hours for complete oxidation (98%).

D-**Mannitol.** The oxidation of D-mannitol to D-fructose was reported in many ketogenic acetic acid bacteria, as in the case of D-sorbitol oxidation. Fulmer and Underkofler[195] examined the possibilities of utilizing this oxidation for the industrial production of D-fructose using A. suboxydans. Interestingly, Hermann and Neuschul[123] reported that A. pasteurianus produced D-mannose as well as D-fructose from D-mannitol, and that A. kützingianus produced only D-mannose from D-mannitol. Isbell and Karabinoos[196] obtained D-fructose-1,6-^{14}C from the oxidation of D-mannitol-1,6-^{14}C by A. suboxydans.

According to Frush and Tregoning,[197] oxidation by A. suboxydans of D-mannitol-2-^{14}C was slower than that of D-mannitol-5[14]C.

Recently Terada et al.[198] found that A. suboxydans, A. oxydans, A. gluconicus, A. melanogenus, G. roseus and G. cerinus oxidized D-fructose with 5-ketofructose as an intermediate.

Other **hexitols.** Brown[138] reported the inability of A. aceti to oxidize dulcitol and Bertrand[157] also reported that A. xylinum, which oxidized sorbitol, failed to oxidize dulcitol. Hermann and Neuschul[123] found that A. gluconicus oxidized dulcitol to galactose and Asai,[199] using the acetic acid bacteria isolated from fruits, found the oxidation of dulcitol and identified the product as galactose. Steiger and Reichstein[200] reported the oxidation of meso-allitol to L-allulose (L-psicose) by A. xylinum.

Zhdan-Pushkina and Kreneva[191] reported that cells of G. suboxydans

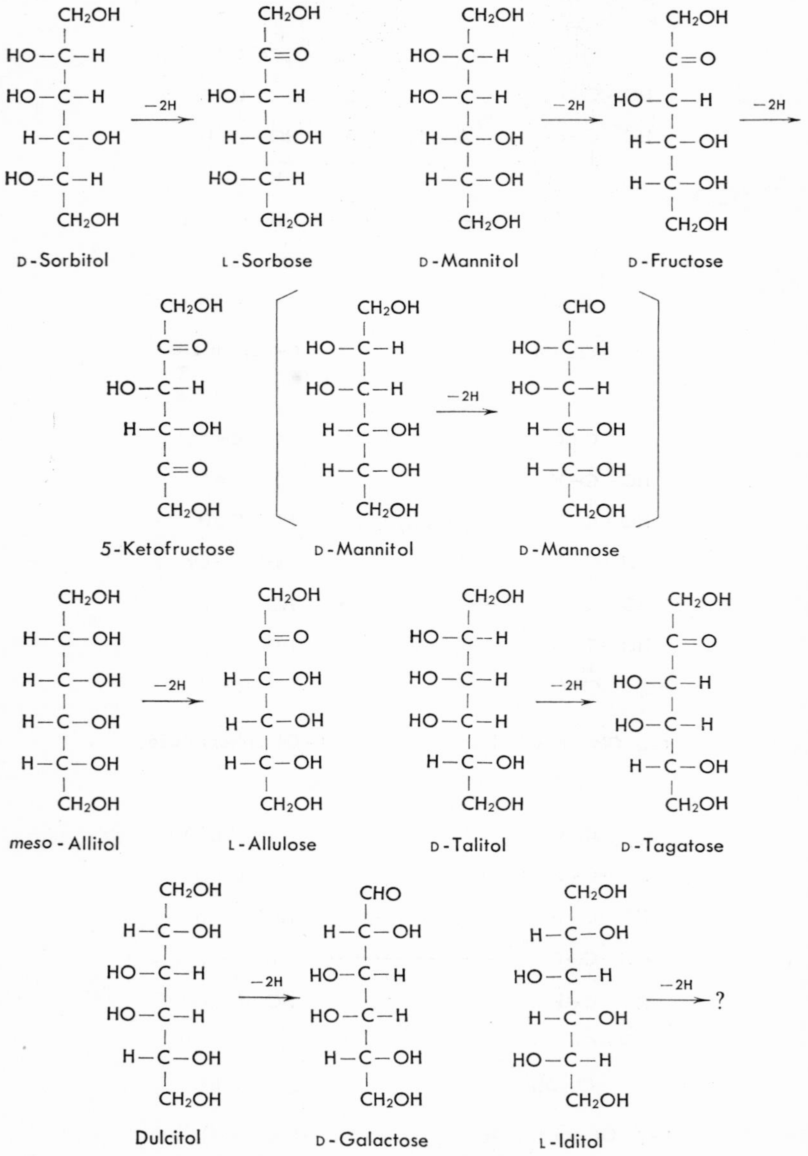

Fig. 7. Oxidation of hexitols.

```
        CH₂OH                                    CH₂OH
         |                                        |
   HO — C — H                                     C = O
         |                                        |
   HO — C — H        A. suboxydans          HO — C — H
         |                                        |
    H — C — OH        A. xylinum             H — C — OH
         |                                        |
    H — C — OH      ──────────────→          H — C — OH
         |                                        |
   HO — C — H             −2H             OH — C — H
         |                                        |
        CH₂OH                                    CH₂OH

     D- Perseitol                           L - Perseulose
```

```
        CH₂OH                                    CH₂OH
         |                                        |
   HO — C — H                                     C = O
         |                                        |
   HO — C — H        A. suboxydans          HO — C — H
         |                                        |
    H — C — OH        A. xylinum             H — C — OH
         |                                        |
   HO — C — H      ──────────────→         HO — C — H
         |                                        |
   HO — C — H             −2H              HO — C — H
         |                                        |
        CH₂OH                                    CH₂OH

   D- α- Glucoheptitol                     L - Glucoheptulose
```

```
        CH₂OH                                    CH₂OH
         |                                        |
   HO — C — H                                     C = O
         |                                        |
   HO — C — H         A. xylinum            HO — C — H
         |                                        |
    H — C — OH      ──────────────→          H — C — OH
         |                                        |
   HO — C — H             −2H              HO — C — H
         |                                        |
    H — C — OH                              H — C — OH
         |                                        |
        CH₂OH                                    CH₂OH

   D- β- Glucoheptitol                      D- Idoheptulose
                                              (Predicted)
```

Fig. 8.   Oxidation of heptitols

```
      CH₂OH                    CH₂OH                        CH₂OH
        |                        |                            |
      C=O                   HO—C—H                       HO—C—H
        |          A. xylinum    |        A. suboxydans       |
   HO—C—H    ◄─────────    HO—C—H    ─────────►        HO—C—H
        |                        |                            |
    H—C—OH          -2H     H—C—OH          -2H          H—C—OH
        |                        |                            |
    H—C—OH                   H—C—OH                       H—C—OH
        |                        |                            |
    H—C—OH                   H—C—OH                        C=O
        |                        |                            |
      CH₂OH                    CH₂OH                        CH₂OH

   Sedoheptulose             D - Volemitol              D - Mannoheptulose
```

```
      CH₂OH                    CH₂OH
        |                        |
   HO—C—H                      C=O
        |                        |
   HO—C—H                   HO—C—H
        |      A. suboxydans      |
    H—C—OH    ─────────►      H—C—OH
        |                        |
   HO—C—H                   HO—C—H
        |                        |
   HO—C—H                   HO—C—H
        |                        |
   HO—C—H                   HO—C—H
        |                        |
      CH₂OH                    CH₂OH

 D - α,β- Glucooctitol     L - Altro - L - sorbo - octose
                                  (Predicted)
```

```
      CH₂OH
        |
   HO—C—H
        |
    H—C—OH
        |
    H—C—OH              ─────────►       No oxidation
        |
   HO—C—H
        |
   HO—C—H
        |
    H—C—OH
        |
      CH₂OH

 D - α,α - Galaoctitol
```

and octitols.

D-Talitol, according to Totton and Lardy,[201] is oxidized by *A. suboxydans* to D-tagatose with a yield of 75–84%. This reaction is used for the production of D-tagatose. In order to have a good yield the concentration of D-talitol should be below 5%.

L-Iditol, like D-sorbitol, is present in the berry juice of the mountain ash. This was not oxidized by *A. xylinum* in Bertrand's experiments,[202] but recently Arcus and Edson[173] reported the oxidation of L-iditol by cell-free extracts of *A. suboxydans* in alkaline conditions (Fig. 7).

Heptitols and octitols.   The oxidation of perseitol to L-perseulose by *A. xylinum* was first reported by Bertrand.[203] Hann *et al.*[172] confirmed the oxidation in *A. suboxydans*. The industrial production of perseulose with this organism was studied by Tilden.[204]

Earlier, Bertrand[157] had observed the oxidation of volemitol to a ketoheptose by *A. xylinum*. Stewart *et al.*[205] identified it as sedoheptulose. According to Ettel *et al.*,[206] however, the oxidation product of volemitol by *A. suboxydans* is a mixture of sedoheptulose and D-mannoheptulose.

The oxidation of D-α-glucoheptitol to L-glucoheptulose by *A. xylinum* was reported by Bertrand and Nitzberg.[207] Hann *et al.*[172] confirmed the oxidation in *A. suboxydans*. Cozic[208] found that not only D-α-glucoheptitol but also β-glucoheptitol was oxidized by *A. xylinum* to ketoheptose. This was reported also by Khouvine and Nitzberg.[209] Moore *et al.*[210] examined various yeasts and bacteria for the oxidation of heptitol and found that only *Acetobacter* spp. could oxidize the compound.

Hann *et al.*[172] reported the oxidation of D-gluco-L-*talo*-octitol (D-α,β-glucooctitol) by *A. suboxydans* to L-altro-L-*sorbo*-octose (predicted), but D-*gala*-L-*gala*-octitol (D-α,α-galaoctitol) was not oxidized (Fig. 8).

Glycerol- and Other Polyol-Dehydrogenases

As stated previously acetic acid bacteria oxidize not only ethanol but also glycerol and other polyols. The enzyme systems involved in the oxidation of polyols have been studied rather recently. Goldschmidt and Krampitz[156] reported the oxidation of glycol to the corresponding ketol by an NAD-linked dehydrogenase. Most studies are concerned with the oxidation of glycerol, mannitol, sorbitol, and cyclic polyalcohols.

Glycerol dehydrogenase.   King and Cheldelin[94] proposed the presence of two enzyme systems for glycerol oxidation in the cell-free extracts of *A. suboxydans*, one being a pyridine nucleotide independent glycerol dehydrogenase with maximum activity at pH 6.0 and the other being most active at pH 8.5 and converting glycerol to α-glycerophosphate with ATP, then oxidizing it to dihydroxyacetone phosphate by an NAD-linked α-glycerophosphate dehydrogenase (Hauge et al.[163]). The details of these systems were discussed in a previous section.

De Ley and Dochy[171] studied the localization of oxidase systems in the cells of G. liquefaciens (see Fig. 6) and found that the oxidation of glycerol and meso-erythritol by particles A and by "cell débris" was carried out only to the corresponding keto-derivatives stage with the final uptake of 0.5 mole O_2/mole substrate. The " ghosts " and the " protoplast débris " were scarcely able to carry out this oxidation.

Treatment of particles A for 3 hours at 37°C remarkably decreased both oxidations, and it may be assumed that during the preparation of the " protoplasts " at 37°C, both oxidase systems, which are probably located on the " ghosts," were almost completely destroyed.

Mannitol dehydrogenase. Müller[211] reported the presence of mannitol dehydrogenase in the cells of acetic acid bacteria. A detailed study of the polyol dehydrogenase of these organisms was carried out by Arcus and Edson.[173] According to this study the cell-free extracts of A. suboxydans ATCC 621 have an NAD- or NADP-independent particulate dehydrogenase active on mannitol and other polyols at pH 5, the substrate specificity following Bertrand-Hudson's rule. This enzyme is associated with cytochrome-containing particles and is called cytochrome-linked D-mannitol dehydrogenase or " acid enzyme." The organism also contained a soluble NAD-linked polyol dehydrogenase which was most active at pH 8. This enzyme is called NAD-linked D-mannitol dehydrogenase or " alkaline enzyme " (Table 12). Its substrate specificity does not follow a simple rule. The presence of NADP-linked D-mannitol dehydrogenase in A. suboxydans was also reported by other workers. (See Sasajima and Isono's work[215] which is introduced later. See also the Addendum at the end of the text.)

Sorbitol dehydrogenase. According to the studies of Cummins et al.,[183] the cell-free extracts of A. suboxydans contain the enzymes participating in three pathways of sorbitol oxidation. In addition to a particulate dehydrogenase (Widmer et al.[212]) catalyzing one-step oxidation of sorbitol (presumably to hexose), there are two alternative enzyme systems in soluble portions of the cells. One forms sorbose in the presence of NADP and the other fructose in the presence of NAD.

Fructose is phosphorylated with ATP and is further oxidized via the pentose cycle. This was presumed from the identification of oxidation

Table 12. Substrate specificity of "acid" and "alkaline" dehydrogenases of
A. suboxydans ATCC 621 (after Arcus and Edson).

Substrate	"Acid" dehydrogenase without NAD	"Alkaline" dehydrogenase with NAD
Erythritol	Attacked rapidly	Not attacked
Ribitol	„ „	Attacked slowly
D-Arabitol	„ „	„ „
L-Arabitol	Not attacked	„ „
Xylitol	Attacked slowly	Attacked rapidly
Allitol	„ „	—
D-Mannitol	Attacked rapidly	Attacked rapidly
D-Gulitol	Not attacked	Not attacked
Sorbitol	Attacked rapidly	Attacked rapidly
Dulcitol	Not attacked	Not attacked
D-Talitol	Attacked slowly	Attacked slowly
D-Iditol	—	„ „
L-Rhamnitol	—	Not attacked
Perseitol	Attacked slowly	„ „
Voleimitol	„ „	Attacked slowly
β-Sedoheptitol	„ „	Attacked rapidly
L-*gulo*-D-*gala*-Heptitol	Not attacked	„ „
meso-Inositol	„ „	Not attacked
(+)-Inositol	„ „	„ „
(−)-Inositol	„ „	„ „

products in the presence of NAD.   Sorbose, however, cannot be further
phosphorylated or oxidized with cell-free extracts and can be dissimilated
only in whole cells in the presence of an energy source.   Widmer *et al.*[212]
purified the NAD-enzyme of the two dehydrogenases about 16-fold and
studied its characteristics.   The NADP-enzyme was so fragile that it was
destroyed almost completely during fractionation.   The NAD-linked sorbitol dehydrogenase was resistant against heat denaturation in the presence
of substrate and pyridine nucleotides.

The optimum pH of both enzymes was about the same (8.0 to 8.5),
but the activity range for the NAD-enzyme was much broader. Mg^{++} or
Mn^{++} stimulated the activity of both enzymes, Zn^{++} or Ca^{++} did not.
Sulfhydryl reagents at high concentrations inhibited the purified NAD-
linked sorbitol dehydrogenase.   Inhibition by *p*-chloromercuribenzoate

and $HgCl_2$ in particular required high concentrations. The purified NAD-linked sorbitol dehydrogenase was specific for sorbitol and did not oxidize mannitol, ribitol, dulcitol, perseitol, glycerol, ethanol, acetaldehyde, or 2-butene-1,4-diol.

Elsaesser et al.[213] studied sorbitol dehydrogenase activity in relation to the growth phase of A. melanogenus in shaking culture and found that the cells harvested after 11 hours ' culture, corresponding more or less to the first half of the logarithmic phase, showed maximum dehydrogenating activity. Quantities up to one half of the sorbitol supplied were dehydrogenated during the logarithmic phase, and the remainder was dehydrogenated in the stationary phase. They also studied the influence of heavy metals and found that the highest concentration of each metal at which growth was not inhibited was 0.005 mg/ml for Cu^{++}, 0.01 mg/ml for Ni^{++}, 0.5 mg/ml for Fe^{++} and 0.1 g/ml for Mn^{++}. Cu^{++} blocked the respiratory enzyme, whereas Ni^{++}, Fe^{++} and Mn^{++} had no effect on the respiration at concentrations where growth was inhibited. Huber et al.[214] also studied the influence of heavy metals on sorbitol dehydrogenase activity in A. suboxydans and reported that Cu^{++} (8 μg/ml) and Ni^{++} (20 μg/ml) showed 60% inhibition, whereas Fe^{++} (4 μg/ml) and Hg^{++} (10 μg/ml) showed complete inhibition. NaCN stimulated the activity in parallel with the concentration up to 40 μg CN^-/ml; a higher concentration resulted in inhibition. NaN_3 had no effect up to a concentration of 300 μg/ml. The inhibitory action of Cu^{++} and Ni^{++} was fortified by the addition of oxytetracycline and EDTA, but not by dehydrostreptomycin, whereas that of Hg^{++} and Fe^{++} was not influenced by the addition of these compounds.

Sasajima and Isono[215] investigated the polyol dehydrogenase system using the soluble and particulate fraction of the cells of G. melanogenus IFO 3292 grown in sorbitol medium. NAD-linked mannitol dehydrogenase, NADP-linked mannitol dehydrogenase and NAD-linked sorbitol dehydrogenase were obtained from the soluble fractions by fractionation with ammonium sulfate and DEAE-cellulose column chromatography. NAD-linked mannitol dehydrogenase was specific for the interconversion between D-mannitol and D-fructose and was very unstable in alkaline conditions (inactivated at pH>8.5), while its optimum pH was 8.9. NADP-linked mannitol dehydrogenase reduced 5-keto-D-fructose to fructose in the presence of $NADPH_2$. This enzyme was stable against heat and the activity did not decrease even after the incubation period of 30 min at 50°C, while that of the NAD-linked enzyme decreased considerably in the course of heat treatment. NAD-linked sorbitol dehydrogenase reduced

5-keto-D-fructose to L-sorbose in the presence of $NADH_2$. The pH optima were 9.4 for both NAD-linked sorbitol and NADP-linked D-mannitol dehydrogenase.

The pH optima in the reverse reaction were 6.5 for NAD-linked D-mannitol dehydrogenase, between 5.0 and 7.5 for NAD-linked sorbitol dehydrogenase and 7.5 for NADP-linked D-mannitol dehydrogenase. All three enzymes were inhibited with sulfhydryl reagents such as p-chloromercuribenzoate, Cu^{++} or Zn^{++}. Thioglycollate, however, did not activate nor reactivate the enzymes. Metal chelating reagents such as ethylene-diamine-tetraacetate, o-phenanthroline and 8-hydroxyquinoline had no effect on any of the enzyme activities.

The particulate fraction contained the oxidative enzyme system catalyzing the oxidation of mannitol, sorbitol, ribitol, D-arabitol, erythritol, and glycerol, probably caused by a single enzyme not specific to the substrate.

Addendum.   Oxidation of Disaccharide Alcohols

French et al.[216] reported that A. suboxydans could not oxidize melibiitol and maltitol, but oxidized epimeliitol (1-galactosyl mannitol) to a ketose with an R_f value corresponding to that of planteobiose. It was suggested that this oxidation can be used for the preparation of ketodisaccharides or higher oligosaccharides, provided that the corresponding sugar alcohols are available.

Chapter 6

STEREOCHEMICAL CHARACTERISTICS OF
THE OXIDATION OF POLYALCOHOLS

Bertrand[217] studied the oxidation of various sugar alcohols by his sorbose bacterium (*A. xylinum*). From his observations he concluded that *A. xylinum* oxidizes only those sugar alcohols having the *cis* arrangement of the two secondary alcohol groups contiguous to the primary alcohol group. This general rule is called Bertrand's rule.

$$
\begin{array}{ccc}
\underset{\text{Attacked}}{\underset{\displaystyle \substack{|\ \ |\\ \text{OH OH}}}{\overset{\displaystyle \substack{\text{H H}\\ |\ \ |}}{\text{CH}_2\text{OH}-\text{C}-\text{C}-\text{R}}}}
& \xrightarrow{\text{oxidized to}}
& \underset{\displaystyle \substack{|\\ \text{OH}}}{\overset{\displaystyle \substack{\text{O H}\\ \|\ \ |}}{\text{CH}_2\text{OH}-\text{C}-\text{C}-\text{R}}}
& \underset{\underset{\text{Unattacked}}{\displaystyle \substack{|\ \ |\\ \text{H}\ \ \text{OH}}}}{\overset{\displaystyle \substack{\text{OH H}\\ |\ \ |}}{\text{CH}_2\text{OH}-\text{C}-\text{C}-\text{R}}}
\end{array}
$$

The formation of L-sorbose from D-sorbitol, D-fructose from D-mannitol, D-xylulose from D-arabitol, L-erythrulose from *meso*-erythritol, and L-perseulose from D-perseitol follows Bertrand's requirement of configuration.

L-Rhamnitol and β-rhamnohexitol, however, are not oxidized although they too follow Bertrand's rule. The oxidation of dulcitol reported by Hermann and Neuschul[123] and Asai[199] is an example of cases not following the rule. The oxidation of gluconic acid to 2-ketogluconic acid is another example, although gluconic acid is not a sugar alcohol.

Hann *et al.*[172] studied the oxidation of a number of sugar alcohols and two pairs of enantiomorphs, using *A. suboxydans*, and found that D-arabitol and D-perseitol were readily oxidized, but their enantiomorphs were not attacked. They properly concluded that the oxidation by the organism is so specific that the *cis* pair of secondary alcohol groups must have the D-configuration. The fact that only ketoses of L-configuration are obtained in high yields from *meso*-sugar alcohols, which was previously shown to be the case in the formation of L-erythrulose from *meso*-ery-

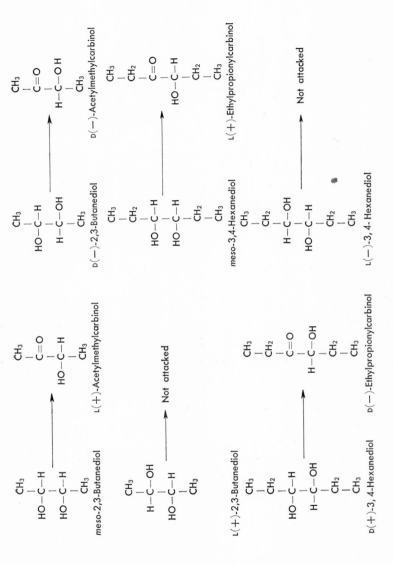

Fig. 9. Oxidation of glycols.

thritol, L-adonulose from *meso*-adonitol and L-allulose from *meso*-allitol, confirms this generalization. This is known as Hudson's rule.

meso-2,3-Butanediol, which is similar to *meso*-erythritol except that it has two terminal methyl groups, is oxidized to L(+)-acetoin by *A. suboxydans*. D(−)-2,3-Butanediol is also oxidized to D(−)-acetoin by the same organism, while L(+)-,2,3-butanediol is scarcely oxidized (Underkofler et al.[218]).

Bertrand's rule is followed only in the case of the *cis* secondary alcohol group with D-configuration. With glycols, as in the case of D(−)-2,3-butanediol and of D(+)-3,4-hexanediol, only the secondary alcohol group with D-configuration is oxidized, but the *cis* arrangement is not required (Fig. 9).

In the case of deoxy sugar alcohols, however, there are cases, as in L-fucitol oxidation, which do not follow either Bertrand's rule or the requirement of D-configuration.

Fulmer and Underkofler,[219] after studying the oxidation of 31 straight chain sugar alcohols, excluding the exceptional case of L-fucitol, summarized the substrate specificity of *A. suboxydans* as follows.

1. Only 2-keto compounds are formed from sugar alcohols having a terminal primary alcohol group.

2. The secondary alcohol group oxidized must possess the D-configuration.

3. Polyalcohols having more than two secondary alcohol groups must have *cis* arrangement as well as D-configuration.

The same authors also classed polyalcohols into the following four groups according to the way in which they are oxidized by *A. suboxydans*.

1. Oxidized at high concentrations (25% or above)—sorbitol and D-mannitol.

2. Oxidized at relatively low concentrations—glycerol and *meso*-erythritol.

3. Require additional assimilable substrate for subculture—*meso*-inositol, *meso*-2,3-butanediol, and D(−)-2,3-butanediol.

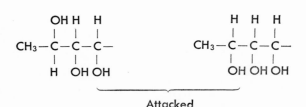

Attacked

4. Not oxidized even in the presence of an additional assimilable substrate—dulcitol, L-rhamnitol, and L(+)-2,3-butanediol.

Richtmyer et al.,[178] studying the oxidation of ω-deoxy sugars by A. suboxydans, suggested that Bertrand's rule is applicable to those sugars if the CH₃CHOH group is considered as the primary alcohol group stated in the rule.

Chapter 7

OXIDATION OF CYCLIC POLYALCOHOLS WITH REFERENCE TO STEREOCHEMICAL CHARACTERISTICS

A. suboxydans oxidizes chain polyalcohols readily to the corresponding ketogenic compounds. Dunning *et al.*,[220, 221] however, were among the first to study the oxidation of cyclic polyalcohols. They inoculated a *meso*-inositol medium supplemented with a small amount of sorbitol with *A. suboxydans* and identified a ketoinositol purified from the fermentation mixture as diketo-*meso*-inositol, from the analytical results of acetyl derivatives and other analytical procedures. Previously, however, Kluyver and Boezaardt[222] had identified the product obtained from the fermentation of *meso*-inositol by the same organism as scyllo-*meso*-inosose. Dunning *et al.* attributed the apparent contradiction to differences in strains, cultural conditions, *etc.*

Later, however, Posternak[223] and Chargaff and Magasanik[224] also identified the oxidation product as scyllo-*meso*-inosose, *i.e.* myo-(*meso*)-inosose, and Carter *et al.*[225] confirmed their results.

myo-(*meso*)-Inositol   myo-(*meso*)-Inosose-2

Magasanik and Chargaff[226] studied the oxidation of L-, D-, and *epi*-inositol and proposed the following theoretical generalization.

Inositol takes the form of a cyclohexane having a stable " chair " form. There are two possible types of bonding of the C and OH of the ring, resulting in their pointing in different directions with regard to the rotatory axis of the ring.   One is a bonding parallel to the axis and is called (a) bonding (axial).   The other has an angle of 109°28′ to the axis and is called (e) bonding (equatorial).   At every carbon atom one bond extends above the horizontal plane of the ring and another below.   The situation is reversed at the neighboring carbons.   The stereochemical structure is shown in Figure 10.

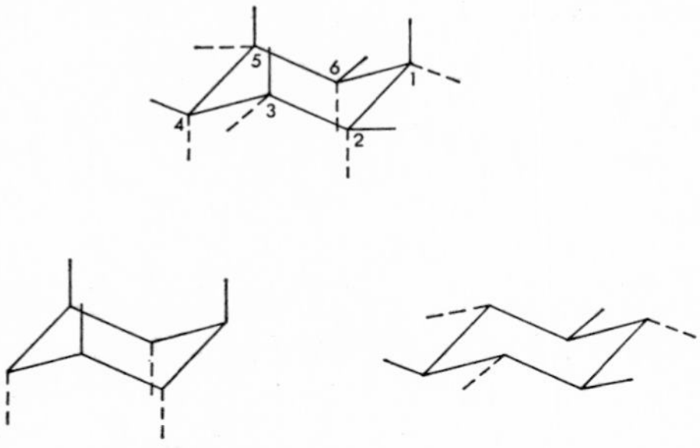

(a) bonding   (e) bonding

Fig. 10.   Structure of cyclohexane
(chair form).

There are eight known isomers of inositol.   The stereochemical con-formation and the oxidation products of the isomers used by Magasanik *et al.*[227] are shown in Figure 11.

Thus the specific steric requirements for oxidation could be defined as follows: Only axial hydroxy groups are oxidized (rule 1), and the carbon atom in *meta* position to the one carrying the axial hydroxy group (in coun-terclockwise direction, if north axial; clockwise, if south axial) must carry one equatorial hydroxy group (rule 2).   This generalization is known as Chargaff's rule.   It was further suggested that an additional rule might also apply, *viz.*, there must be an equatorial hydroxy group in *para* position to the one carrying the axial hydroxy group (rule 3).

Later, however, Posternak and Reymond[228] pointed out exceptions to the rule, especially among triols and tetrols.   Anderson *et al.*,[229] using *A. suboxydans* ATCC 621, studied the oxidation of cyclitols and their derivatives not examined by Magasanik *et al.* and obtained the results shown in Figure 12 (the oxidation products were not identified).

It was concluded that Chargaff's rule is a valid description of the minimal steric requirements for higher cyclitol oxidation, but that the cyclic polyalcohols which fulfil these requirements are not necessarily oxidized.

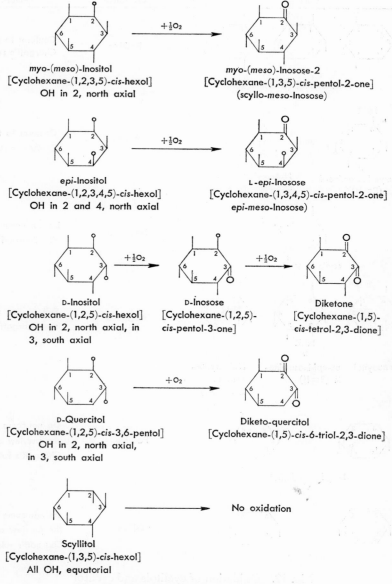

Fig. 11. Oxidation of inositol isomers by
A. suboxydans ATCC 621.

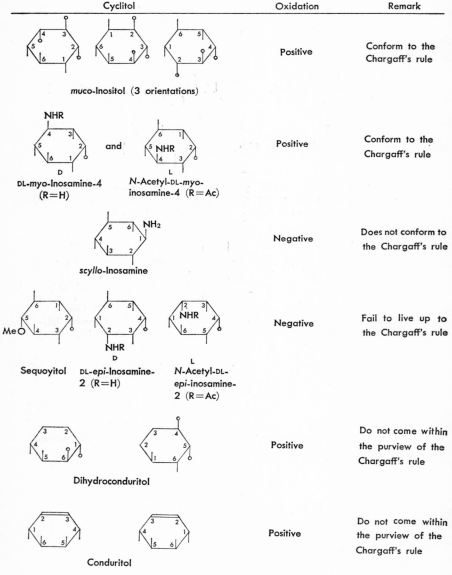

Fig. 12. Oxidation of cyclitols and cyclitol
derivatives by *A. suboxydans* ATCC 621
(after Anderson *et al.*).

In another experiment with 1,-*cis*-2,*cis*-3-cyclohexanetriol (*γ*-cyclo-hexanetriol) as the substrate, oxidation occurred, contradictory to the rule. The enzyme which oxidizes the triol may be different from that which oxidizes higher cyclitols, since the triol tested was not attacked in cell-free extracts of the organism.

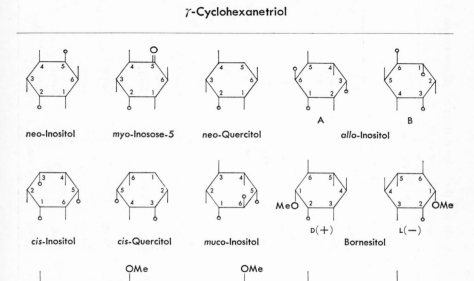

γ-Cyclohexanetriol

Fig. 13.   Formulas of the cyclitols and some derivatives.

Table 13. Oxidation of cyclitols and their derivatives by *A. suboxydans* ATCC 621 (after Anderson *et al.*).

Substrate	Position of axial hydroxyls	Predicted oxidizable position after rules		Position attacked	Oxidation product
		1, 2 and 3	1 and 2		
neo-Inositol	2, 5	None	2, 5	2, 5	2, 5-Diketo-*myo*-inositol
myo-Inosose-5	2	None	2	2	2, 5-Diketo-*myo*-inositol
neo-Quercitol	2	None	2	2	5-Deoxy-*myo*-inosose-2
allo-Inositol, form A	1, 3, 6	1	1, 6	1 (and?)	{D-*allo*-Inosose-1
allo-Inositol, form B	1, 3, 6	None	3, 6	?	(+Diketone)
cis-Inositol	1, 3, 5	None	None	1	*cis*-Inosose
cis-Quercitol	3, 5	None	None	3 or 5	3-Deoxy-*cis*-inosose
muco-Inositol	1, 5, 6	1, 6	1, 6	1(?)(and?)	Not identified
D(+)-Bornesitol	2	2	2	2	D-1-O-Methyl-*myo*-inosose-2
L(−)-Pinitol	2, 3	3	2, 3	3	D-5-O-Methyl-*myo*-inosose-1
L(−)-Bornesitol	2	2	2	None	—
Sequoyitol	2	None	2	None	—
Dambonitol	2	2	2	None	—
Quebrachitol	2, 3	2	2	None	—
D(+)-Pinitol	2, 3	None	2	None	—
5,6-Di-O-Methyl-L-inositol	2, 3	None	2, 3	None	—
D-1-O-Methyl-*allo*-inositol	2, 4, 5	None	2	None	—
L(+)-Ononitol	2	None	None	None	—

Recently Anderson *et al.*[230] studied the oxidation of several rare inositols and quercitols, inositol methyl ethers and two cyclohexanetetrols by *A. suboxydans* ATCC 621. Formulas of the cyclitols studied and the results of their oxidation are shown in Figure 13 and Table 13 respectively.

Dambonitol and quebrachitol, with axial hydroxyl groups conforming to Chargaff's rules 1, 2, and 3, were not oxidized. D(+)-Pinitol, 5,6-di-*O*-methyl-L-inositol, and D-1-*O*-methyl-*allo*-inositol were not oxidized although their structures conformed to rules 1 and 2. L(+)-Ononitol and 2-*O*-methyl-*myo*-inositol did not conform to any rule and were not oxidized.

From these results Anderson *et al.* proposed amending Chargaff's rule 3 as follows: The oxidation of an axial hydroxyl in a higher cyclitol proceeds best when there is an equatorial hydroxyl in the *para* position, and when there is a choice, the axial hydroxyl satisfying this rule is attacked preferentially.

However, when different strains of bacteria are used the above rules do not necessarily apply. Posternak *et al.*[231] reported that only rule 1 applied to the oxidation of cyclitols by *A. suboxydans* Kluyver.

Posternak[232] reported that *A. suboxydans*, following Chargaff's rule, oxidized D-viburnitol and L-viburnitol (deoxy-inositol) to the corresponding monoketones.

Posternak and Ravenna[233] studied the oxidation of cyclohexanetriols-1,2,3 by *A. suboxydans* and *A. xylinum*. For oxidation at least two neighboring OH groups with *cis* relation are necessary. When they tested triols with three neighboring OH groups with *cis* relation they confirmed the oxidation of 1, *cis*-2, *trans*-3-cyclohexane triol (β-cyclohexane triol) and 1, *cis*-2,*cis*-3-cyclohexane triol (γ-cyclohexane triol) to the corresponding monoketones.

Posternak and Reymond[228] reported that cyclitols with five or six hydroxyls or related substituents were oxidized according to Chargaff's rule, but the oxidation of those with a smaller number of hydroxyls or substituents did not follow the rule.

Chapter 8

OXIDATION OF MONOSACCHARIDES

The oxidation of ethyl alcohol to acetic acid characterizes the biochemical activity of acetic acid bacteria. As stated previously, however, various other aliphatic (chain and cyclic) and aromatic compounds with alcohol radicals (primary and secondary alcohol radicals) or aldehyde radicals are also widely oxidized by acetic acid bacteria depending on their configurations.

The second important biochemical activity of acetic acid bacteria is the oxidation of glucose to gluconic acid. Both of these activities have been recognized for a long time.

Since the oxidative metabolism of carbohydrates is an important aspect of bacterial metabolism, there have recently been an increasing number of studies on carbohydrate metabolism using acetic acid bacteria. In this chapter the oxidation of monosaccharides (aldoses and ketoses) will be discussed.

Oxidation of Aldoses

Many acetic acid bacteria oxidize aldoses to the corresponding onic acids and very often accumulate them. According to Visser't Hooft,[115] *A. xylinum*, *A. suboxydans*, and *A. melanogenus* oxidize glycolaldehyde to glycolic acid. Murooka et al.[234] reported the uptake of 0.5 mole O_2 per 1 mole of glycolaldehyde by intact cells of *G. suboxydans* ATCC 621 and identified the oxidation product as glycolic acid by paper chromatography. Using Calkin's method, 95% conversion of glycolaldehyde to glycolic acid was reported.

Bertrand[235] reported the oxidation by *A. xylinum* of L-arabinose and L-xylose to L-arabonic acid and L-xylonic acid, respectively. Hermann and Neuschul[123] reported the formation of L-arabonic acid from L-arabinose by *A. gluconicus*, *A. xylinoides*, *A. aceti*, and *A. acetosus*. Frateur[236] examined various acetic acid bacteria for pentonic acid forming abilities and

178

$$
\begin{array}{ccc}
\text{CHO} & & \text{COOH} \\
| & & | \\
\text{H}-\text{C}-\text{OH} & & \text{H}-\text{C}-\text{OH} \\
| & & | \\
\text{HO}-\text{C}-\text{H} & +\tfrac{1}{2}\text{O}_2 & \text{HO}-\text{C}-\text{H} \\
| & \longrightarrow & | \\
\text{H}-\text{C}-\text{OH} & & \text{H}-\text{C}-\text{OH} \\
| & & | \\
\text{H}-\text{C}-\text{OH} & & \text{H}-\text{C}-\text{OH} \\
| & & | \\
\text{CH}_2\text{OH} & & \text{CH}_2\text{OH} \\
\text{D-Glucose} & & \text{D-Gluconic acid}
\end{array}
$$

found the formation of arabonic acid from arabinose by *A. rancens* and *A. xylinum*, and arabonic and xylonic acids from arabinose and xylose, respectively, by *A. suboxydans*, *A. suboxydans* var. *muciparum*, *A. melanogenus*, *A. melanogenus* var. *maltovorans*, *A. suboxydans* var. *biourgianum* and *A. suboxydans* var. *hoyerianum*. Bernhauer and Riedl-Tůmová[237] also reported the oxidation of L-arabinose and D-xylose to corresponding pentonic acids by *A. suboxydans* var. *muciparum* and *A. melanogenus*. D-Arabinose was oxidized, but L-xylose was not. Frateur[238] found that *A. melanogenus* formed a small amount of ketopentonic acid in addition to D-arabonic acid from D-arabinose. Liebster *et al.*[239] obtained 4-keto-D-arabonic acid from D-arabonic acid with a strain of *A. suboxydans*. Aida *et al.*[240] observed that *G. liquefaciens*, in shake-culture with $CaCO_3$, formed xylonic acid from D-xylose as well as a reducing substance (ketopentonic acid?) and unknown substances with R_f values of 0.28 and 0.37 in *n*-butanol-formic acid-water (4: 1.5: 1). L-Arabonic acid was obtained from L-arabinose in a yield of 40%, as well as an unknown substance with an R_f value of 0.23.

With a stationary culture of *A. aceti*, De Ley and Schell[241] observed the formation from D-ribose and L-xylose of corresponding pentonic acids.

Federico and Gobis,[166] however, made the interesting observation that *A. aceti* formed a small amount of acetoin from xylose. Kaushal *et al.*[242] verified the formation of glycolaldehyde from D-xylose and L-arabinose by *A. acetigenus*.

In an early paper Fred *et al.*[243] reported the formation of acetone as well as ethanol and CO_2 from xylose by an old culture of *A. xylinum*. This observation suggests that acetic acid bacteria, under weak oxidizable or anaerobic conditions, can carry out anaerobic metabolism.

Oxidation of D-glucose

Production of D-gluconic acid.   The oxidation of glucose to gluconic acid by acetic acid bacteria was first noted by Brown and was later confirmed by numerous workers on various strains.   Among these strains *A. industrius*, *A. oxydans*, and *A. xylinum* were especially noted for their high gluconic acid production.   Yagi and Hashitani[244] isolated acetic acid bacteria with a high gluconic acid accumulation from the Dutch strawberry.   Hermann[245] found a strain in " alga-tea " that accumulated gluconic acid in quantities up to 60 to 80% of the glucose added, and named the strain *Bact. gluconicum* (*A. gluconicus*).   These studies paved the way for the industrial production of gluconic acid using acetic acid bacteria. Both gluconic acid fermentation and acetic acid fermentation are regarded as important industrial applications of acetic acid bacteria.

Asai[246] isolated a number of acetic acid bacteria from various kinds of fruits.   Most of the strains isolated had strong glucose oxidizing activities and accumulated gluconic acid nearly to theoretical yield.   He proposed a new genus name, *Gluconobacter*, based on the oxidative behaviors of these strains towards ethanol and glucose, as stated earlier.   These *Gluconobacter* strains carry out active ketogenic oxidation where gluconic acid is further oxidized to 5-ketogluconic acid or 2-ketogluconic acid, and polyalcohols to corresponding ketoses.   The ethanol oxidizing activity, however, is generally weak.

Frateur,[236] however, reported on strains of two species of the " Peroxydans group " and of *A. ascendens* which did not form any acid from glucose.   For these organisms, and others which form gluconic acid from glucose but not acetic acid from ethanol, see Part 1, Chapter 7 (page 42–43).

The optimal temperature and pH for gluconic acid production are, according to Asai,[247] about 25°C and pH 4.0–6.0.   Addition of $CaCO_3$ increases the amount of gluconic acid produced at an early phase, but later the gluconic acid is oxidized to 5-ketogluconic or 2-ketogluconic acid and the yield decreases.   Thus the non-addition of $CaCO_3$ to a culture favors gluconic acid production.   Substrate concentration should be between 10 and 15%.   Hermann[245] reported the accumulation of gluconic acid up to a concentration of 23% by growing *A. gluconicus* in a medium containing 40% glucose.   According to Tanaka,[248] the optimal pH for glucose oxidation by *A. aceti* lies between 5 and 6 (manometric study).   Tanaka also observed that the carbon source of the growth medium had a marked influence on the glucose oxidation of harvested cells of *A. aceti*.   As carbon sources for growth he used ethanol, ethanol plus glucose, and glucose, and studied the glucose oxidizing ability of the three kinds of cells obtained. Cells obtained from the ethanol medium showed the strongest activity and those grown on glucose the weakest. When both glucose and ethanol were present in the medium, ethanol was first oxidized and then glucose. Tanaka further found that the intact cells of *A. aceti* could utilize methylene blue and quinone as hydrogen acceptors for glucose oxidation, and concluded that glucose oxidation in this organism is carried out by a typical dehydrogenase system. Glucose oxidation with oxygen was inhibited by KCN, CO and toluene, but not the oxidation with quinone. Thus the oxidation with oxygen was apparently mediated by a cytochrome system, but the inhibition by KCN and CO was less marked than in the case of ethanol oxidation. Therefore a possibility remained that a cytochrome-independent enzyme system was partly responsible for the glucose oxidation.

Production of ketogluconic acids and aldehydogluconic acid. Acetic acid bacteria, particularly *Gluconobacter*, can oxidize glucose and gluconic acid to ketogluconic acids (5-ketogluconic acid, 2-ketogluconic acid, and 2,5-diketogluconic acid). This property can be considered one of the important characteristics of *Gluconobacter*.

According to Asai's studies[249] in 1935, all strains of genus *Gluconobacter* examined accumulated 5-ketogluconic acid as well as gluconic acid, with the exception of some strains of *G. liquefaciens* and *G. nonoxygluconicus*. Among strains of *Acetobacter*, however, only *A. xylinum* Brown accumulated 5-ketogluconic acid (Asai[250]). *A. aceti* Hansen, *A. aceti* Henneberg, *A. pasteurianus*, *A. kützingianus*, *A. rancens*, *A. acetosus* Henneberg, *A. acetosus* (Institut fur Gärungsgewerbe, Berlin), *A. ascendens* Henneberg, *A. ascendens* (Institut fur Gärungsgewerbe, Berlin), and *A. viniacetati* Henneberg did not produce 5-ketogluconic acid. *A. ascendens* Henneberg did not form gluconic acid from glucose (without the addition of $CaCO_3$).

Kulka and Walker,[251] studying the ketogenic activity towards glucose of various acetic acid bacteria, found the formation of both 2-ketogluconic acid and 5-ketogluconic acid by *A. acetigenus*, *A. gluconicus*, *A. kützingianus*, *A. orleanense*, *A. suboxydans*, *A. turbidans*, *A. viscosus*, and *A. xylinum*. Those which did not form 5-ketogluconic acid were *A. acetosus* NCIB 2224, *A. ascendens* NCIB 4937 (isolated from African vinegar, 1944), *A. capsulatus* NCIB 4943, *A. acetosus* NCIB 6425, *A. acidum-mucosum* NCIB 6429, *A. acetosus* (isolated from beer), *A. ascendens* NCIB 4937, and *A. aceti* NCIB 6423.

A. capsulatus did not form 5-ketogluconate at first, but acquired the ability after repeated transfers. Bernhauer and Görlich[252] reported that *A. gluconicus*, which had originally actively formed 2-ketogluconic acid, changed into a predominantly 5-ketogluconic acid-forming variant.

According to Frateur,[236] *A. ascendens* may be distinguished from other species by reason of its failure to produce acid from glucose. Kulka and Walker,[251] however, observed the formation of a small amount of Ca gluconate in a $CaCO_3$-glucose medium by *A. ascendens* NCIB 4937, although there was no formation of gluconic acid in the absence of $CaCO_3$. Generally the species of acetic acid bacteria which accumulated a large amount of Ca gluconate from glucose formed little reducing acid or Ca 5-ketogluconate, and those which formed little Ca gluconate accumulated a large amount of 2-ketogluconate or 5-ketogluconate. It should be pointed out, however, that the strains of *A. ascendens* used by these workers had very weak gluconate-forming and reducing acid-forming activity.

Kondo and Ameyama[253] classified acetic acid bacteria into the following seven groups according to the kind of acids formed from glucose.

1) Gluconic acid not formed: *A. ascendens* Henneberg.
2) Gluconic acid formed: *A. aceti* Beijerinck, *A. pasteurianus* Hansen, *A. kützingianus* Hansen, *A. rancens* Beijerinck.
3) Gluconic acid and 2-ketogluconic acid formed: *A. acetosus* Henneberg, *A. dioxyacetonicus* Asai, *A. orleanense* Henneberg.
4) Gluconic acid, 2-ketogluconic acid and 2,5-diketogluconic acid formed: *A. melanogenus* Beijerinck, *A. rubiginosus* nov. sp.
5) Gluconic acid, 2-ketogluconic acid, 5-ketogluconic acid and 2,5-diketogluconic acid formed: *A. aurantius* nov. sp.
6) Gluconic acid, 2-ketogluconic acid, and 5-ketogluconic acid formed: *G. cerinus* Asai, *A. gluconicus* Hermann, *A. xylinum* Brown, *A. industrius* Henneberg, *A. oxydans* Henneberg, *G. roseus* Asai, *A. albidus* nov. sp., *A. suboxydans* Kluyver.
7) Gluconic acid and 5-ketogluconic acid formed: *A. suboxydans α* nov. var.

2,5-Diketogluconic acid was first reported by Katznelson *et al.*[254] as being produced by old cells and cell-free extracts of *A. melanogenus* from glucose, gluconate or 2-ketogluconate. Aldehydogluconic acid was first reported by Takahashi and Asai[255] in a glucose-$CaCO_3$ culture of *A. industrius* var. *hoshigaki*. These workers tentatively identified the compound as L-glucuronic acid from the results of the naphthoresorcinol test, melting point and nitrogen content determinations of the phenylosazone, and optical rotation measurement. They obtained 25 g of Ca glucuronate from 100 g of Ca gluconate.

Later, Bernhauer and Irrgang[256] found that a reducing uronic acid other than 5-ketogluconic acid was formed from glucose by *A. gluconicus* and *A. xylinum*, and identified the acid as D-aldehydogluconic acid (L-guluronic acid) from its reduction of Fehling's solution, positive naphthoresorcinol reaction, and green coloration with orcinol-HCl, as well as its melting point and the nitrogen content of the phenylosazone. An *A. gluconicus* culture in 5% calcium gluconate gave calcium D-aldehydogluconate with a yield of 52%. They stated that the formation of L-glucuronic acid was not likely in view of its chemical configuration.

$$
\begin{array}{ccc}
\text{COOH} & \text{COOH} & \text{COOH} \\
| & | & | \\
\text{H}-\text{C}-\text{OH} & \text{H}-\text{C}-\text{OH} & \text{H}-\text{C}-\text{OH} \\
| & | & | \\
\text{HO}-\text{C}-\text{H} & \text{HO}-\text{C}-\text{H} & \text{HO}-\text{C}-\text{H} \\
| & | & | \\
\text{H}-\text{C}-\text{OH} & \text{H}-\text{C}-\text{OH} & \text{H}-\text{C}-\text{OH} \\
| & | & | \\
\text{H}-\text{C}-\text{OH} & \text{C}=\text{O} & \text{H}-\text{C}-\text{OH} \\
| & | & | \\
\text{CH}_2\text{OH} & \text{CH}_2\text{OH} & \text{CHO}
\end{array}
$$

D-Gluconic acid   D-5-Ketogluconic acid   D-Aldehydogluconic acid (L-Guluronic acid)

Riedl-Tůmová and Bernhauer[257] grew *A. melanogenus* in a yeast-extract medium containing 5% calcium gluconate with shaking and obtained, in addition to 2-ketogluconate, the soluble calcium salt of a reducing acid. The acid was thought to be D-aldehydogluconic acid, but further attempts to identify it were unsuccessful.

Janke[258] was dubious about the identification because of the failure of earlier workers to find, in fermentation media, saccharic acid which should have been formed easily through oxidation from D-aldehydogluconic acid.

Production of 5-ketogluconic acid. In 1866 Boutroux[259] reported that a bacterium called *Micrococcus oblongus* (later identified as an acetic acid bacterium) formed oxygluconate (5-ketogluconate) from glucose in the presence of $CaCO_3$. Bertrand[217] also reported the formation of 5-ketogluconate by *A. xylinum*. Similar reports of 5-ketogluconate formation by acetic acid bacteria were published by Visser't Hooft[115] on *A. rancens*, by Hermann and Neuschul[123] on *A. gluconicus*, *A. xylinoides*, *A. orleanense*, and *A. aceti*, by Bernhauer and Knobloch[260] on *A. suboxydans* and *A.*

suboxydans var. *muciparum*, and by Takahashi and Asai[261] on acetic acid bacteria isolated from fruits. Kulka and Walker observed the accumulation of 5-ketogluconic acid by *A. turbidans*, *A. acetigenus*, and *A. viscosus*.

$$
\begin{array}{ccc}
\text{COOH} & & \text{COOH} \\
| & & | \\
\text{H—C—OH} & & \text{H—C—OH} \\
| & & | \\
\text{HO—C—H} & \xrightarrow{\;-2H\;} & \text{HO—C—H} \\
| & & | \\
\text{H—C—OH} & & \text{H—C—OH} \\
| & & | \\
\text{H—C—OH} & & \text{C=O} \\
| & & | \\
\text{CH}_2\text{OH} & & \text{CH}_2\text{OH} \\
\text{D-Gluconic acid} & & \text{D-5-Ketogluconic acid}
\end{array}
$$

Hermann called these 5-ketogluconate-forming bacteria " ketogenic acetic acid bacteria," while Asai stated that the strong ketogenic activity can be considered a characteristic of *Gluconobacter*.

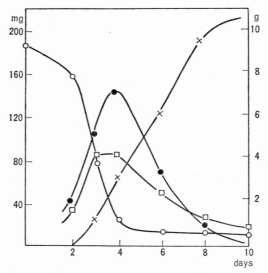

Fig. 14.   Course of glucose oxidation by *A. suboxydans* var. α.

Stationary culture, in presence of 10% glucose and 2.5% CaCO₃ (after Kondo and Ameyama).

○, reducing power (Cu mg/ml); ●, Ca gluconate, g; ×, Ca 5-ketogluconate, g; □, Ca dissolved, g.

For completion of oxidation of glucose to 5-ketogluconate the addition of $CaCO_3$ as a neutralizing agent is required. According to Stubbs et al.[262] and also to Kondo and Ameyama,[263] the oxidation proceeds through two steps. The first step is the oxidation of glucose to gluconate, and the growth of the organism virtually ceases after this step. The second step is the oxidation of gluconate to 5-ketogluconate (see Fig. 14).

The production of 5-ketogluconate from calcium gluconate was first verified by Kluyver and De Leeuw[264] using A. suboxydans. Industrial production of 5-ketogluconate was attempted by Berhauer and Schön[265] with A. xylinum as a test organism. A weak alkaline reaction (pH 7–8) was found favorable for the oxidation. 5-Ketogluconate was obtained from calcium gluconate with a yield of 50%, and from glucose in the presence of $CaCO_3$ and a small amount of butyric acid (in order to prevent fungus infection) with a yield of 65%.

Hermann's claim[266] that A. gluconicum formed 5-ketogluconate from fructose with a yield of 70% is somewhat doubtful.

There are reports by Bernhauer and Knobloch,[267] Teramoto et al.,[268] Sumiki and Hatsuta,[269] Khesghi et al.[270] and Yamazaki[271] concerning the conditions for production of 5-ketogluconic acid.

Khesghi et al., using A. suboxydans NRRL B-72, obtained yields as high as 84.8% of the theoretical yield with shake flask fermentation in 7 days. Addition of catalase, vanadium pentoxide, and borate increased neither the yields of 5-ketogluconic acid nor the rate of fermentation. Optimal concentration of corn steep liquor as a nitrogen source was 0.3% and higher concentrations increased the proliferation of bacteria but lowered the yield of 5-ketogluconate. As antifoam reagents, corn oil and mineral oil were innocuous. In paper chromatograms of fermentation products 5-ketogluconic and gluconic acid as well as tartaric acid were found, but not 2-ketogluconic acid.

Yamazaki, using G. suboxydans ATCC 621, studied 5-ketogluconic acid production with shake flask fermentation. The optimal pH for production was around 4.0, and alkaline pH's decreased the yield. This point should be kept in mind when adding $CaCO_3$. Under favorable conditions of 10% glucose, 0.5% fish extract (as nitrogen source) and 2.45% $CaCO_3$, glucose was converted to calcium 5-ketogluconate with a yield of 90.2%. Yamazaki and Asai[272] found that the parent strain, after repeated transfers on an agar medium containing fish extract or corn steep liquor as a nitrogen source, mutated to a strain with a markedly decreased ability to produce 5-ketogluconate and a greatly increased ability to produce 2-ketogluconate. The coloration of colonies also changed. Addition of 0.5% NaCl, $CaCl_2$, or $CaSO_4$ inhibited the formation of 5-ketogluconate markedly. Mutation of *A. suboxydans* from a 5-ketogluconate producer to a 2-ketogluconate producer was reported also by Kulka and Walker.[251] Foda and Vaughn[273] reported that 5-ketogluconate may be the end-product of maltose oxidation by *A. melanogenus*.

Production of 2-ketogluconic acid. Bernhauer and Görlich[252] reported the formation of 2-ketogluconate from calcium gluconate by *A. gluconicus*. Bernhauer and Knobloch[274] found the production of K 2-ketogluconate from K gluconate by *A. suboxydans* with a yield of 75%. Starting from glucose, however, they could produce 5-ketogluconate up to 37% of the theoretical yield. Kondo and Narita[275] obtained 2-ketogluconate with a newly isolated *Gluconobacter 2-ketogluconicus* in a yield higher than 50%. All these experiments were carried out with surface cultures and the fermentation required more than 10 days for completion. Asai and Ikeda[276] used *G. cerinus* and obtained 2-ketogluconate in high yields, as high as 83% of the theoretical yield using K gluconate as a substrate in surface culture.

The production of 2-ketogluconic acid by *Acetobacter* or *Gluconobacter*, however, gives generally low yields compared to that of *Pseudomonas*, and is often accompanied by 5-ketogluconate production. Thus for industrial production of 2-ketogluconate *Pseudomonas* is more suitable.

According to Knobloch and Tietze,[277] acetic acid bacteria accumulated 2-ketogluconate more often than 5-ketogluconate. Frateur *et al.*[278] reported that all strains of the *Mesoxydans* group and *Suboxydans* group examined and some strains of *A. rancens* formed 2-ketogluconate.

Production of 2,5-diketogluconic acid. Katznelson *et al.*[254] reported the production of 2,5-diketogluconate as well as 2-ketogluconate from glucose or gluconate by old intact cells and by cell-free extracts of *A. melanogenus* forming brownish pigment. A suspension of old cells consumed approximately 1.0 and 0.5 moles of O_2 per mole of gluconate and 2-ketogluconate respectively, without CO_2 liberation, and 2,5-diketogluconate was found as the oxidation product in the reaction mixture. It was thus shown that a conversion occurred with the following equation:

$$\text{Glucose} \xrightarrow{\frac{1}{2}O_2} \text{Gluconate} \xrightarrow{\frac{1}{2}O_2} \text{2-Ketogluconate} \xrightarrow{\frac{1}{2}O_2} \text{2,5-Diketogluconate}$$

Oxidation by young intact cells proceeded further, accompanied by CO_2 evolution.

Kondo *et al.*[279] confirmed the formation of 2,5-diketogluconate in a stationary culture of *A. melanogenus* with addition of $CaCO_3$.

Aida *et al.*[280] incubated Ca 2-ketogluconate with a dried cell preparation of *G. liquefaciens* grown for 48 hours at 30°C on a shaker and obtained
2,5-diketogluconate.   The same authors[281] checked for the acid in a culture of the organism growing with glucose and gluconate and found instead
another reducing acid (R_f 0.06–0.08 in *n*-butanol : acetic acid : water=
4 : 1 : 5).   This acid was not definitely identified.   Oxidation of glucose,
gluconate and 2-ketogluconate was studied manometrically with intact
cells of *G. liquefaciens* (Fig. 15).   Approximately 1.5, 1.0 and 0.5 moles
of O_2 were consumed for each mole of glucose, gluconate and 2-keto-
gluconate, respectively.   When cell-free extracts were used, evolution
of CO_2 did not occur and the oxygen uptake of each substrate was
essentially the same as with an intact cell suspension in the presence of
NAD.

2,5-Diketogluconic acid is unstable at high temperatures and easily

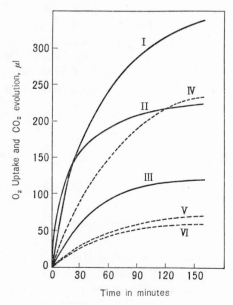

Fig. 15.   Oxidation of glucose, gluconate, and 2-ketogluconate by
intact cell suspensions of *G. liquefaciens*.

Each vessel contained 0.5 ml of the cells, 1 ml of phosphate buffer solution
(M/15, pH 6.0) and 5 μM of the substrate.   Total volume 2.5 ml; temp. 30°C.
Oxygen uptake : I------glucose, II------gluconate, III------2-ketogluconate.
Carbon dioxide evolution : IV------glucose, V------gluconate, VI------2-keto-
gluconate.

polymerizes or decomposes at a pH above 4.5. The brownish color of the fermentation liquor of *A. melanogenus* is probably related to the formation of this acid.

Formation of γ-pyrone compounds and other substances. Takahashi and Asai[282] reported the formation of kojic acid from fructose and comenic acid[283] from galactose by *Gluconobacter*. Since Yabuta's discovery[284] of kojic acid as a metabolic product of *Asp. oryzae* there had been no report on the formation of γ-pyrone compounds from sugars until Takahashi and Asai's.

Aida *et al.*[240] grew *G. liquefaciens* in a glucose-CaCO₃ medium on a shaker and observed the formation of substances with a reddish violet reaction to FeCl₃, one of which was comenic acid. Two unknown substances were isolated and named rubiginol and rubiginic acid. These substances were not produced in the absence of $CaCO_3$ or in surface culture.

Aida[285] later established the chemical structure of rubiginol as 3,5-dihydroxy-1,4-pyrone and that of rubiginic acid as 3,5-dihydroxy-1,4-pyrone-2-carboxylic acid.[286] Thus the following three γ-pyrone compounds are formed from glucose by *G. liquefaciens*. All three compounds are produced from 2,5-diketogluconate by the intact cells of *G. liquefaciens;* 2,5-diketogluconate was also shown to be a precursor of these γ-pyrone compounds in an experiment with 1-¹⁴C-gluconate (unpublished).

Comenic acid   Rubiginic acid   Rubiginol

Datta and Katznelson[287] obtained from 2,5-diketogluconate a pentose-like substance and α-ketoglutarate as metabolic intermediates with cell-free extracts of *A. melanogenus* in the presence of phenazinemethosulfate. The conversion ratio to α-ketoglutarate was calculated as 50%.

In addition to the compounds described above, acetic acid bacteria produce succinic, lactic, acetic and glycolic acids from glucose (Takahashi and Asai[288]). Aida *et al.*[289] reported the formation of aldehyde (acetaldehyde?), formic acid, acetic acid, glycolic acid, and tartronic acid from glucose by *G. liquefaciens*. Asai *et al.*[290] found that *G. cerinus* produced α-ketoglutaric acid and pyruvic acid from glucose in a shake culture.

The formation of tartronic acid was also confirmed by Kulka et al.[291] with *A. acetosus*. In a later publication, however, Hall et al.[292] attributed the formation of tartronate from 2-keto-D-gluconate to a chemical, rather than biochemical, process consisting of the decomposition of 2-ketogluconate in the presence of calcium hydroxide. Interestingly, there is a patent on the production of tartaric acid from glucose through 5-ketogluconate with *A. suboxydans* at pH 5.0–7.0 with the addition of alkali (Kamlet[293]).

Kondo et al.[279] confirmed Aida's report and using *A. melanogenus* obtained aldehyde (acetaldehyde?), formic acid, succinic acid, glycolic acid, tartronic acid, rubiginol, rubiginic acid and comenic acid from glucose. They[294] further reported the accumulation of oxalic acid in the medium in stationary culture with $CaCO_3$ for one and a half months. Oxalic acid production by acetic acid bacteria will be discussed in detail elsewhere.

Ameyama and Kondo[295] reported the production of D-lyxuronic acid from glucose by a growing culture of *A. melanogenus*, and suggested a possible pathway for the formation of this compound, which will be also discussed later.

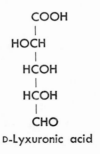

COOH
|
HOCH
|
HCOH
|
HCOH
|
CHO
D-Lyxuronic acid

Oxidation of other aldohexoses

With respect to the oxidation of D-mannose to D-mannonic acid and D-galactose to D-galactonic acid, Hermann and Neuschul[296] reported that they had noted these reactions in *A. gluconicus*. Takahashi and Asai[283] found the production of galactonic acid and comenic acid (5-hydroxy-1,4-pyrone-2-carboxylic acid) from galactose by *G. liquefaciens*, *G. roseus*, and *G. cerinus*.

Frateur[297] reported the formation of galactonic acid from galactose by the following organisms: *A. rancens Beijerinck*, *A. aceti* (Pasteur) Beijerinck No. 2, *A. xylinum* (Brown) Beijerinck No. 2, *A. xylinum* var.

xylinoides No. 1, *A. mesoxydans* var. *saccharovorans*, *A. mesoxydans* var. *lentum*, *A. mexosydans* var. *lentum-saccharovorans*, *A. suboxydans* var. *muciparum*, *A. melanogenus* var. *maltovorans*, and *A. melanogenus* Beijerinck. Ettel *et al.*[206] found the formation of 2-ketogalactonic acid from D-galactonic acid by a strain of *A. suboxydans*.

Aida *et al.*[289] also reported the production by *G. liquefaciens* of mannonic acid and galactonic acid from mannose and galactose, respectively.

Recently Terada *et al.*[298] found the production of 5-ketomannonic acid and 2-ketogluconic acid as well as mannonic acid from D-mannose.

Stereochemical characteristics of the oxidation of aldoses and aldonic acids

Little systematic work has been done to discover whether acetic acid bacteria follow Bertrand's rule, Hudson's rule, or Bertrand-Hudson's rule of polyol oxidation in the oxidation of aldoses and aldonic acids. Bertrand's rule can be applied to the oxidation of gluconic acid to 5-ketogluconic acid but not to 2-ketogluconic acid.

Recently, Komatsu[299] studied the oxidation of various aldoses by growing cells of *A. suboxydans* ATCC 621 and obtained the results shown in Table 14.

Table 14.   Oxidation of aldoses by *A. suboxydans* (after Komatsu).

Substrate	Result	Substrate	Result
D-Ribose	Not oxidized	L-Idose	Not oxidized
D-Arabinose	,,	D-*glycero*-D-*gulo*-Heptose	Oxidized
D-Xylose	Oxidized	3-*O*-Methyl-D-glucose	Not oxidized
L-Arabinose	Not oxidized	6-*O*-Methyl-D-glucose	,,
D-Glucose	Oxidized	Maltose	,,
D-Mannose	Not oxidized	Lactose	,,
D-Galactose	,,	D-Glucosamine	,,
D-Gulose	Oxidized		

D-Xylonic acid was formed from D-xylose, D-gluconic acid from D-glucose, and D-*glycero*-D-*gulo*-heptonic acid from D-*glycero*-D-*gulo*-heptose.

Komatsu assumed that aldose has a pyranose structure and concluded that the rule of cyclitol oxidation is not applicable, and that the C-1 hydroxyl is oxidized when the C-3 and C-4 positions bear equatorial-equatorial or axial-axial hydroxyls, and that there are no axial-equatorial or

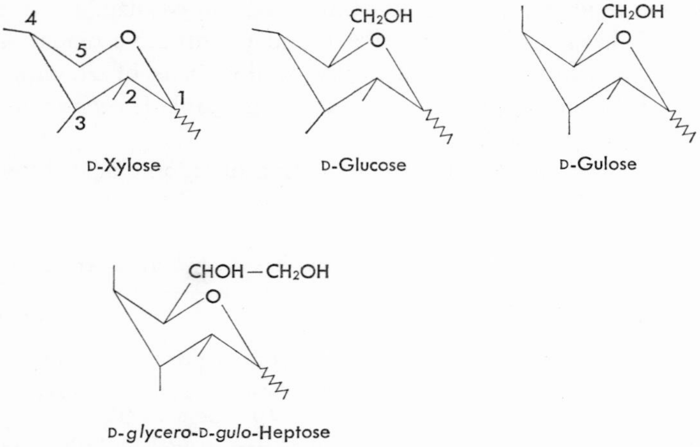
4
5
O
2
1
3
D-Xylose
CH2OH
O
D-Glucose
CH2OH
O
D-Gulose
CHOH—CH2OH
O
D-*glycero*-D-*gulo*-Heptose

Fig. 16. Structural formulas of the oxidizable substrates.

equatorial-axial hydroxyls in C-2 and C-5 positions, as shown in Figure 16.

King and Cheldelin,[300] on the other hand, using the same species, reported the oxidation of D-galactose by a cytochrome-linked glucose dehydrogenase prepared from the solubilized particulate fraction. Komatsu attributed this discrepancy to the cell-wall permeability of substrates which would affect the results in growing cell experiments. Production of D-galactonic acid from D-galactose and of L-arabonic acid from L-arabinose by growing cells was also reported by other workers, as stated previously. Hence Komatsu's generalization may have to be revised in the future.

Table 15. Oxidation of glucosamine, galactosamine and *N*-acetylglucosamine by acetic acid bacteria (after Takahashi and Kayamori).

Strain	O_2 uptake μl/min/mg (cell dry wt.)		
	D-Glucosamine	Galactosamine	*N*-Acetylglucosamine
G. melanogenus	84	0	0
G. suboxydans	91	0	0
A. rancens	118	0	0
G. roseus	6	0	0

Cells grown in glucosamine were used. Substrate : 50 μM.
Cells grown in glucose medium were also capable of actively oxidizing glucosamine.

Komatsu reported that D-glucosamine was not oxidized by *A. suboxydans*, but Takahashi and Kayamori[301] found quantitative oxidation of D-glucosamine to D-glucosaminic acid by resting cells of *A. melanogenus* Beijerinck.　In their experiments, *A. suboxydans* and *A. rancens* also oxidized glucosamine (Table 15).

Komatsu[302, 303] also studied the oxidation of aldonic acids by growing cells of *A. suboxydans* (Table 16).

Table 16.　Oxidation of aldonic acids by *A. suboxydans* (after Komatsu).

Substrate	Result	Substrate	Result
D-Erythronic acid	Not oxidized	L-Idonic acid	Not oxidized
D-Ribonic acid	,,	D-*glycero*-D-*gulo*-Heptonic acid (II)	Oxidized
D-Arabonic acid	,,		
D-Xylonic acid	,,	D-*glycero*-D-*ido*-Heptonic acid	Not oxidized
L-Arabonic acid	,,	D-*glycero*-D-*galacto*-Heptonic acid	,,
D-Gluconic acid (I)	Oxidized		
D-Mannonic acid	Not oxidized	D-*erythro*-L-*galacto*-Octonic acid	,,
D-Galactonic acid	,,	D-*erythro*-L-*talo*-Octonic acid (III)	Oxidized
D-Gulonic acid	,,		

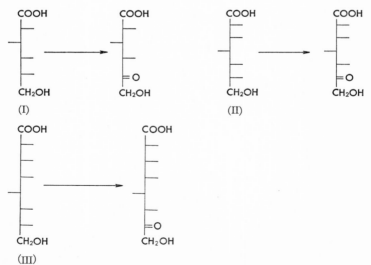

Fig. 17.　Structural formulas of oxidizable substrates and products.

Horizontal bars indicate secondary alcohol radicals.

5-Keto-D-gluconic acid was obtained from D-gluconic acid, 6-keto-D-*glycero*-D-*gulo*-heptonic acid from D-*glycero*-D-*gulo*-heptonic acid, and 7-keto-D-*erythro*-L-*talo*-octonic acid from D-*erythro*-L-*talo*-octonic acid. The structural formulas of the substrates and products are shown in Figure 17. It was thereby concluded that the rule of polyol oxidation can not be applied to the oxidation of aldonic acids by *A. suboxydans*. It was also assumed that the secondary alcohol adjacent to the primary alcohol is oxidized when the acid, viewed from the primary alcohol group, possesses D-*gluco*-type structure in the molecular terminal, and does not possess a *trans*-hydroxyl between its D-*gluco*-type structure and carboxyl group.

Since Kulhánek[304] reported the production of 4-keto-D-arabonic acid from D-arabonic acid by the same species, however, these conclusions may also have to be revised later.

Oxidation of Ketoses

Ketoses are not so readily attacked by acetic acid bacteria as aldoses. Glycerol is oxidized to dihydroxyacetone by many strains, but the further oxidation of dihydroxyacetone is difficult.

Visser't Hooft[115] found consumption of oxygen by resting cells of *A. rancens* with dihydroxyacetone as a substrate, and Cozic[133] noted the same phenomenon in *A. xylinum*. More detailed studies on the metabolism of glycerol and dihydroxyacetone have been carried out recently by King, Cheldelin and their coworkers (for references see the section on glycerol metabolism). Ikeda[159] suggested that reductone formation from fructose may result from the oxidation of dihydroxyacetone after aldolase-splitting of the fructose molecule, in addition to a possible direct oxidative degradation of fructose. Visser't Hooft[115] reported the oxidation of L-erythrulose by *A. rancens*.

Oxidation of D-fructose.   Bertrand[217] reported a complete oxidation of fructose by the sorbose bacterium after a long incubation. Henneberg found that many acetic acid bacteria formed acids from fructose.

Hermann and Neuschul,[123] using *A. gluconicus* and other acetic acid bacteria, identified one of the acids as acetic acid. Bernhauer and Knobloch[274] reported the accumulation of a small amount of succinic acid as well as a considerable amount of acetic acid by *A. gluconicus* in the presence of $CaCO_3$. Hermann and Neuschul,[123] using the same organism, found that a large amount of D-5-ketogluconic acid was formed from D-fructose.

One of the important oxidation products of D-fructose is kojic acid. Its occurrence was first reported by Takahashi and Asai[282] with *G. lique-faciens*. Aida *et al.*[280] reported the formation of a new γ-pyrone compound, iso-kojic acid (6-hydroxy-2-hydroxymethyl-1, 4-pyrone). Terada *et al.* later corrected the identification to 5-oxymaltol. Ikeda[305] studied the mechanism of the formation of kojic acid from D-fructose. He grew *G. roseus* on fructose in a shake culture and confirmed the formation of glucosone, 2-ketogluconic acid and rubiginol. He then theorized that fructose was converted to kojic acid *via* the enol-form of fructose and glucosone according to the following equations:

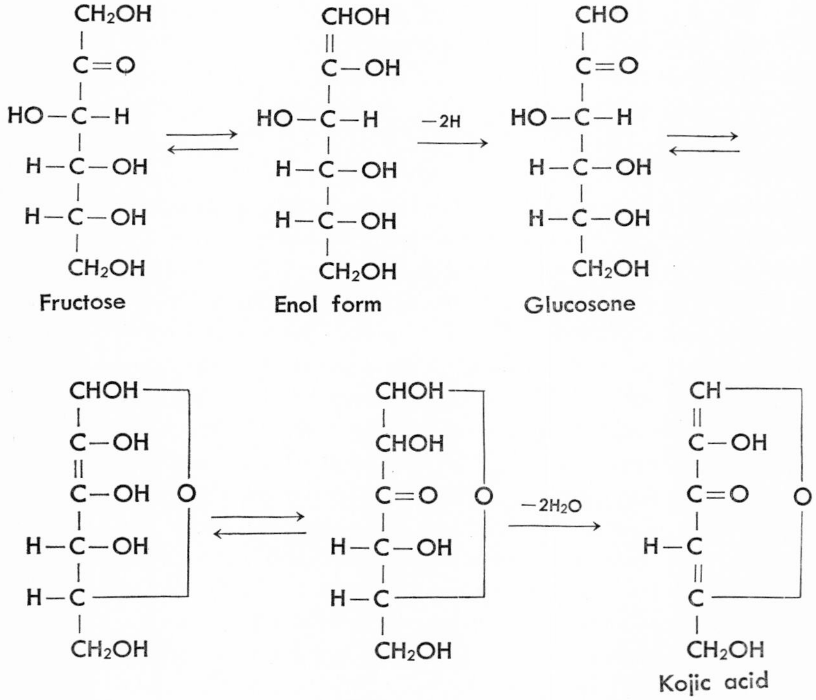

$$
\begin{array}{ccc}
\text{CH}_2\text{OH} & \text{CHOH} & \text{CHO} \\
| & || & | \\
\text{C}=\text{O} & \text{C}-\text{OH} & \text{C}=\text{O} \\
| & | & | \\
\text{HO}-\text{C}-\text{H} & \text{HO}-\text{C}-\text{H} & \text{HO}-\text{C}-\text{H} \\
| & | & | \\
\text{H}-\text{C}-\text{OH} & \text{H}-\text{C}-\text{OH} & \text{H}-\text{C}-\text{OH} \\
| & | & | \\
\text{H}-\text{C}-\text{OH} & \text{H}-\text{C}-\text{OH} & \text{H}-\text{C}-\text{OH} \\
| & | & | \\
\text{CH}_2\text{OH} & \text{CH}_2\text{OH} & \text{CH}_2\text{OH} \\
\text{Fructose} & \text{Enol form} & \text{Glucosone}
\end{array}
$$

(Fructose ⇌ Enol form, −2H → Glucosone ⇌)

$$
\begin{array}{ccc}
\text{CHOH} & \text{CHOH} & \text{CH} \\
| & | & || \\
\text{C}-\text{OH} & \text{CHOH} & \text{C}-\text{OH} \\
|| & | & | \\
\text{C}-\text{OH}\quad\text{O} & \text{C}=\text{O}\quad\text{O} & \text{C}=\text{O}\quad\text{O} \\
| & | & | \\
\text{H}-\text{C}-\text{OH} & \text{H}-\text{C}-\text{OH} & \text{H}-\text{C} \\
| & | & || \\
\text{H}-\text{C} & \text{H}-\text{C} & \text{C} \\
| & | & | \\
\text{CH}_2\text{OH} & \text{CH}_2\text{OH} & \text{CH}_2\text{OH} \\
 & & \text{Kojic acid}
\end{array}
$$

(⇌, −2H₂O →)

Aida *et al.*[280] also supported this pathway for the formation of kojic acid.

Earlier Sakaguchi *et al.*[306] isolated and identified dihydroxyacetone as a metabolic product from D-fructose in a stationary culture of *G. opacus*. They proposed the following scheme for its formation:

This scheme is interesting in relation to the isolation and identification of 5-ketofructose from fructose by Terada *et al.*[198] with gluconobacters.

```
      CH₂OH                CH₂OH                CH₂OH
        |                    |                    |
      C=O                  C=O                  C=O      Dihydroxyacetone
        |                    |                    |
   HO—C—H              HO—C—H               CH₂OH
        |                    |
    H—C—OH    ⟶      H—C—OH    ⟶        CHO
        |                    |                    |
    H—C—OH                C=O                  CHOH     Glyceraldehyde
        |                    |                    |
      CH₂OH                CH₂OH                CH₂OH
     Fructose           5-Ketofructose
```

Ikeda[159] observed that *G. roseus* produced from fructose a substance which rapidly reduced 2,4-dichlorophenolindophenol, and identified the substance as reductone, a compound first named by Euler.

Weidenhagen and Bernsee[307] isolated a crystalline dicarbonylhexose (6-aldehyde-D-fructose?) as a product in the oxidation of D-fructose by *A. suboxydans*. A similar compound was obtained and was confirmed to be 5-keto-D-fructose by Avigad and Englard[308] in a growing culture of *G. cerinus*. The latter found the presence of $NADPH_2$-linked 5-keto-D-fructose reductase in cell-free extracts of this organism.

Terada et al.[198] grew *G. cerinus* and other *Gluconobacter* strains in a yeast extract medium containing D-fructose, with shaking, and observed the formation of a strongly reducing substance which was not glucosone. The substance was identified as 5-ketofructose. From their observation that *G. cerinus* forms kojic acid from 5-ketofructose as well as from fructose, they determined that 5-ketofructose is the precursor of kojic acid.

According to recent experiments by Terada et al.,[309] *G. cerinus*, when grown on fructose with $CaCO_3$, accumulated 5-ketofructose first and then, as growth proceeded, accumulated glyceric acid, glycolic acid and succinic acid with concomitant disappearance of 5-ketofructose. From these observations they determined that these organic acids are derived at least partly from the splitting of 5-ketofructose. Formation of succinic acid from fructose has been known for many years, since the work of Takahashi and Asai.[282] Terada et al.[298] also found, in a culture with $CaCO_3$, reducing organic acids which reacted positively with *p*-anisidine. These were identified as 5-ketomannonic acid and 2-ketogluconic acid. The formation of mannonic acid was also verified. A small amount of reducing organic acid with a color reaction similar to that of 5-ketogluconic acid was obtained but was not definitely identified.

Since fructose cannot be directly converted to these reducing keto-acids, Terada *et al.* showed that there was an isomerization reaction of fructose to mannose.   They further verified the formation of 5-ketoman-nonic acid and 2-ketogluconic acid from mannose, using intact cells.   Thus they proposed that the oxidation of D-fructose by *Gluconobacter* can proceed through the pathway shown below (Fig. 18) in addition to one going through the γ-pyrone compounds.

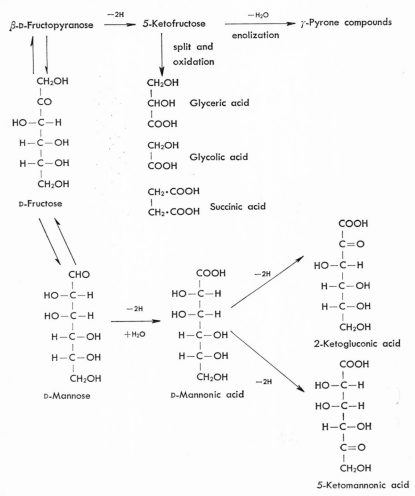

Fig. 18.   Oxidation of D-fructose by *G. cerinus*
(after Terada *et al.*).

According to Terada and his coworkers, 5-ketofructose is formed by a wide variety of acetic acid bacteria and can be isolated in crystalline form in a medium containing 5–10% fructose and 0.3–0.5 % yeast extract. Some of the strains producing it are *A. suboxydans, A. oxydans, A. gluconicus, A. melanogenus, G. cerinus,* and *G. roseus.*

Carr *et al.*[310] examined the metabolic products of 38 strains of *Acetomonas (Gluconobacter)* using paper chromatography, with fructose as the carbon source, and found that all but 6 of the *Acetomonas* gave 2,5-D-*threo*-diketofructose (5-ketofructose) as the main product from fructose.

Recently Terada *et al.*,[311] working on *G. cerinus* var. *ammoniacus* Asai IFO 3267, isolated and identified a new γ-pyrone compound 3-oxykojic acid from fructose in addition to kojic acid. This compound melted at 187°C and showed a reddish purple color with ferric chloride reagent.

The same authors,[312] from melting point determination, infrared and ultraviolet absorption spectra, and elemental analysis proved that the compound previously called isokojic acid by Aida *et al.* did not actually have any alcoholic hydroxyls, but had instead two phenolic hydroxyls ; they

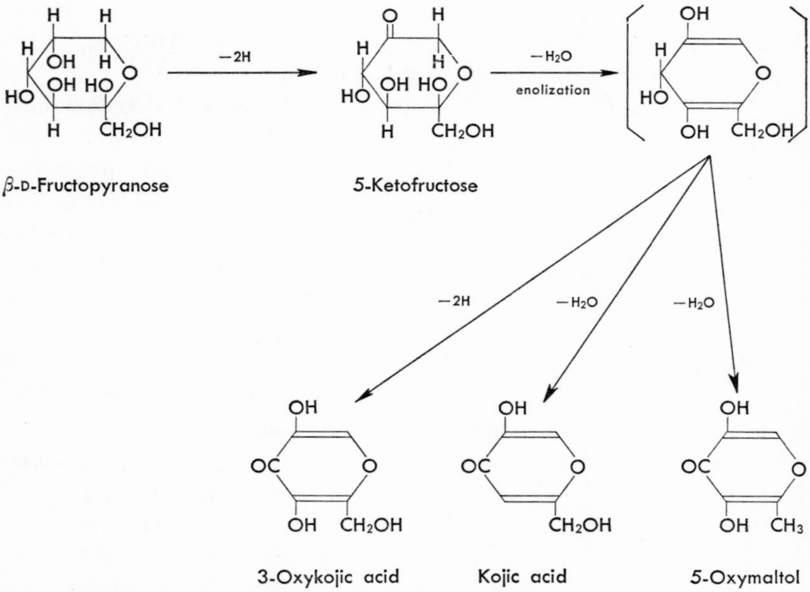

Fig. 19. Presumable pathways through which three different
γ-pyrone compounds are formed from D-fructose
(after Terada *et al.*).

identified the compound as 5-oxymaltol (3,5-dihydroxy-2-methyl-1,4-pyrone, 3-oxyallomaltol or 2-methyl-rubiginol). This is also a new γ-pyrone compound formed by *Gluconobacter*. The chemical pathway for the formation of γ-pyrone compounds from D-fructose was presented as shown in Figure 19.

Recently Aida and Yamada[313] found a new enzyme, "5-ketofructose reductase," catalyzing the reduction of 5-ketofructose to fructose, in fructose-grown cells of *G. albidus* IFO 3250. The reaction proceeds as follows:

$$5\text{-Ketofructose} + \text{NADPH}_2 \rightleftharpoons \text{Fructose} + \text{NADP}$$

The partially purified enzyme preparation was specific for 5-ketofructose and reduced neither 5-ketogluconate nor 2-ketogluconate. A reverse reaction could not be detected at pH 7.0. NADH_2 showed only one thirty-fifth the activity of NADPH_2. Optimum pH for the oxidation of NADPH_2 was about 7.0 and the Michaelis constant (Km) for 5-ketofructose was $6.7 \times 10^{-3}\text{M}$. The reaction was strongly inhibited by 10^{-5}M p-chloromercuribenzoate and 10^{-5}M phenylmercuric nitrate.

Englard and Avigad[314] also reported that NADPH_2 was oxidized by cell-free extracts of *G. cerinus* in the presence of 5-ketofructose, but they did not describe the enzyme involved.

In subsequent experiments Aida et al.[315] purified the enzyme preparation about 200-fold, and found that it still retained a weak activity for the reduction of D-fructose and 2-keto-D-gluconate. But the activities were only 1.7 and 6%, respectively, of that of 5-keto-D-fructose. SH-reagent strongly inhibited the enzyme activity. Monoiodoacetate, phenylhydrazine and arsenite had appreciable inhibitory effects. The enzyme was not influenced by L-cysteine and EDTA (10^{-4}M each). It was inhibited by Cu^{++} and Ni^{++} at a concentration of 10^{-3}M, but was not inhibited by other metals such as Ba, Zn, Cd, and Mg. The enzyme activity was maintained for six months at $-20°C$ without noticeable decrease. It was concluded that the enzyme differs from the 5-ketogluconate reductase of De Ley[316] and Okamoto[317] and also differs from the 5-keto-D-glucono-L-idono-reductase of Takagi,[318] because of its substrate specificity. The reaction proceeds strongly to reduction and the reverse reaction was not detected. Therefore, it was presumed that the enzyme is not involved in the metabolic oxidation of fructose and functions most probably only in the reduction of 5-ketofructose, which has been formed from fructose by another particulate-linked fructose dehydrogenase.

Yamada et al.[319] further reported the positive effects of growth substrate on the activity of 5-ketofructose reductase. The activity was intensified about four-fold when the organism was grown on fructose. The same effect was seen with galactose, sorbose or maltose. 5-Keto-fructose reductase is widely distributed in the genus *Gluconobacter*, especially in the strains belonging to *G. suboxydans*. A very weak activity was seen in *Acetobacter* and no activity was seen in *Pseudomonas*.

The particulate-linked D-fructose dehydrogenase was detected by Yamada *et al.*[320, 321] in the cells of *G. cerinus* IFO 3267. The enzyme preparation, purified about fifty-fold, strongly oxidized D-fructose and the product was confirmed to be 5-ketofructose. The preparation did not oxidize L-sorbose, 2-ketogluconate and polyols, but oxidized glucose and gluconate. The latter activities were attributed to glucose and gluconate dehydrogenases existing in the preparation. The optimum pH for the oxidation of D-fructose was 5.0. Oxidation was strongly inhibited by 10^{-4}M *p*-chloromercuribenzoate and 10^{-4}M phenylmercuric nitrate, but inhibition by atebrine was not strong, only up to 30% at a concentration of 10^{-4}M. The preparation was completely inactivated by 10^{-4}M Ag^+ or 10^{-4}M Hg^{++} and partially by 10^{-3}M Cu^{++} or 10^{-3}M Zn^{++}. The pattern of inhibition indicated that the enzyme is most likely an SH-enzyme. Neither NAD nor NADP was required for the oxidation of D-fructose and 2,6-dichlorophenolindophenol was found to be a most effective electron acceptor.

From these results the enzyme was identified as " D-fructose dehydrogenase," present in the particulate fraction of the cell, and differing from both the glucose and gluconate dehydrogenases of the gluconobacters. An initial metabolic pathway for the oxidation of D-fructose by this organism was also presented, as shown in Figure 20.

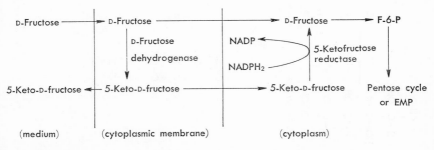

Fig. 20. A presumable pathway for the oxidation of D-fructose initiated by *G. cerinus* (after Yamada *et al.*).

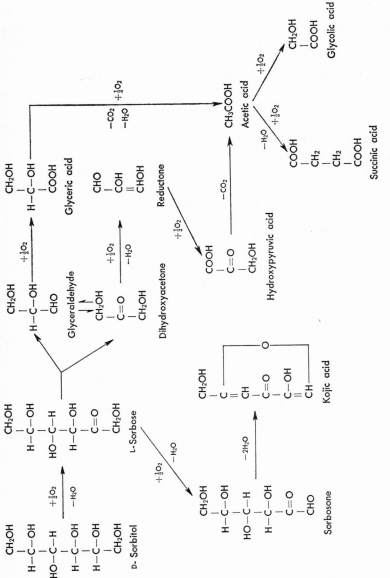

Fig. 21.   Presumable pathway of sorbose oxidation by *A. suboxydans* var. α (after Kondo and Ameyama).

In recent investigations by the same authors[322] on the electron-transfer system in fructose oxidation, the presence of c-type cytochrome was indicated, in addition to the possible participation of cytochrome b, cytochrome o, and $CoQ_9(?)$, while the presence of cytochrome a was disproved. The following electron transfer chain was suggested:

D-Fructose dehydrogenase$\rightarrow CoQ_9(?)\rightarrow$Cyt. $b\rightarrow$Cyt. $c544\rightarrow$Cyt. $o\rightarrow O_2$

The participation of quinone in electron transfer in fructose oxidation was also presumed, because in the presence of the particulate fraction of the cells and of fructose, O_2 uptake was markedly diminished by irradiation with a " black light lamp " which specifically inactivates quinone.
Oxidation of L-sorbose.   Kondo and Wada[323] reported the formation of kojic acid from L-sorbose by some cultures of acetic acid bacteria. There was concomitant formation of glyceric acid, hydroxypyruvic acid, dihydroxyacetone, glycolic acid and succinic acid. Later Kondo and Ameyama[324] observed that only 5-ketogluconate-forming strains oxidized sorbose rapidly, and they isolated and identified as the oxidation products glyceraldehyde, oxymethylene-glycolaldehyde (reductone) and sorbosone as well as kojic acid and related compounds. They proposed a metabolic pathway for sorbose oxidation by *A. suboxydans* var., as shown in Figure 21.

Recently Terada *et al.*[325] confirmed the formation of kojic acid from sorbose and the concomitant formation of 5-ketofructose with a culture of *A. suboxydans* var. *non-acetum* IFO 3254. Kojic acid was also produced from 5-ketofructose by a growing culture of this organism, and interconversion between sorbose and fructose was noted with intact cells under anaerobic conditions. The resulting ketoses were identified chromatographically.

Clearly, then, 5-ketofructose is an intermediate in the formation of kojic acid from sorbose. It is not yet certain, however, whether 5-ketofructose is formed directly from sorbose or from fructose after isomerization of sorbose. A possible pathway for the formation of kojic acid is shown below.

Sorbose $\rightarrow$
$\uparrow \downarrow$ 5-Ketofructose $\dashrightarrow$ Kojic acid
Fructose $\rightarrow$

The transformation of sorbose to fructose is a non-epimeric isomerization of sorbose at C-5. This is the first known case of a non-epimeric transformation of sugars by microorganisms.

Recently, Sato *et al.*[326] demonstrated the formation of three γ-pyron compounds from D-sorbitol *via* 5-keto-D-fructose by a growing culture of *A. suboxydans* "Daiichi No. 1." L-Sorbose and 5-keto-D-fructose were confirmed to be intermediates and the products were identified as kojic acid, 3-oxykojic acid and 5-oxymaltol. The chemical pathway was assumed to be as follows:

$$\text{D-Sorbital} \longrightarrow \text{L-Sorbose} \longrightarrow \text{5-Keto-D-fructose} \left\{ \begin{array}{l} \longrightarrow \text{Kojic acid} \\ \longrightarrow \text{3-Oxykojic acid} \\ \longrightarrow \text{5-Oxymaltol} \end{array} \right.$$

In another report the same authors[327] investigated the particle-bound L-sorbose dehydrogenase in *A. suboxydans* "Daiichi No. 1" which catalyzes the following reaction:

$$\text{L-Sorbose} + \text{Acceptor} \longrightarrow \text{5-Keto-D-fructose} + \text{Reduced acceptor}$$

The particulate fraction obtained from sonicated cells of the organism showed an activity toward L-sorbose and the oxidation product was identified as 5-keto-D-fructose. The optimum pH of the enzyme activity was 6.4 in the presence of 2,6-dichlorophenolindophenol. NAD or NADP did not stimulate the rate of L-sorbose oxidation. The Km value for L-sorbose was 4.2×10^{-3}M (pH 6.5, 15°C).

p-Chloromercuribenzoate and phenylmercuric nitrate as well as silver and mercuric ions inhibited the enzyme activity, suggesting that the enzyme possesses an active SH group. In addition to L-sorbose, the enzyme preparation oxidized D-glucose, D-mannose, D-xylose, L-arabinose, D-sorbitol, D-mannitol, *i*-erythritol, glycerol, D-gluconate, and 2-keto-D-gluconate. D-Fructose, D-arabinose, dulcitol, 5-keto-D-gluconate, trehalose, and sucrose were not oxidized. The enzyme was very unstable and the activity for L-sorbose was completely lost at 40°C for 5 minutes or below pH 4.0 for 20 minutes, whereas that for aldoses, polyols, D-gluconate or 2-keto-D-gluconate was retained to some extent under the above conditions. The enzyme was distinguishable from any other glucose dehydrogenase, ketogluconate dehydrogenase, mannitol dehydrogenase and even from the *Trametes* L-sorbose oxidase. From these experimental results, the enzyme was identified as "L-sorbose dehydrogenase," and Bertrand-Hudson's rule was considered inapplicable to it.

The formation of 2-keto-L-gulonic acid from sorbitol and sorbose has recently been reported by Isono *et al.*[328] In addition to this keto-acid, L-sorbose, D-fructose, L-idonic acid, D-mannonic acid, 2-keto-D-gluconic acid and 5-keto-D-mannonic acid were detected and identified in a shake culture of *G. melanogenus* IFO 3292 when D-sorbitol was supplied as the carbon source. A possible pathway for the formation of these compounds is shown in Figure 22.

Moore et al.[329] studied the oxidation of various heptoses by four strains of *Acetobacter* sp. and found that sedoheptulose was the only sugar metabolized.

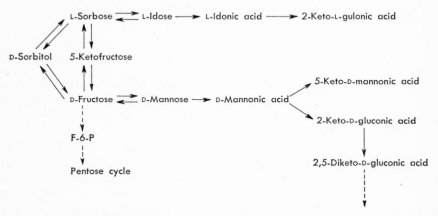

Fig. 22. Pathway for the degradation of D-sorbitol and
L-sorbose by *G. melanogenus* IFO 3292
(after Isono *et al.*).

Chapter 9

BREAKDOWN OF DI- AND OLIGO-SACCHARIDES

The direct oxidation of disaccharides, for instance the conversion of maltose to maltobionic acid seen in *Pseudomonas graveolens*, does not take place in acetic acid bacteria. Disaccharides are oxidized after decomposition to monosaccharides by hydrolyzing enzymes.

Sucrase was found in *A. aceti* and *A. xylinum* by Hoyer.[1] Takahashi[330] also confirmed sucrase in some varieties of *A. aceti* and *A. xylinoides* isolated from Japanese vinegar. Later, the existence of sucrase was reported in many acetic acid bacteria by Hermann and Neuschul,[123] Frateur,[236] and others. Henneberg[331] found that *A. xylinum* produced acid from sucrose. Janke[4] observed a strong acidification of sucrose by *A. lafarianus*. Frateur[236] confirmed that *A. aceti, A. xylinum* var. *xylinoides, A. suboxydans* var. *muciparum, A. suboxydans* var. *biourgianum, A. melanogenus* var. *maltosaccharovorans, A. mesoxydans* var. *saccharovorans* and *A. suboxydans* var. *lentum-saccharovorans* produced acid from sucrose in the presence of $CaCO_3$. In most cases, 5-ketogluconic acid was produced in addition to gluconic acid. He also observed 5-ketogluconic acid production from maltose by *A. suboxydans* No. 1, *A. suboxydans* var. *muciparum, A. xylinum* var. *maltovorans* and *A. melanogenus* var. *maltovorans* in the presence of $CaCO_3$. The acidification of maltose was first reported by Henneberg[139] with strains of *A. oxydans* and *A. industrius*. It was also observed later with some varieties of *A. aceti* Brown and *A. xylinum* by Janke,[4] with *A. rancens* by Visser't Hooft[115] and with *A. suboxydans* by Kluyver and de Leeuw.[264]

Foda and Vaughn[273] investigated the oxidation of maltose by cell suspensions of four maltose-utilizing strains of *A. melanogenus* and concluded that maltose was hydrolyzed to glucose prior to oxidation and then converted to gluconic and 5-ketogluconic acids. This was verified by O_2-uptake and also by paper chromatography of the reaction products. They concluded that in maltose utilization phosphorus uptake could not be demonstrated; maltobionic acid was not found as an end product.

204

It was also observed that cell suspensions of maltose-oxidizing cultures did not utilize this substrate unless the cells had been grown previously in its presence. Later, Tanenbaum and Katznelson[332] reported the formation of 2,5-diketogluconic acid from maltose by two strains of *A. melanogenus* grown previously in maltose. They suggested that the oxidation proceeded in the following way: 1/2 maltose—→glucose—→ gluconic acid—→2-ketogluconic acid—→2,5-diketogluconic acid. However, no evidence was offered for the formation of 5-ketogluconic acid. Oxidation with intact cells in the presence of 2,4-dinitrophenol, as with cell-free extracts, demonstrated that this process is adaptive and not constitutive.

Acidification of lactose was early demonstrated by Henneberg[139] with *A. xylinum*, by Hermann and Neuschul[123] with *A. vini acetati* and by Janke[4] with *A. viscosus*. Asai[333] also reported acidification with two strains of *G. cerinus*. On the acidification of trehalose and raffinose there are few reports. De Ley and Schell[241] observed acid formation from trehalose in a shaking culture of *A. aceti*. Raffinose is attacked by some strains, according to Henneberg[139] and Hermann and Neuschul.[123]

Asai[333] observed acidification of raffinose by some strains of *Gluconobacter*, and Frateur[236] reported that *A. mesoxydans* var. *saccharovorans* produced a remarkable viscid substance accompanied by a small amount of acid.

Henneberg[139] reported that acid was produced from dextrin by *A. industrius* and Takahashi[330] reported the same with some *Acetobacter* strains. Acid production from inulin by *A. xylinoides*, and from pectin by *A. aceti* was reported by Takahashi,[330] and Pitman and Cruess[334] respectively. We have no detailed information, however, on the hydrolyzing enzymes involved in polysaccharide decomposition.

Chapter 10

OXIDATION OF LACTATE AND PYRUVATE

All the strains of genus *Acetobacter* oxidize D-lactate and acetate nearly to completion. This oxidative ability is used as one of the criteria for differentiating between *Acetobacter* and *Gluconobacter* (*Acetomonas*).

The oxidation of lactate was early reported by Hoyer[1] and Seifert.[335] According to Visser't Hooft,[115] *A. rancens* oxidized lactate completely to CO_2 and H_2O while *A. suboxydans*, *A. xylinum*, and *A. melanogenus* oxidized it to the acetate level. Hermann and Neuschul[336] reported that 13 species of acetic acid bacteria gave acetic acid, CO_2 and acetoin as metabolic products of lactic acid. Frateur[236] also reported that most strains, except those of the *Suboxydans* group, exhibited complete oxidation. From enzymatic studies on lactate oxidation, De Ley and Schell[337] found that resting cells of DL-lactate grown *A. peroxydans* NCIB 8618 oxidized D-lactate about four times as fast as L-lactate.

The presence of D-lactate oxidizing enzymes is unusual, since the enzymes responsible for the production and decomposition of lactate in microorganisms are usually specific for the L-configuration. The particulate fractions from crude cell-free extracts of *A. peroxydans* NCIB 8618 contained all the enzymes for the oxidation of D-and L-lactate, pyruvate, ethanol, and acetaldehyde to form acetate with a concomitant O_2 uptake, requiring no additive cofactors. The particles also contained pyruvate decarboxylase, cytochromes and cytochrome oxidase. Both lactate isomers were oxidized ultimately by way of the cytochromes and cytochrome oxidase, as reported by Chin.[338] The optimum pH for the oxidation of D- or L-lactate was approximately 6.0, and no evidence was found for a pyruvate oxidase. The soluble fraction from crude cell-free extracts contained D-lactate and L-lactate dehydrogenase. Both required N-methylphenazium methosulfate as an artificial carrier, but the latter enzyme was very weak. The pH optimum for soluble D-lactate dehydrogenase was 5.5–6.0. The soluble fraction also contained pyruvate decarboxylase and coenzyme-linked ethanol and acetaldehyde dehydrogenases. NADP- and

NAD-linked lactate dehydrogenases and lactate racemase were not detected. Acetate oxidation did not occur and was not stimulated by the carriers or cofactors tested. The properties of the lactate-oxidizing enzymes of *A. peroxydans* are shown in Table 17.

Table 17. Cofactors, inhibitors and optimum pH of lactate-oxidizing enzymes of *A. peroxydans* NCIB 8618 (after De Ley and Schell).

Localization	Enzymes	Opt. pH	End product	Cofactors	Inhibitors
Particulate fraction	D-Lactate oxidase L-Lactate oxidase	About 6.0	Pyruvate	Not required	CO, KCN, BAL (5×10^{-3} M), arsenite (10^{-2} M), H_2O_2 (40 μM), *p*-benzoquinone (M/640), dimedon
Soluble fraction	D-Lactate dehydrogenase L-Lactate dehydrogenase	5.5–6.0	Pyruvate	*N*-Methylphenazinium methosulfate, methylene blue.* NADP and NAD were not required.	Atebrin H_2O_2 (40 μM), *p*-benzoquinone (M/640), dimedon

BAL : British anti-lewisite (2, 3-dimercaptopropanol).
* In the case of L-lactate dehydrogenase methylene blue does not act as cofactor.
D- and L-lactate oxidases are not inhibited by atebrin (3.3×10^{-4} M), urethane (5×10^{-3} M), amytal (5×10^{-3} M) and *o*-phenanthroline (10^{-2} M). D- and L-lactate dehydrogenases are not inhibited by arsenite (10^{-3} M). D-Lactate dehydrogenase is not inhibited by EDTA (10^{-2} M), KCN (10^{-3} M) and *p*-chloromercuribenzoate (10^{-3} M).

Tanenbaum[101] reported, however, that pyruvate was rapidly oxidatively decarboxylated by suspensions of *A. peroxydans* NCIB 8618 and that the reaction in the cell-free system was linked only to NADP reduction. Lactate behaved in the same manner as pyruvate.

De Ley and Vervolet[339] looked for a possible role of peroxidase in the oxidation of D-lactate by *A. peroxydans* NCIB 8618. Intact resting cells showed marked peroxidase activity for D-lactate, acetaldehyde, several primary alcohols and aromatic diamines (*o*-, *p*- and dimethyl-*p*-phenylenediamine). However, it was confirmed that oxidation of D-lactate by resting cells in the presence of M/640 *p*-benzoquinone ceased when 0.5 mole O_2 was consumed per mole of D-lactate oxidized and one mole of pyruvate accumulated. The same concentration of *p*-benzoquinone inhibited neither the spontaneous nor the enzymatic peroxidation of pyruvate. Further, the assumption that D-lactate would be oxidized with production of H_2O_2 was invalidated since the respiration of D-lactate was not noticeably enhanced in the presence of p-phenylenediamine. Thus they concluded that peroxidase did not play a physiological function in the oxidation of D-lactate.

The probable mechanism of D-lactate oxidation by both soluble and oxidosome-linked enzymes of *A. peroxydans* has been presented as follows:

$$D\text{-Lactate} + 0.5\ O_2 \longrightarrow Pyruvate + H_2O$$

$$Pyruvate \longrightarrow CH_3CHO + CO_2$$

$$CH_3CHO + 0.5\ O_2 \longrightarrow Acetate$$

$$\text{Sum : } D\text{-Lactate} + O_2 \longrightarrow Acetate + CO_2 + H_2O$$

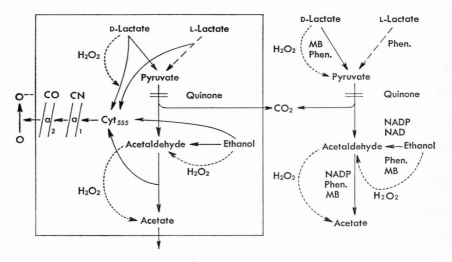

Fig. 23.   Two pathways for the formation of acetate from either lactate or ethanol by *A. peroxydans* (after De Ley).

The rectangle represents an ultramicroscopic particle (oxidosome), the enclosed reaction being linked to it. The other reactions are carried out by soluble enzymes in the cytoplasm. The broken lines show weak activities. The dotted lines may have no physiological functions. Abbreviations: MB, methylene blue; Phen., *N*-methyl-phenazinium methosulfate.

Figure 23 shows the dual pathways for oxidation of lactate or ethanol by *A. peroxydans* that appeared in a publication by De Ley.[340]

Later experiments by De Ley and Schell[341] on other acetic acid bacteria revealed that *A. aceti* strain Ch 31, *A. aceti* (rancens) strain 23, *A. aceti* (pasteurianus) strain 11, *A. aceti* (liquefaciens) strain 20 (=*G. liquefaciens* or "Intermediate" Strain IAM 1834), *Gluconobacter oxydans* (suboxydans) strain 26, and *G. oxydans* (melanogenus) NCIB 8086 contained a constitutive pyruvate decarboxylase that required thiamine pyrophosphate and Mg^{++}. A pyruvate decarboxylase system was also found in the particulate fraction when the organisms were grown on lactate. With most of the strains D-lactate was oxidized faster than L-lactate.

King and Cheldelin[342] reported that pyruvate oxidation to acetate in *A. suboxydans* was carried out in two steps: first, the pyruvate is decarboxylated to acetaldehyde by a yeast-type decarboxylation, and then the latter is oxidized to acetate. But, according to Rao's data[343] with a strain of *A. aceti*, acetaldehyde was not an intermediate in pyruvate breakdown, and lipoic acid did not play a role. The existence of a pyruvate oxidase system requiring CoA, glutathione, thiamine pyrophosphate and NAD was reported in *A. pasteurianus* by King et al.[344] The oxidizing system was fully confirmed with extracts of glycerol-grown cells of *A. aceti* Ch 31 by De Ley and Schell,[341] but the data indicated that acetaldehyde is an intermediate and that a pyruvate oxidase system is not involved. The following pathway was suggested.

$$Pyruvate \xrightarrow[\substack{Mg^{++} \\ ThPP}]{CO_2} CH_3CHO \xrightarrow[NAD]{\substack{CoA \\ GSH}} Acetyl\text{-}CoA + NADH_2$$

Since it was proved that the same extracts contained an NAD-linked acetaldehyde dehydrogenase which is activated by CoA and glutathione, De Ley and Schell demonstrated that oxidation involving a pyruvate decarboxylase and NAD-linked acetaldehyde dehydrogenase should operate without participation of a separate pyruvate oxidase system. It was noted that particles from lactate-grown cells displayed a pyruvate decarboxylase activity, but the particulate pyruvate decarboxylase is not very tightly bound to the particles and could easily be detached by washing with buffer, in contrast to the oxidase systems which could not be removed thus.

According to De Ley and Dochy,[171] cell débris and small particles of cells of *G. liquefaciens* oxidized D-lactate to the acetate stage, while intact cells oxidized D-lactate, acetate, glycerol and ethanol almost to completion. Dupuy and Maugenet[345] studied the aerobic oxidation of lactate under non-proliferating conditions with *A. rancens* grown at pH 3.8 and 6.0. With Ca D-lactate the oxidation rate was substantially constant between pH 3.8 and 7.0 for both the cells grown at pH 3.8 and 6.0. CO_2 evolution was slightly greater in both cases than would correspond to the O_2 absorption. When respiration ceased the total O_2 absorption was greater in the cells grown at pH 6.0 (about 75% of that required for total oxidation) than in the cells grown at pH 3.8 (about 50% of that required for total oxidation). The unoxidized substrate was tested with cells grown at pH 3.8 by using DL-lactate-U-[14]C. At pH 6.0, 80% of the radioactivity appeared in the CO_2 evolved, while the remainder was fixed by the cells. At pH 3.8 the lactate also disappeared but 20% of the radioactivity appeared in solution as acetate while a smaller proportion than before was absorbed by the cells. It was shown in the aerobic oxidation of pyruvate and acetaldehyde that the acetaldehyde oxidation stage tends to be delayed relative to the pyruvate decarboxylation stage by an increase in pH. It was also shown that decarboxylation of pyruvate under anaerobic conditions was much greater in the cells grown at pH 3.8 than in the cells grown at pH 6.0. Other experiments using cell-free extracts showed that the former had much more pyruvate decarboxylase in their cytoplasm than the latter.

Chapter 11

ACETOIN FORMATION

Formation of acetoin by acetic acid bacteria was first reported by Kitasato[346] with acetone-treated cells of *A. ascendens*, using pyruvate as the substrate. Lemoigne[347] also reported formation of acetoin from pyruvate by a strain of *A. xylinum*. Hermann and Neuschul[336] studied the decomposition of DL-Na lactate by 13 species of *Acetobacter* and found that all but *A. gluconicus* were able to produce acetoin. *A. rancens* and *A. ascendens* were the most active acetoin-producers, giving 31–36% of the theoretical amount. Later, Frateur[348] reported that strains of the *Oxydans* group decomposed Ca lactate and produced acetoin. Recently, De Ley[349] studied aerobic fermentation of DL-lactate with 44 strains of *Acetobacter*. Most strains produced only limited amounts of acetoin, but some strains belonging to *A. rancens*, *A. pasteurianus* or *A. ascendens* could convert DL-lactate into acetoin up to 74% of the theoretical yield. These may be of importance in the industrial production of acetoin.

The presence of pyruvate decarboxylase was confirmed by manometric experiments with intact cells, and the formation of acetoin from pyruvate was also verified with crude cell-free extracts of *A. rancens* strain 23. Pyruvate decarboxylase and the acetoin-forming enzyme were present most actively in the soluble fractions of crude extracts. Thiamine pyrophosphate (ThPP) stimulated both CO_2 production and acetoin synthesis from pyruvate. Mg^{++} and Mn^{++} had no noticeable effect, but when the soluble fraction was dialyzed against 0.6% versene at pH 7.5, it became inactive unless ThPP and Mg^{++} were added. Acetoin synthesis was active over a pH range of 5–8. Addition of acetaldehyde to pyruvate under anaerobic conditions resulted in a considerable increase of acetoin with crude cell-free extracts, but the rate of CO_2 production was hardly affected, supporting the theory that acetoin could be formed by the reaction between acetaldehyde and ' aldehyde-ThPP.' In addition, acetoin was also formed from α-acetolactate accompanied by decarboxylation when crude cell-free extracts of *A. rancens* strain 23, *A. pasteurianus* strain 11,

211

or *A. ascendens* strain A were used. This supports the existence of another pathway involving an active α-acetolactate decarboxylase. The two proposed pathways of acetoin formation by these organisms are:

1)   2 Pyruvate+2 ThPP $\longrightarrow$ 2 'Aldehyde-ThPP '+2 CO_2
     'Aldehyde-ThPP ' $\longrightarrow$ $CH_3 \cdot CHO$+ThPP
     'Aldehyde-ThPP '+$CH_3 \cdot CHO$ $\longrightarrow$ Acetoin+ThPP
2)   'Aldehyde-ThPP '+Pyruvate $\longrightarrow$ α-Acetolactate+ThPP
     α-Acetolactate $\longrightarrow$ Acetoin+CO_2

Stouthamer[350] examined *Acetobacter* for acetoin formation with lactate as the substrate, and confirmed it in strains of the *Rancens* and *Mesoxydans* groups, and especially in *A. rancens* strain 24.

Acetoin formation from 2,3-butanediol has already been described (p. 145).

Oxidation of acetoin to diacetyl by the *Oxydans* group of *Acetobacter* is unlikely, according to De Ley,[349] in view of the almost quantitative recovery of acetoin from the medium.

Cheldelin and co-workers[351] supported the presence of two acetoin-forming enzymes in soluble extracts of *A. suboxydans:* one employing α-acetolactate as the preferred substrate, and the other employing acetaldehyde.

According to recent investigations by Wixom,[352] four *Acetobacter* strains, *A. rancens* NCIB 6430, *A. rancens* NCIB 4148, *A. kützingianus* NCIB 3294, and *A. acetosus* NCIB 2224 produced acetoin in a medium containing lactate-ammonium sulfate as the sole carbon and nitrogen source (other than vitamins), while two strains of *Gluconobacter* (*Acetomonas*), *G. melanogenus* NCIB 8084 and *G. suboxydans* ATCC 621 failed to grow in this medium and even cells grown in glucose-lactate-ammonium sulfate medium could not produce acetoin. Dehydrating activity of ultrasonically disrupted cells against DL-α,β-dihydroxyisovalerate was demonstrated in both *Acetobacter* and *Gluconobacter* strains. The specific dehydration activity from organisms grown on valine- and isoleucine-deficient media, such as an ammonium sulfate medium, was greater than those grown on complete casein hydrolysate medium.

Not only the valine intermediate, α,β-dihydroxyisovalerate, but also the isoleucine intermediate, α,β-dihydroxy-β-methyl-*n*-valerate, was dehydrated by extracts of both *Acetobacter* and *Gluconobacter* strains. Since the existence of acetolactate-forming enzyme(s) and acetolactate decarboxylase was already confirmed in *Acetobacter* strains by De Ley,[349] as mentioned

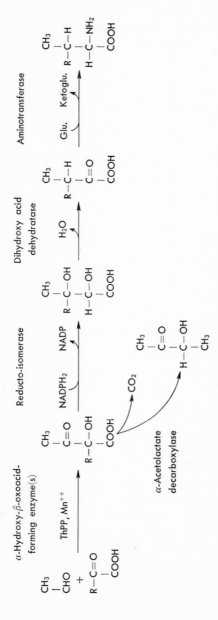

Fig. 24. The metabolic pathway of α-acetolactate or its homologue in *Acetobacter*.

If R=CH₃, acetoin and valine are the end products.

If R=CH₂CH₃, acetyl-ethylcarbinol and isoleucine may be formed.

earlier, Wixom suggested that *Acetobacter* strains at least can carry out both valine synthesis and acetoin formation with α-acetolactate as the intermediate. The proposed metabolic pathway of α-acetolactate or its homologue in *Acetobacter* is shown in Figure 24.

Chapter 12

OXIDATION OF ACETATE

The strains belonging to the genus *Acetobacter* oxidize acetate into carbon dioxide and water. This " over-oxidation " had been considered one of the undesirable phenomena in vinegar production. Visser't Hooft[115] reported that almost all acetic acid bacteria except the *Suboxydans* group are able to oxidize acetate. Tanaka[353] also determined by manometric experiments with intact cells that various acetic acid bacteria showed O_2 uptake with an R.Q. of 1. Sodium bisulfite and HCN strongly inhibited oxidation. Intact cells of *A. aceti* could utilize neither methylene blue nor quinone as hydrogen acceptors in acetate oxidation. Oxygen uptake was inhibited by monoiodoacetate, carbon monoxide, and KCN.

Over-oxidation occurred most markedly at a low concentration of acetic acid, and with strains tolerant against a low pH, according to Wiame and Lambion.[354] Dratwina[355] pointed out that a limited pH range specific for the strains exists in over-oxidation of acetic acid. Tanenbaum[101] obtained disrupted cell particles maintaining acetate oxidizability from a catalase-negative *A. peroxydans* NCIB 8618, whereas the cell-free extracts were devoid of enzyme activity. The R.Q. for acetate oxidation with one-day-old cells of this organism was close to 1.0, indicating that the following conversion occurred:

$$CH_3COOH + O_2 \longrightarrow CO_2 + (CH_2O) + H_2O$$

Acetate oxidation was almost completely inhibited by $10^{-4}M$ dinitrophenol or by a similar concentration of monoiodoacetate. Intact fresh cells oxidized formate quantitatively to carbon dioxide and water. De Ley and Schell,[341] however, using the same organism, showed that acetate was not oxidized by soluble enzymes either in the absence of carriers or in the presence of phenazinemethosulfate or methylene blue. The particulate fraction was also incapable of oxidizing acetate. Federico and

215

Gobis[356] observed that certain organic acids involved in the TCA cycle stimulated acetic acid oxidation of *A. aceti*, while oxaloacetic acid affected it weakly and citric acid had no effect. Antoniani *et al*.[357] reported that pyruvic acid behaved like oxaloacetic acid but that DL-glyceric acid stimulated the oxidation of acetic acid, most probably accompanied by the formation of fumaric acid. The slow effect of added L-glutamic acid was thought to be due to oxidative deamination to α-ketoglutaric acid prior to active stimulation. Later Federico[358] found that addition of malic or succinic acid strongly promoted the oxidation of acetic acid by *A. aceti*. Formation of citric acid in the reaction mixture was also confirmed in subsequent experiments by Antoniani *et al*.[359] Tošić[126] observed that addition of formate, lactate, succinate, malate, pyruvate, glycerol, glucose or ATP shortened the lag period of acetate oxidation by intact cells of *A. turbidans*.

As to the metabolic pathway of acetate oxidation, Butlin[360] viewed it as the conversion of acetate into glyoxylate and formate. Tanaka,[361] on the other hand, suggested the operation of the Wieland-Thunberg pathway in *A. aceti*, *i.e.* 2 moles of acetate convert stepwise into one mole of succinate, fumarate, malate, oxaloacetate and pyruvate; the latter may convert into one mole each of acetate and carbon dioxide, and this acetate participates again and the cycle is oxidized to completion. According to Tanenbaum,[101] members of the dicarboxylic acid cycle such as succinic, oxaloacetic, malic and fumaric acids were oxidized, whereas citric, isocitric and α-ketoglutaric acids were not attacked either by sonically disrupted homogenates which oxidized acetate, or by cell-free extracts of *A. peroxydans* NCIB 8618. Therefore, he supposed that acetate oxidation by this organism might take place *via* a dicarboxylic cycle rather than the TCA cycle. Other studies of acetate oxidation by this particular organism will be introduced later.

Recently Loitsyanskaya and Mamkaeva[362] reported that *A. aceti*, in cultures with acetic acid as the carbon source, is capable of using acetic acid not only in energy metabolism but also in anabolism. An increase in the mass of cells was always accompanied by the intensive consumption of acetic acid.

During its development in the nutrient media with ethanol, *A. aceti* used the acetic acid formed from the ethanol for anabolism. When the total concentration of ethanol and acetic acid in the medium was 5% or somewhat higher, relatively large quantities of acetic acid were accumulated in the logarithmic growth phase and depression of the pH limited further multiplication of the bacteria. After a lag, active growth of the cells resumed, accompanied by a decrease in the amount of acetic acid and a rise in pH.

Chapter 13

OXIDATION OF ORGANIC ACIDS OTHER THAN TCA CYCLE MEMBERS

Formic acid. An early observation by Müller[116] on the oxidative break-down of formic acid with intact cells of *A. pasteurianus* was re-confirmed by Tanaka[353] with, in addition to *A. pasteurianus*, *A. ascendens*, *A. peroxydans*, *A. aceti* Pasteur, *A. rancens*, and *A. zythi*. The R.Q. was about 2, indicating that the following reaction might be taking place:

$$HCOOH + 0.5\ O_2 \longrightarrow CO_2 + H_2O$$

The reaction was inhibited by HCN. Methylene blue was utilized as a hydrogen acceptor in contrast to the case of acetate oxidation. Intact cells of *G. suboxydans* and *A. xylinum* could not oxidize formate.

Propionic, butyric, and isobutyric acids. According to Hoyer,[1] a strain of *A. aceti* oxidized Ca propionate to Ca carbonate, whereas *A. pasteurianus* and *A. rancens* did not. Visser't Hooft[115] reported oxidation of Ca propionate, Ca butyrate and Ca isobutyrate by *A. rancens* in yeast extract medium.

Early experiments by Banning[363] showed oxidation of isobutyric acid to oxalic acid by *A. rancens*, while oxidation of butyric acid by *A. lovaniense*, *A. ascendens*, *A. rancens*, *A. rancens* var. *filamentosum*, *A. rancens* var. *turbidans*, *A. rancens* var. *saccharovorans*, *A. rancens* var. *vini*, *A. aceti* and *A. mesoxydans* was reported by Frateur.[348] Only *A. aceti* oxidized propionic acid.

Malonic acid. Banning[363] reported the formation of oxalic acid from malonic acid by various acetic acid bacteria. Tanaka[353] observed a noticeable oxidation of malonic acid by several species of *Acetobacter*.

Glyceric acid. In addition to Tanaka's experiments[353] with intact acetobacter cells, Frateur[348] confirmed oxidation of Ca glycerate by the *Oxydans* and *Mesoxydans* groups of *Acetobacter*.

218

Visser't Hooft[115] reported oxidation of tartrate to carbonate by *A. rancens*.

Glutaric acid. Hoyer[1] and Visser't Hooft[115] reported oxidation of glutarate to carbonate by *A. rancens*. However, Tanaka's data[353] with intact cells gave negative results with various acetic acid bacteria including *A. rancens*.

Aconitic acid. According to Visser't Hooft,[115] the acid was oxidized to carbonate by *A. rancens*.

Table 18 shows the oxidizability of several organic acids by intact cells of various acetic acid bacteria.

Table 18. Oxygen uptake of acetic acid bacteria for several organic acids (Warburg's manometric experiments, ml/hr/mg, after Tanaka).

Organism / Substrate	*A. aceti* Pasteur	*A. ascendens*	*A. pasteurianus*	*A. peroxydans*	*A. rancens*	*A. suboxydans*	*A. xylinum*	*A. zythi*	*A. aceti*
Oxalic acid	0	0	0	0	0	0	0	0	0
Fumaric acid	5–8	25	3	25	10	0	10–15	30–50	50
Maleic acid	2	2	0.3	2	2		0.3	8	3
Malic acid	5–8	28	7	25–30	10	0	10–15	30–50	50
Tartaric acid	0	0	0	0	0		0	0	0
Citric acid	2	0.5	1	0.5	1		0	1	0
Aconitic acid	1.5	0.3	0.4	0.6	0.5		0	1	0
Glyceric acid	2–5	5–9	3	2	1		3	10–15	5

D-*Quinic acid and related acids.* Whiting and Coggins[364] reported that growing cells of a few strains of *Acetomonas oxydans* oxidized D-quinate to 5-dehydroquinate, and D-shikimate to 3-dehydroshikimate. D-Dihydroshikimate was also oxidized to the corresponding 5-dehydro compound but epidehydroshikimate was not (see Figure 25). Cell-free extracts oxidized D-quinate to 5-dehydroquinate with the consumption of the stoichiometric amount of oxygen, but the oxidation of shikimate and dihydroshikimate did not go to completion. Quinate was oxidized by a constitutive particulate enzyme probably localized in the cytoplasmic membrane. No evidence was presented for the participation of NAD, NADP, or free flavins in electron transport and the system was presumed to be cytochrome-linked.

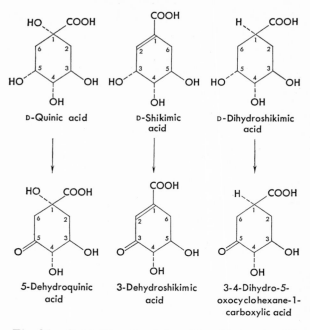

Fig. 25.   Oxidation of quinic acid and related acids.

Chapter 14

CYTOCHROMES AND FLAVIN-ENZYMES

Cytochromes. In acetic acid bacteria the cytochromes, which act as electron-transferring carriers in terminal oxidation, have been studied primarily with *A. pasteurianus*, *A. peroxydans*, and *G. suboxydans* strains. Detailed studies on the cytochromes of the latter two species will be introduced later in the text (p. 286, 259). Here we will discuss the cytochromes of the other species of *Acetobacter*, and especially *A. pasteurianus*.

Smith,[365] using intact cells of *A. pasteurianus*, and King and Cheldelin[300] using the solubilized particles obtained by deoxycholate-treatment, found the cytochromes shown in Table 19.

Table 19.   Absorption bands (mμ) of the reduced cytochromes in *A. pasteurianus*.

Preparation	Cytochrome *a* group (a₁)		Cytochrome *b* group		
Intact cells	588	445	554	523	428
Solubilized particles	590–595	440–445	558–560	528–532	430

Fujita and Kodama[366] found a cytochrome *a*-group in both *A. pasteurianus* and *A. aceti*; cytochrome a_1 was considered to participate most probably in the terminal oxidation.   Chance,[367] and Castor and Chance[368] supported this view.   Cytochrome a_2 (625–635 mμ), which was found in *A. pasteurianus* by Fujita and Kodama,[366] and by Kubowitz and Haas,[369] was not likely to participate in the terminal oxidation, according to Smith.[365] Cytochrome a_3, which is known to function in yeast and animal tissues, was confirmed spectroscopically in *A. pasteurianus* by Warburg and Negelin[370] but its active participation in terminal oxidation was denied by Smith.[371]

Kasai *et al.*[372] reported finding in the solubilized particulate fraction

221

of A. *dioxyacetonicus* strain A 15, cytochrome c_1 which is identical to the cytochrome c_1-552 or cytochrome c_1-554 of G. *suboxydans* found by Iwasaki;[373] it was concluded that the electron released in the oxidation of glyoxylate to oxalate is most probably transferred through this cytochrome to atmospheric oxygen.

Flavin-enzymes.   Warburg and Christian[374] were the first to report the presence of " old yellow enzyme," *i.e.* flavin-enzymes, in A. *pasteurianus.* The literature on the existence and function of flavin-enzymes is very slight compared to that on the pyridine nucleotide-enzymes and cytochromes. It is presumed, however, that the flavin-enzymes probably act together with the cytochrome system in terminal oxidation, taking a role in transporting activated hydrogen or electrons to atmospheric oxygen to form H_2O.

Tanenbaum[101] reported the participation of a flavin-enzyme associated with diaphorase and a reduced NADP cytochrome c-reductase in cell-free extracts of A. *peroxydans.*

According to De Ley and Dochy,[171, 375] flavin-enzymes seem to be tightly bound to the protoplasmatic membrane of acetic acid bacterial cells and therefore the protoplast freed from the cell wall possesses enzymatic activities.   The participation of flavin-enzyme was suggested in the direct oxidation of glucose by the particulate fraction of G. *suboxydans.*

Chapter 15

OXALIC ACID FORMATION

Oxalic acid production from glucose by *Acetobacter* was discovered in 1900 by Zopf[376] in three species of *Acetobacter, viz. A. aceti, A. pasteurianus,* and *A. kützingianus.* Later, *A. xylinum, A. acetigenus, A. ascendens,* and *A. acetosus* were added to the above species by the same author. Banning[363] reported that *Termobact. aceti, A. oxydans, A. industrius,* and *A. aceti oxalici* formed oxalate not only from sugars but also from alcohols and organic acids. Henneberg[331] reported that *A. curvus* and *A. schützenbachii* also formed oxalic acid. Frateur[377] found oxalic acid formation from D-fructose by *G. melanogenus* and *G. suboxydans* var. *biourgianum.*

As is evident from these early reports, many species of acetic acid bacteria can form oxalic acid from various carbonaceous substrates. The substrates and the species producing oxalic acid therefrom were summarized by Tanaka and Masumoto[378] as shown in Table 20.

The mechanism of oxalate formation has been studied primarily in fungi because of their ability to accumulate high amounts of oxalic acid from sugars. Recently, Hayaishi and his associates[379] found hydrolysis of oxaloacetate to acetate and oxalate by cell-free extracts of *Aspergillus niger.*

$$\begin{array}{ccc}
\text{COOH} & & \text{COOH} \quad \text{Oxalic acid}\\
| & & |\\
\text{C=O} & \xrightarrow[\text{Mn}^{++}]{+\text{H}_2\text{O}} & \text{COOH}\\
| & & \\
\text{CH}_2 & & \text{CH}_3\\
| & & | \qquad\quad \text{Acetic acid}\\
\text{COOH} & & \text{COOH}\\
\text{Oxaloacetic acid} & &
\end{array}$$

Challenger *et al.*[380] had earlier proposed a direct oxidation mechanism of acetate to oxalate and confirmed the formation of Ca oxalate from

223

Table 20. Oxalic acid production from various carbonaceous substrates by *Acetobacter* (after Tanaka and Masumoto).

Carbon source	A. xylinum	A. ascendens (Inst. f. Gärungsgewerbe, Berlin)	A. aceti Pasteur	A. acetosus Henneberg	A. pasteuri-anus	A. aceti Henneberg	A. rancens	A. acetosus (Inst. f. Gärungsgewerbe, Berlin)	A. xylhi	A. aceti
Glycol	≣	≣	≣	≣	≣	≣	≣	≣	≣	≣
Glycolic acid	−	−	≣	≣	≣	≣	≣	≣	−	≣
Butyric acid	+	+	≣	+	+	+	−	+	(+)	+
Glycerol	≣	≣	+	≣	≣	≣	−	+	−	−
Glyceraldehyde	≣	≣	≣	≣	≣	≣	−	−	−	−
Ethanol	≣	≣	≣	−	+	+	−	−	−	−
Glucose	≣	≣	+	≣	−	−	−	−	−	−
Gluconic acid	≣	≣	(+)	(+)	(+)	−	−	−	−	−
Lactic acid	≣	≣	+	−	−	−	−	−	−	−
Pyruvic acid	≣	≣	+	−	−	−	−	−	−	−
Mannitol	≣	≣	+	−	−	−	−	−	−	−
Methyl glyoxal	≣	+	−	−	−	−	−	−	−	−
Acetic acid	(+)	(+)	−	−	−	−	−	−	−	−
Succinic acid	(+)	+	−	−	−	−	−	−	−	−
Fumaric acid	(+)	(+)	−	−	−	−	−	+	−	−
Malic acid	−	(+)	−	−	−	−	−	−	−	−
Propyl alcohol	−	−	+	−	−	−	−	(+)	−	−
Propionic acid	−	−	+	−	−	−	−	+	−	−

Medium: peptone 1%, meat extract 1%, carbon sources 0.5%, $CaCl_2$ trace, pH 6.0, agar added. Ca oxalate which had crystallized out was observed microscopically.

Ca acetate.   They identified glycolic acid as one of the intermediates of acetate oxidation in *Asp. niger*, and reported the isolation of NH_4 oxalate from NH_4 glycollate.   Although they could not prove direct formation of oxalate from glyoxylate, they isolated glyoxylate as an aminoguanidine derivative from culture fluid containing Ca acetate.   Thus the following hypothetical scheme was proposed.

$$
\begin{array}{cccccccc}
CH_3 & & CH_2OH & & CHO & & COOH \\
| & \longrightarrow & | & \longrightarrow & | & \longrightarrow & | \\
COOH & & COOH & & COOH & & COOH \\
\text{Acetic acid} & & \text{Glycolic acid} & & \text{Glyoxylic acid} & & \text{Oxalic acid}
\end{array}
$$

No evidence, however, has yet been offered for the formation of oxalic acid from glyoxylic acid.

Tanaka and Masumoto[378] observed that glycollate was the best substrate for oxalate formation and emphasized the possibility of oxalate formation *via* glycollate as a side reaction for the active acetate oxidation.

Recently Asai *et al.*, in unpublished experiments, confirmed the presence of the TCA cycle and glyoxylate by-pass as the terminal oxidation system of *A. dioxyacetonicus* strain A 15, and also confirmed oxidation of glyoxylate to oxalate by intact cells of the organism.   A possible pathway for oxalate formation is shown in Figure 26.

Kasai *et al.*[141] reconfirmed the oxidation of glyoxylic acid to oxalic acid.   On paperchromatograms of oxidation products from glyoxylic acid-1,2-^{14}C by either intact cells or cell-free extracts, oxalic acid was the only radioactive compound other than the substrate.   Since the oxidation

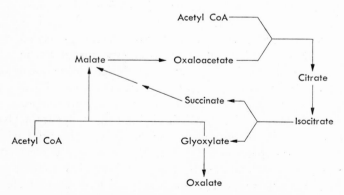

Fig. 26.   A possible pathway for oxalate formation by
A. dioxygluconicus strain A 15.

was not inhibited by 2,4-dinitrophenol it was concluded that it does not involve a reaction requiring resynthesis of acetyl CoA, which would have been a prerequisite for the condensation of glyoxylic acid and acetyl CoA to form malic acid.   Moreover, since oxalic acid was not formed from malic acid, the system reported by Hayaishi *et al.* in *Asp. niger*, *i.e.* the hydrolytic formation of oxalic acid from oxaloacetic acid, apparently does not exist here.

The cell-free extracts oxidized glyoxylic acid as well as acetaldehyde and hypoxanthine with consumption of oxygen.   The oxidation of glyoxylic acid was strongly inhibited by NaN_3, while that of acetaldehyde and hypoxanthine was not.   The extracts oxidized glyoxylic acid and acetaldehyde rapidly with 2,6-dichlorophenolindophenol as a hydrogen acceptor, but not hypoxanthine.   Kasai and his coworkers separated a particulate fraction and a soluble fraction at $100,000 \times g$ for 60 min.   The oxidative activity for acetaldehyde was concentrated in the soluble fraction, glyoxylic acid in the particulate fraction.   The particulate fraction from a strain with a decreased activity for acetaldehyde oxidation lost the ability to oxidize acetaldehyde completely when washed with a Tris-buffer of pH 7.0 and oxidized only glyoxylic acid.   The soluble fraction, after dialysis, also oxidized only glyoxylic acid.

The glyoxylic oxidase, unlike acetaldehyde dehydrogenase, possessed an optimum pH range around 6 and did not oxidize acetaldehyde, glyceraldehyde, benzaldehyde, glucose, glucuronic acid or ribose.   Cell-free extracts did not oxidize oxalic acid and glycolic acid.   There was, in addition, an $NADH_2$-linked glyoxylate reductase which catalyzed the reduction of glyoxylic acid to glycolic acid.

It was concluded that oxalic acid formation by *A. dioxyacetonicus* proceeds at least partly from the direct oxidation of glyoxylic acid, which is in turn derived from isocitric acid through the glyoxylate by-pass. This oxidase shows a high substrate specificity for glyoxylic acid and differs in many characteristics from acetaldehyde dehydrogenase or xanthine oxidase.   The oxidase is located in the particulate fraction, can use both oxygen and 2,6-dichlorophenolindophenol as electron acceptors, and is independent of NADP or NAD, which distinguishes it from the glyoxylate dehydrogenase found in *Pseudomonas oxalatics* by Quayle and Taylor.[381]

In subsequent experiments Kasai *et al.*[382] studied the electron-transfer system in glyoxylate oxidation by this organism.   The enzyme preparation, obtained from the particulate fraction by solubilizing with deoxycholate and then salting out with 15–45% ammonium sulfate, carried out a stoichiometric conversion of glyoxylate to oxalate.   Phenazine-methosulfate, 2,6-dichlorophenolindophenol and cytochrome c were used as electron acceptors, while methylene blue was weakly reduced and methyl viologen, riboflavin and triphenyltetrazolium chloride were not reduced. The oxidation of glyoxylate was not inhibited by cyanide at a concentration of $4 \times 10^{-4} M$, which completely inhibited the acetate oxidation, suggesting the formation of cyanhydrine in the glyoxylate oxidation. Carbon monoxide inhibited oxidation with 2,6-dichlorophenolindophenol. This inhibition was restored by light, indicating that carbon monoxide inhibited cytochrome oxidase. These observations strongly suggested that the cytochrome system is associated with the terminal electron transfer system of glyoxylate oxidation. Kasai *et al.* also observed an absorption band of cytochrome c_1 type in the solubilized particulate fraction, which coincided with the cytochrome c_1-552 or c_1-554 of *G. suboxydans*(?) reported by Iwasaki.[373] The enzyme preparation dialyzed against sucrose-tris-buffer showed a reduced form of this cytochrome having an α-band at 535.5 mμ, β-band at 522 mμ and γ-band at 417 mμ. Most probably the electron

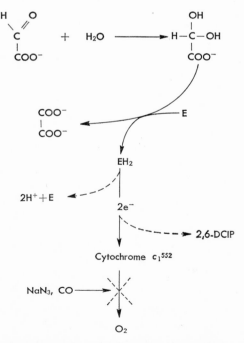

Fig. 27. A presumed mechanism for glyoxylate oxidase in
A. dioxyacetonicus strain A 15.

released in the reaction is transferred *via* this cytochrome to oxygen. A possible mechanism for the glyoxylate oxidase system is shown in Figure 27.

New findings were also presented on the distribution of the glyoxylate oxidase system in acetobacters and gluconobacters. The four tested strains of *Acetobacter, i.e. A. aceti, A. rancens, A. dioxyacetonicus* and *A. pasteurianus*, showed remarkable oxidative activities for glyoxylate, while *G. suboxydans* and *G. roseus* did not. The " intermediate " strain IAM 1834 (*G. liquefaciens*) oxidized glycollate like the acetobacters. This also indicated that the glyoxylate oxidizing system, as well as the TCA cycle and glyoxylate by-pass, all of which are lacking in the genus *Gluconobacter*, may have some physiological role in *Acetobacter*.

Chapter 16

DEAMINATION AND TRANSAMINATION
OF AMINO ACIDS

In an earlier work Miyaji[383] reported that tyrosine was reductively deaminated to *p*-oxyphenyl lactic acid by *A. rancens* and *A. xylinoides*. *A. xylinoides* also converted glycine to acetic acid. These results were obtained with cultures grown in synthetic media, after 30 days' incubation. Formation of L-isopropyl lactic acid from L-leucine by *A. rancens*, *A. schützenbachii*, and *A. xylinoides*, and L-β-imidazolelactic acid from L-histidine by *A. xylinoides* was also observed.

Oxidative deamination of glutamic acid followed by decarboxylation was also demonstrated with *A. aceti*, *A. schützenbachii*, *A. xylinoides* and *A. xylinum*. In this case α-ketoglutaric acid and succinic acid were found. The same observations were made by Antoniani *et al.*[384] Federico and Gobis[385] reported the formation of pyruvic acid from α-alanine by *A. aceti*, and the formation of acetoin as an additional product derived from pyruvic acid.

Stokes and Larsen,[58] in studies of the amino acid requirements of *A. suboxydans*, found that this organism deaminated the following amino acids (the products were not identified): 1) glutamic acid, aspartic acid, alanine, lysine, arginine, tryptophan, histidine, serine, proline and oxyproline——deaminated 50–100%; 2) cysteine, methionine, tyrosine, phenylalanine, threonine, norleucine and glycine——deaminated 10–50%. Leucine, isoleucine and valine were not deaminated.

According to Joubert *et al.*,[386] the strains in the *Peroxydans*, *Rancens*, and *Mesoxydans* groups (genus *Acetobacter*) attacked and deaminated L-alanine, L-glutamic acid and L-aspartic acid. Many strains also attacked and deaminated serine, threonine and proline. The strains of the *Suboxydans* group (genus *Gluconobacter*), however, were powerless against these amino acids. The authors suggested that these characteristics might be useful to differentiate between *Acetobacter* and *Gluconobacter*. Other researchers, however, have shown that this distinction does not necessarily hold.

Janke *et al.*[387] examined the breakdown of amino acids by four different strains of acetic acid bacteria: *A. aceti* Beijerinck, *A. rancens* NRRL B-65, *G. suboxydans* 1 and *G. liquefaciens* ("intermediate" strain IAM 1834). The amino acids supplied were α-DL-alanine, L-glutamic acid and L-aspartic acid. *G. suboxydans* deaminated DL-alanine oxidatively with formation of pyruvate in experiments using cell suspensions, but deamination of aspartate and glutamate did not occur. *A. rancens* and *G. liquefaciens* deaminated these two dibasic amino acids but *A. aceti* did not. The deamination of alanine by *G. suboxydans* and *A. rancens* occurred not only with the L-isomer but also with the D-isomer. The inability of *G. suboxydans* to deaminate glutamic acid was attributed to a lack of the TCA cycle involved in the oxidation of α-ketoglutarate; hence the oxidative deamination to α-ketoglutarate would be inactivated.

Acetic acid bacteria also possess transaminase activities. According to Cooksey and Rainbow,[388] cell extracts of lactaphilic strains such as *A. acidum-mucosum*, *A. mobile*, *A. rancens*, *A. ascendens* and *A. oxydans* could transaminate glutamate and aspartate. But it was also shown that the conversion of L-aspartate to α-alanine by cell extracts of lactaphiles was not due to transamination but due to β-decarboxylation. This process, to the author's knowledge, has never been reported in aerobic bacteria.

By contrast, reversible transamination from glutamate to aspartate was feeble or nonexistent in glycophilic strains such as *A. capsulatus*, *A. gluconicus*, *A. turbidans* and *A. viscosus*.

Addendum. Röhr[389] found that *G. suboxydans* could grow in a medium containing glucose and ammonium sulfate as the sole carbon and nitrogen sources, and that pyruvate and α-ketoglutarate, in addition to gluconate and 2- and 5-ketogluconate, were accumulated in the broth. Growth was enhanced about ten-fold when glycerol was supplied instead of glucose, suggesting that some organic compounds with less carbon than C_4 may condense and give rise ultimately to glutamic acid. Possibly a condensation reaction of pyruvic acid with an active C_2 compound (active acetaldehyde?) occurs. Röhr and Chiari[390] also observed that *G. suboxydans* formed serine from oxypyruvic acid in the presence of alanine. Other observations by Sekizawa *et al.* will be discussed in the section on carbohydrate metabolism. The inactivating mechanism of aspartate deamination is still unclear. Intact cells of *A. aceti* were also incapable of deaminating aspartate and glutamate. Again, the reason is unclear, but perhaps coenzymes related to the dehydrogenic system are washed away in the preparation of resting cells.

Klungsöyr and Aasheim[391] have recently reported that a strain of *A. xylinum* produced indole-3-acetic acid and tryptophol when grown in a medium containing peptone, or when the washed cells were incubated with DL-tryptophan. In the presence of $NaHCO_3$ the yields of these substances decreased, but small amounts of indole-3-acetaldehyde were found, suggesting that the latter may be an intermediate metabolite of DL-tryptophan.

Chapter 17

BIOSYNTHESIS OF CELLULOSE AND
OTHER POLYSACCHARIDES

Among the earlier reports of Brown[121] there are observations of cellulose formation by *A. xylinum* from carbon sources added to the medium. Thereafter, various organisms active in the formation of cellulose or " near-cellulose " were reported, *viz. A. xylinoides* by Hibbert and Barsha,[392] *A. acetigenus* by Kaushal and Walker[142] as well as by Barcley *et al.*,[393] *A. pasteurianus* and *A. rancens* by Tarr and Hibbert,[394] and *A. kützingianus* NCTC 3924 by Kaushal and Walker.[395] Steel and Walker[66] also reported cellulose formation by *A. acetigenus* NCIB 8132 and Carr[396] observed a positive cellulose reaction with the pellicle of *A. estunense*; Shimwell and Carr[397] reported cellulose-forming strains of *A. aceti, A. lovaniense, A. rancens,* and *A. orleanense.*

Of the organisms mentioned above, all strains of *A. xylinum* and some strains of *A. xylinoides* produce leathery, hard and tough pellicles on liquid media; the rest make soft slimy pellicles. Shimwell and Carr called the former " hard cellulose " and the latter " soft cellulose."

In the microorganisms synthesizing polysaccharides, pure cellulose producers are limited to the genus *Acetobacter*. Shimwell and Carr listed the cellulose (and " near-cellulose ")-forming acetobacters as follows:

A. xylinum (Frateur's)	hard cellulose
A. xylinum NCIB 8623	,,
A. xylinum NCIB 8034	,,
A. xylinoides	hard (and soft?)
A. orleanense	,,
A. acetigenus	soft
A. aceti (new isolates)	soft
A. estunense	,,
A. rancens (new isolates)	,,

The cellulose-forming ability of *A. xylinum* is not absolutely stable, as the organism often mutates to a non-cellulosic form, as previously described (p. 59, 60).

Characteristics of Hard Cellulose Produced by *Acetobacter xylinum*

The first chemical studies on the hard cellulose formed by *A. xylinum* were carried out by Hibbert and Barsha,[392, 398] who demonstrated that it is a polymer of β-1: 4 linked glucose residues. Its derivatives, tri-*O*-acetate and tri-*O*-methyl ether, were identical with those from cotton cellulose. Aschner and Hestrin[399] isolated cellulose as a fibrillar structure from *A. xylinum*.

Mühlethaler[400] isolated it as homogeneous slime and determined by electron microscopy that the diameter of the fibre was approximately 25 mμ, almost identical with that of cell walls of higher plants. Other papers related to the cellulose structure of *A. xylinum* were presented by Franz and Schiebold,[401] Rånby,[402] and Mühlethaler.[403]

Husemann and Werner[404] reported recently that the degree of polymerization (D.P.) of synthesized cellulose by *A. xylinum* ATCC 12733 rose rapidly during the first few days of cultivation, reaching a maximum of about 6000 after 5–6 days, and then decreasing slowly to <4000 after 25–30 days. The molecular weight distribution showed two distinct maxima during the first 6 days, at D. P. 2000 and 6000, but longer incubations produced a gradual flattening of the curve to form one maximum at D.P. 2500, probably due to a cellulose in the organism.

The cellulose produced by *A. xylinum* from glucose, fructose, glycerol, galactose and mannitol gave X-ray diagrams similar to those of cotton cellulose (Hibbert and Barsha[392, 398]).

Synthesis of Hard Cellulose by *Acetobacter xylinum*

Cellulose formation by *A. xylinum* is remarkably influenced by the type of substrate used. Hexoses such as glucose and fructose and polyalcohols such as glycerol and mannitol are the most suitable. Addition of ethanol promotes growth and increases cellulose production.

In experiments by Dillingham *et al.*[405] the maximum yield of cellulose synthesized from glucose was nearly 40% of the substrate. This was obtained when *A. xylinum* ATCC 12733 was grown in glucose-yeast extract medium containing 0.7% glucose, 2% yeast extract and 0.1% KH_2PO_4 (pH=6.0, surface/volume=0.7) for 20 days at 30°C. Khouvine[406] reported that mannitol gave a higher yield of cellulose than sorbitol; he[407] also tested the suitability of C_3-, C_4- C_5-, C_6- and C_7-polyalcohols for cellulose synthesis. Resting cells, according to Hestrin et al.,[408] could also synthesize cellulose in phosphate buffer with glucose or other monosaccharides or polyalcohols added. The optimum temperature was 37°C and the optimum pH 5.5. Synthesis was inhibited by cyanide, fluoride and nitrogen gas.

Schramm and Hestrin[409] observed that cells of A. xylinum in stationary liquid cultures were swept to the surface by flotation of the submerged cellulose net; a high oxygen pressure at the surface was thought to mediate rapid cellulose production. Cellulose production in submerged cultures was less than in stationary cultures and the shape of the cellulosic fibre changed to a joined spherical form. Submerged cultures also tended to produce mutants deficient in cellulose-forming ability.

Hestrin and Schramm[410] succeeded also in separating cells of A. xylinum from the cellulosic pellicle and proved that cellulose was synthesized from glucose-U-^{14}C, using suspensions of freeze-dried cells. The optimum pH for synthesis was 5–7 and the optimum temperature 30–37°C. There was a loss of up to 90% cell viability in freeze-dried cell preparations, although the cellulose-synthesizing activity remained 75% that of viable cells. It was suggested that cellulose is not formed within the cell or by extracellular enzymes, but rather that the terminal catalyst of the enzyme complement concerned is situated on the outer bacterial surface.

Webbs and Colvin[411] reported that lysozyme lysates of A. xylinum produced cellulose in a manner similar to that of whole cells. Extracts from a number of plants increased cellulose synthesis by whole cells of A. xylinum, particularly with the supernatants obtained by centrifuging tomato homogenates. This increased cellulose production three-fold. Later these researchers[412] investigated the relationship of oxygen uptake by whole or lysed cells of A. xylinum to cellulose synthesis, using glucose as the substrate, and found that the rate of oxygen uptake at pH 8.0 was only 3% that at pH 6.0. The amount of cellulose synthesized was the same at pH 6.0 and pH 8.0. This led them to the conclusion that not more than 3% of the oxygen uptake of the cells at pH 6.0 is directly linked to cellulose formation, despite the large fractions of available glucose incorporated into cellulose. At pH 8.0 under anaerobic conditions no cellulose synthesis was observed with either whole or lysed cells. Tomato supernatant stimulated cellulose formation by whole cells at pH 6.0, especially when it was the sole source of energy. Oxygen uptake in the presence of tomato supernatant, however, was reduced inversely with increased synthesis of cellulose. The most probable explanation for the small proportion of oxygen required in cellulose synthesis is that the rate-limiting step in this synthesis lies between the point of ATP energy input and the final incorporation of glucose into the insoluble cellulose microfibril. The possibility was not excluded, however, that different pathways of cellulose synthesis may operate at pH 6.0 and pH 8.0.

Minor *et al.*,[413] on the basis of label distribution in cellulose-^{14}C formed with D-glucose-1-^{14}C as the substrate, suggested that a pathway involving C_3 intermediates participates in cellulose synthesis. Later, the same workers[414] found that 84–96% of the total radioactivity occurred in positions 1 and 6 in D-glucose-^{14}C molecules obtained by decomposition of cellulose-^{14}C which was synthesized from a D-mannitol-1-^{14}C substrate, indicating that some of the cleavage products from the original D-mannitol became oriented in the cellulose. Ethanol tended to increase the yield and ^{14}C content of the cellulose but did not affect the distribution of ^{14}C in the glucose molecules making up the cellulose. They[415] also studied the label distribution in glucose-^{14}C molecules obtained by hydrolyzing the cellulose synthesized from a D-glucose-2-^{14}C substrate, and found that approximately 60% of the total radioactivity was in the C_2 position, while a significant amount was distributed in the C_1, C_3, and C_4 positions.

Greathouse[416] showed that 82% of the total activity was concentrated in the C_6 position of the glucose unit of cellulose when it was synthesized from D-glucose-6-^{14}C; the activity was distributed in all the carbon atoms of the glucose units when the cellulose was synthesized from a glycerol-1,3-^{14}C substrate, although most of the activity was found in the C_3, C_4 and C_6 positions. Using cell homogenates of *A. xylinum*, Colvin[417] observed the formation of typical cellulose microfibres when the preparation was incubated with glucose, ATP and buffer. In the absence of ATP cellulose formation did not occur.

According to recent investigations by Everson and Colvin,[418] more than 20% of the C atoms in position 6 of the original D-glucose-6-^{14}C were rearranged to position 1 during the formation of bacterial cellulose by *A. xylinum*.

This rearrangement was not remarkably influenced by the age of the cells, the length of time during which they had been metabolizing, or shaking the culture; equilibrium seems to have been reached in < 10 min of incubation. The radioactivity in C-2 and C-3 was quite low even after a 4 hr incubation, suggesting that the rearrangement did not involve enzymes of the pentose phosphate pathway. D-Fructose was also utilized by D-glucose grown cells to form bacterial cellulose. The distribution of the radio activity among C atoms in the D-glucose during the formation of bacterial cellulose suggested that the rearrangement takes place by the action of aldolase and triosephosphate dehydrogenase on D-fructose diphosphate derived from D-glucose-6-^{14}C and that, where equilibrium is reached, nearly half the D-glucose mols ultimately incorporated into cellulose pass first to the triosephosphate level. Half the D-glucose mols incorporated into cellulose did not undergo any intermolecular rearrangement.

Schramm et al.[419] confirmed cellulose synthesis by intact cells from glycerol, dihydroxyacetone, gluconate, 2- and 5-ketogluconates and, after adaptation, fructose. Acetate, pyruvate and other TCA cycle intermediates were oxidized to carbon dioxide but not converted to cellulose. Glucose-6-phosphate, α-glucose-1-phosphate and β-glucose-1-phosphate were not synthesized to cellulose either by fresh or freeze-dried cells, presumably because such phosphate esters failed to penetrate into the cells. Cellulose formation from glucose with intact cells was inhibited by fluoride, cyanide, p-chloromercuribenzoate, N-ethylmaleiamide, 2,4-dinitrophenol, thymol and azide, and also by oxidizable metabolites such as acetate and pyruvate. Arsenate retarded formation but this was reversed by the addition of ortho-phosphate. Fluoro-acetate (10 mM) did not inhibit but rather enhanced CO_2 production. 2,4-Dinitrophenol (4 mM) did not interfere with ketogluconate production but inhibited cellulose formation and retarded CO_2 evolution from glucose.

In experiments by Gromet et al.,[420] cell-free extracts of A. xylinum oxidized gluconate aerobically in the presence of phenazine methochloride. Ketogluconate, largely 2-ketogluconate, was obtained when NAD was added. In the presence of ATP, glucose was oxidized aerobically to carbon dioxide. The existence of glucokinase, gluconokinase, phosphoglucomutase, all the enzymes of the pentose cycle, and all the enzymes by which triosephosphate is converted into pyruvate was verified.

Anaerobic cleavages of fructose-6-phosphate and xylulose-5-phosphate, leading to the formation of acetyl phosphate with uptake of inorganic phosphate, were also demonstrated by Schramm and Racker[421] in extracts of a celluloseless mutant of A. xylinum.

Schramm et al.[422] re-examined the distribution of radioactivity in cellulose monomers synthesized from glucose-1-^{14}C, glucose-2-^{14}C, glucose-6-^{14}C and fructose-1-^{14}C with washed cells and verified that the C_6 of the cellulose monomer was derived exclusively from the C_6 of the added glucose, and to some extent from the C_2 and the C_1 of the exogenous hexose. They suggested that hexose-phosphate arose directly by phosphorylation of the exogenous hexose and indirectly from pentose cycle intermediates, and that it gave rise to cellulose. The polymerization step in the cellulose synthesis was assumed to be catalyzed by the cell surface or to occur spontaneously in the medium from a diffusible precursor containing the carbon skeleton of hexose phosphate but in which the phosphate group is either modified or replaced, since hexose phosphate itself is probably unable to penetrate or leave the cell. Ketogluconate was assumed to be reduced to gluconate and then to enter into the pentose cycle, leading to cellulose synthesis *via* hexose phosphate, while glycerol and dihydroxy-acetone might be synthesized to cellulose *via* a pathway in which they are converted to triose phosphate, then to fructose-1,6-diphosphate and finally to hexose phosphate.

Koshio and Katayama[423] reported that cell-free extracts of *A. xylinum* synthesized cellulose more efficiently from β-glucose-1-phosphate than from α-glucose-1-phosphate, and that cellulose was not formed from β-cellobiose-1-phosphate either by cell-free extracts or washed cells.

Using labeled glucose as the substrate and freeze-dried cells of *A. xylinum* ATCC 10245 as the inoculum, Weigl[424] paperchromatographed the reaction mixture separated from the cell-mass and cellulosic substance. He detected strong radioactive substances and, after treatment by cellulase, was able to identify radioactive cellobiose and a small amount of radioactive cellotriose, as well as radioactive glucose.

Glaser[425] studied cellulose synthesis using uridine diphosphate glucose (UDPG) as the substrate. When the cell-free particles were incubated with ^{14}C-glucose-labeled UDPG, a radioactive, water-insoluble, alkali-insoluble substance was formed; this was identified as cellulose since it gave rise to cellobiose of constant specific activity after partial hydrolysis. With ^{14}C-labeled glucose-1-phosphate and ^{14}C-labeled glucose, no cellulose was synthesized under the same conditions. The cellulose formation was remarkably stimulated by addition of soluble cellodextrins. This supported the theory that cellulose synthesis is most probably achieved by the utilization of UDPG as substrate and cellodextrins as acceptors. This was verified by Klungsöyr,[426] who obtained an increased yield of alkali-insoluble carbohydrate in a reaction mixture of UDPG and sonic homogenates of *A. xylinum* NCIB 8246 with added cellodextrins. Zeigler and Weigl[427] also reported the stimulative effect of UDPG on cellulose synthesis, using intact cells of *A. xylinum* ATCC 10245 grown in *p*-amino-benzoic acid free medium. They supported the view that glucose-6-phosphate is an initial intermediate in cellulose synthesis from glucose.

The presence of UDPG in $HClO_4$ extracts of *A. xylinum* cultures has recently been confirmed by Enevoldsen.[428] Nucleotides were isolated from four-day-old cultures and were adsorbed on Norite, and UDPG was eluted with 50% ethanol. The amount of UDPG present in five pellicles varied from 0.04 to 0.16 μmole. UDPG was also detected in the liquid phase but the concentration was about one-tenth of that found in the pellicles.

By contrast, Colvin[429] reported that neither UDPG nor short β-glucosan chains are intermediate precursors of bacterial cellulose. He found a compound in ethanol extracts of *A. xylinum* cells which, after immersion in water, rapidly changed into cellulose microfibre. The formation of this microfibre was attributed to a heat-unstable, extracellular enzyme.

Benziman and Burger-Rachamimov[430] found that succinate-grown cells of *A. xylinum* could convert pyruvate to cellulose whereas glucose-grown cells could not. The isotope distribution of cellulose monomers synthesized from pyruvate-1-, 2-, or 3-^{14}C by cell suspensions suggested that the cellulose monomer could be formed *via* a condensation involving two molecules of a three carbon compound, most likely triose phosphate formed from phosphoglycerate. The latter in turn could arise from pyruvate *via* phosphoenolpyruvate or *via* still unknown intermediates. According to Gromet-Elhanan and Hestrin,[431] washed cell suspensions prepared from cultures grown on ethanol, acetate or succinate were able to transform TCA cycle intermediates into cellulose, in contrast to cells grown on glucose. This variation in cellulose-synthesizing activity was attributed to differences in the oxidative capacities of the cells. Growth on glucose resulted in loss of ability to oxidize acetate and in a delay in the onset of oxidation of dicarboxylic acids. In these respects, ethanol-grown cells occupied an intermediate position between glucose- and acetate-grown cells.

Janke[432] presented the following scheme for the pathways of cellulose synthesis in *A. xylinum*: Exogenous glucose penetrates into the cells and is phosphorylated with ATP, and thus forms glucose-6-phosphate. The latter is probably converted to glucose-1-phosphate, which is utilized for the synthesis of UDPG. UDPG, by transglucosidic action, gives rise stepwise to cellobiose, cellotriose, *etc.* In the last stage of extracellular cellulose synthesis some permeable precursor must be assumed. This might be a modified glucose phosphate in which the phosphate group is replaced with a radical causing smooth transportation through the cell walls.

Figure 28 summarizes the probable chemical pathways participating in cellulose formation from various substrates by *A. xylinum*.

Recently Gascoigne[433] presented a discussion of three aspects of cellulose formation in *A. xylinum*: the structure and mode of formation

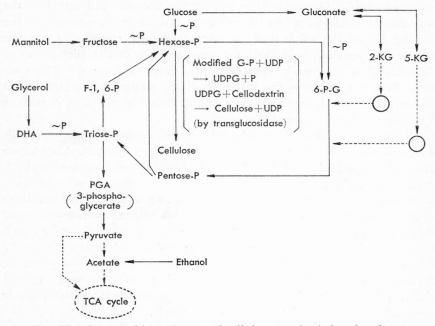

Fig. 28.   Presumable pathways of cellulose synthesis by *A. xylinum* according to Schramm *et al.* and others.

Full lines indicate reactions which have been experimentally demonstrated. Broken lines indicate reactions which are presumed to take place.

The pathways by which TCA cycle members convert into cellulose with ethanol-, acetate- or succinate-grown cells are still obscure. The route to triose-phosphate *via* the Embden-Meyerhof-Parnas pathway or the Entner-Doudoroff pathway has been demonstrated by White and Wang.[426]

of precursors or intermediates, the way in which these precursors are induced to combine, and the changes undergone by the newly-formed polymer to form the tough, insoluble fibrous material.   A slimy, nascent cellulose encasing bacterial cells is commonly produced from sugar, especially from fructose.   Cellulose synthesis is greatly delayed when the cell walls are hydrolyzed by lysozyme, and in this circumstance cello-oligosaccharides frequently appear in the extracellular medium.   *A. xylinum* contains the enzymes responsible for converting UDPG into cellulose, and the building-up of the cellulose polymer takes place entirely outside the cell.   Glucose-1-phosphate, required for the synthesis, is probably passed through the cell wall by temporary attachment to a sugar nucleotide lipid.   Glucoside then crosses the bacterial cell wall into the extracellular medium, where the glucose residue is transferred to the growing tip of the microfibril by an extracellular enzyme.   Once the glucose has been removed the lipid returns to the cell to pick up more glucose, and the process is repeated until cellulose is formed.   This synthesis, however, requires the presence in the medium of certain chain initiators or formers, produced by bacterial cells and probably consisting of a glucose phosphate-lipid complex.

Dennis and Colvin[434] investigated cellulose synthesis using lysed cells of *A. xylinum*.   They found that fructose was a better substrate than glucose for cellulose synthesis, and that glucose-1-phosphate, glucose-6-phosphate, fructose-1,6-diphosphate, and ribose were not efficient as substrates. Lysed cell preparations utilized Ca gluconate as a substrate, but not dihydroxyacetone, in contrast to whole cells which utilized both compounds. Mg, Ca, ATP, and UTP stimulated cellulose synthesis.   Lysed cell preparations in which the cytoplasmic membranes had flowed out of the cell wall and had been precipitated by Mg were active in cellulose synthesis, whereas preparations in which the cytoplasmic membranes were digested with trypsin were inactive.   Hence it was suggested that the cytoplasmic membrane may be the site of cellulose synthesis.

Colvin and his co-workers carried out a series of experiments on the formation of cellulose microfibrils by cellulose-free cells of *A. xylinum*. Colvin and Beer[435] reported that the formation of cellulose microfibrils by incubating cells in a synthetic medium containing glucose takes place away from the cell surface and without an amorphous, intermediate high polymer.   The microfibrils grew only at one or both tips.   The rate of microfibril growth per bacterial cell at 25°C was constant at 0.1μ per minute up to at least 7 minutes.   New microfibrils were produced continously during the incubation period.   According to Millman and Colvin,[436] the formation of extracellular cellulose microfibrils by cells on moist agar surfaces also occurs away from the cell membrane and does not involve an amorphous, high polymeric intermediate, in agreement with conclusions from studies of liquid suspensions.   Individual microfibrils grew at the tip(s) only and the rate of extension was 0.2μ per bacterial cell per minute at 34°C.   The rate of initiation of new microfibrils per 10^3 bacteria per minute at 34°C was roughly 40.   Both rates were constant after an induction period of approximately 30 seconds.   Newly nucleated microfibrils down to a length of about 0.5μ were identified.   The growth of microfibrils on agar surfaces was characterized by the formation of sheaves or bundles, with axes of the microfibrils roughly parallel.   The rate-limiting step in the formation of these microfibrils was suggested to have an activation energy of about 15 kc.

The formation of spherulites in the pellicle of bacterial cellulose produced by static cultures of *A. xylinum* was observed by Colvin.[437] The appearance of these two-dimensional analogues of spherulites were attributed to the radial orientation of bacterial cellulose microfibrils in limited regions of the pellicle plane. These spherulites interacted to form characteristic leaf-like or dendritic forms which resemble leaves of higher plants. Later the same author[438] reported the development of non-spherulitic, thread-like birefringence in the pellicle. This birefringence was attributed to the formation of bundles of parallel cellulosic microfibrils within the pellicle. Its development was often accompanied by the formation of streaks of lipid-like material floating on the surface of the medium.

Most recently Carson *et al.*[439] found that several classes of lipids induced birefringence in the pellicle when they were added in small amounts to the surface of static cultures. The capacity to introduce birefringence was weak or almost absent in lecithins, triglycerides, and Ca stearate. Diglycerides, however, showed a marked capacity to introduce birefringence and the capacity increased with chain length of the particular fatty acid. Cholesterol or its derivatives also induced strong birefringence. This birefringence was attributed to partial orientation of the cellulose microfibrils in the pellicle. However, the orientation of microfibrils was not restricted to those pellicles grown in the presence of lipids. Sections of areas of non-birefringent cultures showed many small regions of highly oriented bundles of microfibrils, though the regions are not outwardly visible. In the presence of lipids some of the regions may overgrow others and produce a macroscopic effect when they reach a suitable size. How the lipids induce alignment of the microfibrils is unknown, but the process appears to involve an interaction between newly formed extracellular microfibrils and the microcrystalline surface of the lipid.

Synthesis of Soft Cellulose by *Acetobacter acetigenus*

Kaushal and Walker[142] reported the formation of a large amount of pellicle material, amounting to 60% dry weight of the initial glucose, in a strain of *A. acetigenus* grown in glucose yeast-water in the presence of Ca carbonate. The pellicle gave a cellulose reaction, and was shown to be identical with cotton cellulose by X-ray data. After hydrolysis it gave glucose and also, by acetolysis, penta-acetyl glucose. Glucose, ethylene glycol, glycerol, mannitol, arabinose, xylose, fructose, galactose, maltose, sucrose, lactose and soluble starch were utilized as carbon sources for cellulose synthesis. In addition to cellulose formation, ethylene glycol and glycerol yielded glycolic aldehyde and dihydroxyacetone respectively, indicating that these two metabolites may not be excluded from participation in cellulose synthesis. Cellulose formation from α- or β-methyl-D-glucoside and erythritol was also observed.[395]

Kaushal et al.[440] determined by electron microscopy that the cellulosic material formed by A. acetigenus was a filamentous crystal of 25–50 mμ width, but that the cellulose formed from ethylene glycol was amorphous. Barcley et al.[393] observed that the formed cellulose consisted of about 600 units of glucose residues in β-glucosidic linkage, and was identical with cotton cellulose in its infrared absorption spectrum.

Creedy et al.[441] detected, by paperchromatography, the formation of D-cellobiose in addition to cellulose in growing cultures of the organism, using ethylene glycol, xylose, glucose or soluble starch as the substrate. Cellulose synthesis was inhibited by the addition of Na arsenite (5×10^{-4}M), which caused accumulation of carbonyl compounds.

Bourne and Weigel[442] obtained radioactive cellulose from a growing culture of A. acetigenus in a synthetic medium composed of $CH_3 \cdot CH$ (OH) [14]COOH (labeled DL-lactic acid), amino acids and vitamins. Strong radioactivity was detected symmetrically in the C_3 and C_4 positions of the cellulose monomer after hydrolysis of the formed cellulose, and the possibility of cellulose synthesis via C_3 fragment binding was pointed out.

Celluloseless mutants of A. acetigenus were reported by Wright and Walker[443] and by Steel and Walker.[66] In the former case a celluloseless mutant arose when the parent strain was cultivated under vigorous aeration and mechanical agitation, while in the latter A. acetigenus NCIB 8132 gave rise to a celluloseless, proteolytic, and non-ethanol oxidizing mutant when it was grown under submerged conditions. Wright and Walker[443] detected the presence of cellobiose, D-melibiose, and cellulose when A. acetigenus NCIB 8132 was grown in glucose-containing synthetic medium in submerged cultures under conditions in which the celluloseless mutant could not proliferate. Chromatographic separation of the products in subsequent experiments[444] revealed fructose, cellobiose, cellotriose and cellotetrose. The latter two were isolated and identified as acetates. The presence of di- and tri-saccharides containing fructose units and of phosphorylated oligosaccharides was demonstrated. According to Steel and Walker,[445] however, neither the celluloseless mutant from cellulose forming A. acetigenus NCIB 8132 or from A. xylinum var. africanus NCIB 7029 could produce oligosaccharides under the same experimental conditions. This suggested that these oligosaccharides may play an important role in cellulose synthesis.

Ramamurti and Jackson[446] reported the formation of dihydroxy-acetone, ribulose (?), xylose, fructose, glucose, cellobiose, cellotriose and cellotetrose in shaking cultures of *A. acetigenus* NCIB 8132 in medium containing glycerol with K lactate as a buffer. The last three substances were not produced by non-cellulosic strains. The presence of ribulose (?) and xylose was attributed, though not exclusively, to the possible operation of a pentose cycle in cellulose synthesis.

Dudman[447] reported that cellulose production in *A. acetigenus* strain EA-I was most effective with the hydrolysates of blackstrap molasses as carbon sources. Ammonium sulfate or a mixture of glutamate and asparagine were suitable sources of nitrogen when glucose was supplied as the carbon source. The addition of acetate, citrate or succinate to glucose medium increased the cellulose production twenty- to thirty-fold.

Synthesis of Polysaccharides by Other *Acetobacter* Species

Frateur[238] found that a strain of *A. pasteurianus* produced an amylose-like polysaccharide in beer when fructose was added. According to Tošić and Walker,[448] *A. acidum-mucosum*, which had been isolated from brewery yeast, could produce a starch-like polysaccharide in malt extract medium at pH 4.5 or lower. The formation of a viscid substance in beer had been reported earlier by Baker *et al.*[449] with *A. viscosus*, and by Shimwell[450] with *A. capsulatus*. The substance was also produced from dextrin. Hehre and Hamilton[451] reported that the viscid substance was immunologically identical with dextran produced by *Leuconostoc mesenteroides* and assumed it to be a polymer of glucopyranose, of largely α-1,6-binding. The enzyme involved was tentatively called " dextrin-dextran-trans-glucosidase " by Hoffmann-Ostenhof.[452] It was thought to be extra-cellular, and the synthesizing activity was not influenced by the addition of streptomycin (1% w/v) at pH 4.6, while in a yeast extract-glucose medium the addition of 1 μg per ml of streptomycin sulfate completely inhibited the growth of the organism, according to Arnold and Hall.[453] They concluded that growth inhibition by streptomycin in *A. capsulatus* was not due to disturbance of its dextrin-dextran synthesizing enzyme system, but probably affected a more fundamental metabolic process. Barker *et al.*[454] described the method for isolating the enzyme.

Walker *et al.*[455] made the interesting observation that *A. aceti* NCIB 7214, *A. acidum-mucosum* NCIB 8133 and *A. acetosus* var. *nariobiense*

NCIB 7212, none of which produced cellulose, could produce gentiobiose (O-β-D-glucopyranosyl-[$1 \rightarrow 6$]-D-glucopyranose), sophorose (O-β-D-glucopyranosyl-[$1 \rightarrow 2$]-D-glucopyranose), and other disaccharides and higher saccharides when grown in defined media containing glucose as the sole source of carbon.

A mucous substance which was formed by the strains of *A. suboxydans* var. *muciparum* in sucrose-containing medium was identified as levan by Loitsyanskaya.[456]   Sucrose was hydrolyzed and liberated fructose was synthesized to levan, while glucose was oxidized to gluconic acid and then to 5-ketogluconic acid.   The bacteria did not grow in culture media containing raffinose or melibiose as the sole carbon source.   However, the addition of 0.2–0.5% glycerol or ethanol permitted growth and levan synthesis.

Chapter 18

METABOLISM OF CARBOHYDRATES WITH REFERENCE TO THE ENZYMES INVOLVED

Much work has been done on the oxidative metabolism of carbo-hydrates by acetic acid bacteria. The overall modes of metabolism and the enzyme systems involved, however, vary widely in the *Acetobacter* and *Gluconobacter*, and even within their species.

On the whole, the *Gluconobacter* (*Suboxydans* group) are better equipped with enzymes for carbohydrate metabolism than the *Acetobacter*. Glucose is readily oxidized to gluconate, 2- and 5-ketogluconates, and 2,5-diketogluconate in *G. melanogenus* or the " intermediate " strain IAM 1834. Polyalcohols are oxidized to the corresponding ketogenic com-pounds, whereas they are not oxidized by most strains of *Acetobacter*. *A. peroxydans* or *A. paradoxus* and some strains of *A. ascendens* cannot even oxidize glucose itself.

The enzyme systems relating to pyruvate or acetate oxidation, how-ever, do occur in the genus *Acetobacter*. Operation of the TCA cycle involving a glyoxylate bypath has been confirmed in many species of *Acetobacter*, but never in *Gluconobacter*.

Table 21 shows the oxidative capacity for glucose, gluconate and ketogluconates of various acetic acid bacteria, as observed in experiments by De Ley.[457]

At present the reaction pathways of carbohydrate metabolism in acetic acid bacteria are generally acknowledged to be initiated *via* the pen-tose cycle, and to a lesser extent *via* the Entner-Doudoroff pathway with a partial participation of the Embden-Meyerhof-Parnas pathway, accom-panied by several side reactions which result in the accumulation of by-products.

An exception to this is *A. peroxydans*, which cannot dissimilate glucose, although it possesses a TCA cycle. Schramm *et al.* have also reported that a " fructose-6-phosphate pathway " exists in *A. xylinum* under anaerobic conditions (see section on *A. xylinum*, pp. 269–270).

Table 21. Oxidation of glucose, gluconate and ketogluconates by some representative strains of acetic acid bacteria (after De Ley).

Group	Strain	Glucose	Na gluconate	Na 2-keto-gluconate	Na 5-keto-gluconate
Peroxydans group	*A. peroxydans* NCIB 8618	0	0	0	0
Oxydans group	*A. ascendens* A	0	0	0	0
	A. rancens 15	0.5	0	0	0
	A. lovaniense 13	0.5	0	0	0
	A. ascendens NCIB 4937	1.0	0	0	0
	A. vini aceti NCIB 4939	1.4	0	0	0
	A. estunensis E	1.2	>0.3	0	0
	A. turbidans NCIB 6424	2.3	>1.6	0	0
	A. mobilis NCIB 6428	2.0	>0.8	0	0
Mesoxy-dans group	*A. aceti* NCIB 8544	1.8	0.2	0	0
	A. mesoxydans NCIB 8622	>3.0	>2.1	0	0
	A. mesoxydans 4	4.9	4.4	>3.1	>3.4
	A. xylinum NCIB 8747	>3.5	>3.3	0.6	>0.4
Suboxy-dans group	*G. capsulatus* NCIB 4943	1.4	0	0	0
	G. suboxydans 26	3.1	2.5	0	0.1
	G. melanogenus NCIB 8086	2.2	1.3	0.5	0
	G. cerinus 21	4.5	4.0	0.3	>1.5
	G. liquefaciens 20	2.2	1.1	0.5	2.6

The results are expressed as mole O_2 uptake per mole substrate with washed cells at the end of the oxidations.

Gluconobacter suboxydans

The aerobic breakdown of glucose by *G. suboxydans* can be divided into two different types: direct (or non-phosphorylative) oxidation and phosphorylative oxidation.

Butlin[458] reported earlier that this organism possesses two separate enzyme systems involved in glucose oxidation. One is acid-labile and splits glucose to CO_2 and water and the other is acid-tolerant and oxidizes glucose to gluconate and ketogluconate. Two kinds of glucose oxidizing systems were thought by Iwasaki[459] to be present in cell-free extracts of

G. suboxydans because of the fact that *p*-chloromercuribenzoate inhibits the reduction of 2,6-dichlorophenolindophenol, while it does not inhibit the oxygen uptake; in addition, a significant difference in the heat-stability of these two activities was observed.

Non-Phosphorylative Oxidation

Oxidation of glucose to gluconate.   Glucose is first oxidized to gluconic acid by two different enzymes.   One is catalyzed by an NAD- or NADP- independent dehydrogenase (oxidase) in the particulate fraction of the cells, and the other is catalyzed by an NADP- or NAD- linked dehydrogenase in the soluble fraction of the cells (King and Cheldelin[460]). Galante *et al.*[461] also reported that cell-free extracts of *G. suboxydans* showed the presence of two kinds of glucose dehydrogenase, one requiring 2,6-dichlorophenolindophenol for the oxidation, the other linked to NADP.

According to Cheldelin,[462] the reaction occurring in the particulate fraction of the cells would proceed as follows:

$$\text{D-Glucose} \xrightarrow{\ -2H\ } \text{D-Glucono-}\delta\text{-lactone} \xrightarrow{\ +H_2O\ } \text{D-Gluconic acid}$$

Flavine ?
Cytochrome ?

This reaction was thought to be similar to that of *Penicillium notatum* or *Aspergillus niger*.   The presence of a lactone-hydrolyzing enzyme in the particulate fraction has been reported by King and Cheldelin.[460] Takahashi[463] confirmed the formation of D-glucono-δ-lactone and γ-lactone from glucose with 72-hour cultured cells of *G. roseus*.   The transformation of the latter to the former was observed non-enzymatically in acidic pH's, while the reverser eaction was very weak.   Takahashi suggested that the following reaction took place.

$$\text{D-Glucose} \longrightarrow \text{D-Glucono-}\delta\text{-lactone} \rightleftarrows \text{D-Glucono-}\gamma\text{-lactone}$$
$$\uparrow \downarrow$$
$$\text{D-Gluconic acid}$$

The optimum pH of the related dehydrogenase (oxidase) was reported to be around 5.5 by King and Cheldelin.[460]   According to De Ley and Dochy,[171] the particulate fraction has been shown not only to

oxidize glucose and gluconate to 2-ketogluconate, but also to oxidize galactose, L-arabinose and xylose to the corresponding aldonic acids.   Mannitol
was oxidized to fructose.   Glycerol, inositol, meso-erythritol, sorbitol,
and adonitol were oxidized with the uptake of 0.5 mole O_2 per mole substrate.   Mannose, maltose, and sucrose were slowly oxidized.   These
particle-linked oxidases, most probably flavin-enzymes, are constitutive and
located on the cytoplasmic membrane.

The other oxidizing system is catalyzed by NADP- or NAD-linked
dehydrogenase in the soluble fraction of the cells.   Phosphate does not
participate in the oxidation.   The pyridine nucleotide-linked glucose
dehydrogenase was first demonstrated by Butlin[360] in washed young cells
of G. suboxydans grown in the presence of $CaCO_3$.   The NADP-linked
enzyme was purified about 100-fold by Cheldelin[462] from the soluble
fraction. The optimum pH was about 8.6 and the purified enzyme oxidized
D-glucose and 2-deoxy-glucose but did not oxidize G-6-P, G-1-P,
gluconolactone, and gluconic acid.   Okamoto[464] also obtained an NADP-
linked glucose dehydrogenase, free from 5-ketogluconate reductase, in the
soluble fraction of cells of G. suboxydans IFO 3432, which oxidized glucose
and mannose stoichiometrically to corresponding aldonic acids.   The
optimum pH was 8.5.

$$\text{Glucose} + \text{NADP} + \text{H}_2\text{O} \longrightarrow \text{Gluconic acid} + \text{NADPH}_2$$

The formation of D-glucono-δ-lactone from glucose was observed,
while the existence of lactonase remained dubious.

According to Fewster,[465] glucose and gluconic acid were oxidized by
ultrasonic extracts to 2- and 5-ketogluconic acid while the keto acids were
not oxidized at pH 7.0.   The oxidation rate of glucose was not increased
by the addition of ATP, Mg^{++}, NAD or cytochrome c, and not appreciably
decreased by 10^{-3}M dinitrophenol or by the absence of added inorganic
phosphate.   Addition of 10^{-2}M dinitrophenol or a rise in pH to 8.0 prevented the oxidation of gluconic acid.   The existence of an NADP-linked
dehydrogenase was supported.   In a subsequent report[19] he emphasized
that the first stage of the oxidation of glucose to gluconate and hence to 2-
and 5-ketogluconate is not likely to involve phosphorylation of the substrate.

Stouthamer[350] detected an NADP-linked glucose dehydrogenase in
cell-free extracts of G. suboxydans strain 18 which rapidly oxidizes D-
glucose and D-mannose, whereas D-galactose, D-xylose, and L-arabinose
are slowly oxidized, and D-ribose and D-arabinose are not oxidized at all.

Oxidation of gluconate to ketogluconates. Gluconate is further oxidized to 2- and 5-ketogluconate. The existence of two different enzymes catalyzing this oxidation has been suggested. NADP-linked 2-ketoglu-conate reductase catalyzing the oxidation of gluconate to 2-ketogluconate was found by De Ley and Stouthamer[83] in soluble and in particulate fractions.

2-Ketogluconate is formed by a particle-linked, possibly cyto-chrome-linked gluconate oxidase (Stouthamer[350]). The oxidation is independent of added NADP or NAD. The optimum pH was shown to be around 5.0. D-Galactonate was oxidized at 1/10, D-gulonate and D-mannonate at about 1/20 of the rate of gluconate. The packed particles after centrifugation are distinctly red, showing the following cytochrome bands: a very strong band at 555 mμ, a strong one between 520–525 mμ, weak ones between 560–570 mμ and 528–534 mμ, and a very weak band at about 515 mμ.

According to Stouthamer,[350] gluconate and galactonate were oxidized by the soluble fraction of *G. suboxydans* strain 18 at pH 10 in the presence of NADP. Gluconate gave rise to 2- and 5-ketogluconate while galacto-nate formed a compound which was thought to be 2-ketogalactonate. Gluconate is oxidized to 5-ketogluconate by a soluble, NADP-linked 5-ketogluconate reductase. The existence of this enzyme in *G. suboxydans* was demonstrated by De Ley and Stouthamer.[83] The reduction of 2,5-diketogluconate in the presence of NADPH$_2$ was also seen in the soluble fraction of *G. suboxydans* var. *biourgianum* strain 2, but it is not clear whether it is caused by an independent 2,5-diketogluconate reductase or by an existing 2- or 5-ketogluconate reductase.

In *G. suboxydans* var. *a*, the optimum pH for the oxidation of gluconate to 5-ketogluconate by washed cells was approximately 4.2, according to Kondo and Ameyama.[466] Murooka[467] confirmed the presence of both NADP-linked 2- and 5-ketogluconate reductase in the dialyzed soluble fraction of *G. suboxydans* ATCC 621. Gluconate oxidation in the presence of NADP proceeded more rapidly at pH 10.0 than at pH 7.0. NAD acted weakly.

Okamoto[317] obtained a crude preparation of NADP-linked 5-keto-gluconate reductase, almost free from 2-ketogluconate reductase, from the soluble fraction of *G. suboxydans* IFO 3432. The optimum pH for gluconate oxidation was about 7.5 and that of 5-ketogluconate reduction was about 9.5. The equilibrium of the reaction was very unfavorable with respect to 5-ketogluconate formation (K=3.5×10^{-12}M); neverthe-less, growing cells produced a copious amount of 5-ketogluconate in the

culture medium. This apparent discrepancy is explained by the fact that
Ca 5-ketogluconate is hardly soluble in water; the removal of 5-ketogluco-
nate from the system pushed the following reaction to the right.

$$\text{Gluconate} + \text{NADP} \rightleftharpoons 5\text{-ketogluconate} + \text{NADPH}_2$$

As to the substrates, D-gluconic, D-mannonic and L-idonic acids were
found to be dehydrogenated by this enzyme.

Okamoto also found a very active $NADPH_2$ oxidizing system, glyoxylate reductase (?), in crude cell-free extracts of this organism (unpublished).
It is presumably coupled with 5-ketogluconate reductase to maintain the
concentration of 5-ketogluconate.

Scalaffa et al.[468] reported that the 5-ketogluconate-producing activity
of G. suboxydans grown in glucose media at pH 6.0 was confined to the
precipitates obtained from cell débris centrifuged at 27,000 ×g. The
activity was somewhat enhanced in the presence of NADP, NAD, FAD,
FMN, thioctic acid, coenzyme A, $CuSO_4$, $ZnSO_4$, $MgCl_2$, $CaCl_2$, and K_2HPO_4
in catalytic quantities. In subsequent experiments by Galante et al.,[469]

Table 22. The non-phosphorylative glucose and

Enzyme	Substrate	Product
Glucose oxidase	Glucose	δ-Gluconolactone
Glucose dehydrogenase	Glucose	δ-Gluconolactone
Gluconate dehydrogenase (2-Ketogluconate reductase)	Gluconate	2-Ketogluconate
Gluconate dehydrogenase (2-Ketogluconate reductase)	Gluconate	2-Ketogluconate
Gluconate dehydrogenase (5-Ketogluconate reductase)	Gluconate	5-Ketogluconate
Gluconate dehydrogenase (5-Ketogluconate reductase)	Gluconate	5-Ketogluconate
Gluconate-oxidizing enzyme	Gluconate	5-Ketogluconate

using subcellular fractions from *G. suboxydans*, oxidation of gluconate to 5-ketogluconate was observed in the heavy particles sedimented at 27,000 × g, in the light particles sedimented at 190,000 × g, and in soluble fractions (supernatant at 190,000 × g). Both particulate fractions formed 5-ketogluconate aerobically without addition of NADP at a pH range of 4 to 6, while the soluble fraction catalyzed the same reaction only in alkaline conditions (pH around 8.5) in the presence of NADP. This soluble fraction is most probably identical with the NADP-linked 5-ketogluconate reductase obtained by De Ley and Stouthamer[316] or by Okamoto.[317]

The enzymes responsible for non-phosphorylative or direct oxidation of glucose and gluconate by *G. suboxydans*, thus far reported, are listed in Table 22, along with their locations, coenzymes, products, and optimum pH's.

Oxidation of ketogluconates. 5-Ketogluconate can be further oxidized by *G. suboxydans*. Washed cells of *G. suboxydans* ATCC 621, however, required the presence of inductive amounts of glucose for steady oxidation (Asai and Murooka[470]). According to Kondo and Ameyama,[466] washed cell suspensions of *G. suboxydans* var. α and *G. suboxydans* NRRL(?)

gluconate oxidizing enzymes of *G. suboxydans*.

Location	Coenzyme	Opt. pH	Ref.
Particulate fraction	None reported	5.5	460 458
Soluble fraction	NADP	8.6	460 459
Particulate fraction	None reported	—	316
Soluble fraction	NADP	—	316
Soluble fraction	NADP	8.5	316 464
Soluble fraction	NADP (NAD, weak)	7.5	317
Particulate fraction	None reported	4–6	464

did not show a noticeable oxygen uptake when 2- or 5-ketogluconate was used as the substrate in the absence of glucose.

Fewster[19] also observed that suspensions of well-washed cells of *G. suboxydans* ATCC 621 did not oxidize fresh solutions of Na 5-ketogluconate unless traces of oxidizable material such as glucose, gluconate, mannitol, fructose, or pyruvate were added, or unless the cell suspensions were incubated in phosphate buffer at pH 6.0 for one hour at 30°C before use. These treatments caused steady oxidation of 5-ketogluconate. This is not always the case, however, with other *G. suboxydans* strains. Resting cells of *G. suboxydans* (isol. 18) were capable of oxidizing 5-ketogluconate with an uptake of 3.1 moles of O_2 per mole of substrate without addition of other oxidizable materials, according to De Ley and Stouthamer.[83]

As to the metabolic products of the oxidation of 5-ketogluconate by washed cells of *G. suboxydans* ATCC 621, Asai et al.[471] reported the formation of α-ketoglutarate, succinate, acetate, glycollate, and pyruvate[234] when a small amount of glucose was added. This finding may support the assumptions of De Ley[316] and of De Ley and Stouthamer[83] that the reduction of 5-ketogluconate to gluconate proceeds most probably prior to the apparent oxidation of 5-ketogluconate; therefore the products seem to be metabolites of gluconate. The TCA cycle does not participate in α-ketoglutarate formation, since it is not present in this organism. A new reaction pathway, leading to α-ketoglutarate formation by way of condensation of glyoxylate and oxaloacetate, proposed by Sekizawa et al.[472] for this organism, will be introduced elsewhere in the text. The mechanism of glycollate formation from 5-ketogluconate is not clear. Glyoxylate reduction seems unlikely, as the organism lacks a glyoxylate bypath in its terminal oxidation system. Murooka et al.[234] originally assumed that 5-ketogluconate is degraded to tartaric semialdehyde and glycolaldehyde after enolization and cleavage between C_4 and C_5 and then converted into tartrate and glycollate, respectively, through oxidation. The experimental results showed only the quantitative oxidation of glycolaldehyde to glycollate; however, glycollate and L-tartrate were not oxidized and no trace of tartrate was formed from 5-ketogluconate.

2-Ketogluconate is also oxidized by this organism. Fewster[19] observed that washed cells of *G. suboxydans* ATCC 621 oxidized 2-ketogluconate to the extent of 0.5 mole O_2 uptake per mole of substrate with an accompanying evolution of 0.5 mole CO_2. The oxidation was not increased by the addition of oxidizable substrates or by the presence of NAD, NADP, ATP or Mg^{++} ion. 2,5-Diketogluconic acid was not produced and the non-reducing product of this oxidative decarboxylation was identified chromatographically as D-arabonic acid. Resting cells of *G. suboxydans* (isol. 18) could oxidize 2-ketogluconate with an uptake of 3.4 moles of O_2 per mole of substrate, according to Stouthamer.[350] De Ley and Stouthamer[83] suggested that the first step in the metabolism of 2- and 5-ketogluconate is a reduction to gluconate, which is phosphorylated and metabolized *via* the pentose cycle. The cells also oxidized 2,5-diketogluconate with an uptake of 2.6 moles O_2 per mole of substrate. Stouthamer showed that 2,5-diketogluconate also is metabolized after an initial reduction to gluconate.

Phosphorylative Breakdown

Initial phosphorylation and oxidation. King and Cheldelin[94] were the first to indicate that phosphorylative breakdown and direct oxidation are both involved in the oxidation of glucose. Kinase activity was found by Fewster[473] in ultrasonic extracts of cells of *G. suboxydans* ATCC 621. Mg^{++} was required. D-Fructose, D-glucose, D-mannose, D-xylose, D-xylulose and L-sorbose were phosphorylated (the latter very weakly), while D-galactose was not phosphorylated. In Fewster's opinion, the phosphorylation of D-xylulose, which may arise by decarboxylation of 5-ketogluconate, provides a substrate for transketolase. Thus a link is established between the early stage of glucose oxidation not involving phosphorylation and the pentose cycle, both of which are active in *G. suboxydans*. Gluconate and 2- and 5-ketogluconate were also found to be phosphorylated. Glucose phosphorylation was completely inhibited by 10^{-4} M phenylmercuriacetate and restored by 5×10^{-3} M cystein, suggesting that kinase most probably contains SH groups essential for its activity. Kinase activity was also demonstrated by Stouthamer.[350]

The breakdown of phosphorylated glucose could take place *via* three different pathways: 1) the Embden-Meyerhof-Parnas pathway, 2) the pentose cycle, or 3) the Entner-Doudoroff pathway.

The Embden-Meyerhof-Parnas pathway Hauge *et al.*[164] confirmed the existence of G-6-P isomerase, phosphofructokinase and aldolase together with triosephosphate isomerase and dihydroxyacetone kinase in cell-free extracts of *G. suboxydans* ATCC 621. Triosephosphate dehydrogenase was also shown to occur in the same organism by Kitos *et al.*[474] Pyruvate decarboxylase was reported by Simon[90] in cells of *G. suboxydans*, as well as methylglyoxal as an intermediate of hexose diphosphate breakdown in an acetone-preparation of the organism. The separation of pyruvate decarboxylase from cell-free extracts of *G. suboxydans* was described by

King and Cheldelin.[342] The purified preparation also decarboxylated α-ketobutyrate and oxaloacetate but had no effect upon α-ketoglutarate, α-ketoisovalerate, α-ketoisocapronate, α-keto-β-methylvarelate or phenylpyruvate.

The results of these enzymatic studies could be taken as support for the operation of glycolysis in this organism. Cheldelin's work,[475] however, suggests that the Embden-Meyerhof-Parnas pathway does not, in fact, operate here, as significant amounts of acetate were not produced from glucose (lactate and pyruvate are converted quantitatively to acetate). The presence of the individual enzymes characteristic of glycolysis does not necessarily mean that glycolytic action occurs, because every enzyme in the scheme, except phosphofructokinase, may also be used in either the pentose cycle or the Entner-Doudoroff pathway; in fact, experiments by Kitos et al.[476] using the radiorespirometric method of Wang et al.[477] argue strongly against glycolysis in whole cells of G. suboxydans ATCC 621.

Cheldelin,[475] in another paper, stated that at least some of the glycolysis reactions can function, if external conditions permit or force their operation.

The pentose cycle. The active participation of a pentose cycle and the presence of a set of enzymes relating to the cycle were verified in G. suboxydans by Hauge et al.[164] Glucokinase was obtained from cell-free extracts and the resulting G-6-P was isolated and identified with its barium salt. The presence of ribokinase was indicated. The reactions of G-6-P and 6-P-G dehydrogenases in the presence of NAD and TTZ (triphenyltetrazolium chloride) were verified manometrically and the products were confirmed as 6-P-G and Ru-5-P respectively by chromatography. These two dehydrogenases are located in the soluble fraction of the cells and have been highly purified. A pH optimum of around 8.0 was reported in each preparation. The presence of hexokinase (glucokinase and fructokinase), gluconokinase, NADP-linked G-6-P and 6-P-G dehydrogenases, transaldolase, and transketolase was also demonstrated by Stouthamer[350] in G. suboxydans (strain 18). The existence of hexokinase, gluconokinase, G-6-P and 6-P-G dehydrogenases together with active oxidation of F-6-P and R-5-P in cell-free extracts of G. suboxydans ATCC 621 was also reported by Murooka.[467]

Figure 29 shows a metabolic map for the oxidation of glucose to the stage entering into the pentose cycle, with reference to the enzyme locations, as presented by De Ley.[340]

G-6-P dehydrogenase (substrate $Km=3.9\times10^{-4}$ for NAD) requires NADP or NAD, with the former slightly more effective, whereas for

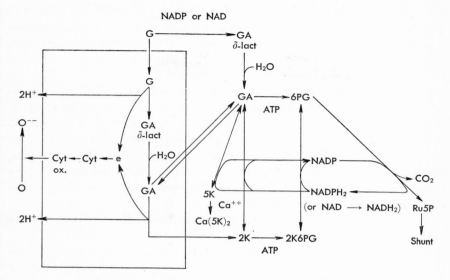

Fig. 29. The two pathways for the oxidation of glucose
by *G. suboxydans* (after De Ley).

The rectangle represents an ultramicroscopic particle (oxidosome) which
carries the enzymes for the reactions included therein.   The other reactions
are carried out by soluble enzymes in the cytoplasm.

Abbreviations : G, glucose ; GA δ-lact, glucono-δ-laction ; GA, gluconate ;
e, electron ; Cyt, cytochrome ; 2K, 2-ketogluconate ; 5K, 5-ketogluconate ;
6PG, 6-phosphogluconate ; Ru5P, ribulose-5-phosphate.

6-P-G dehydrogenase (substrate $Km = 2.3 \times 10^{-4}$ for NAD and 1.3×10^{-3}
for NADP) NAD is more effective than NADP (King and Cheldelin[478]).
Ru-5-phosphate is rapidly oxidized in cell-free extracts with NAD and
TTZ; under non-oxidative conditions it disappears, and sedoheptulose
and triose appear, with the subsequent formation of hexose.   This strongly
indicates the operation of a transketolase-transaldolase system in addition
to pentose phosphate isomerase.   Aldolase, triosephosphate isomerase,
phosphofructokinase and phosphohexoisomerase were also shown to be
present.   King and Cheldelin concluded that the pentose cycle is responsible in this organism for oxidation not only of glucose and ribose but
also of glycerol and dihydroxyacetone.   The oxidation of C_3 compounds
was described previously in the text (p. 209).

In other experiments by Kitos *et al.*[474] the accumulation of 3-phospho-
glycerate from fructose-1,6-diphosphate in cell-free extracts under oxida-

tive conditions with NAD was shown.   Since this organism contains both aldolase and triosephosphate isomerase, it could be deduced that fructose-1,6-diphosphate can serve as a ready source of glyceraldehyde-3-phosphate, from which 3-phosphoglycerate is subsequently formed by the action of triosephosphate dehydrogenase.

A quantitative evaluation of the pentose cycle as a respiratory mechanism in *G. suboxydans* was made by Kitos *et al.*[476] using glucose (or gluconate)-1-, -2-, -3, 4-, -6-, and glucose-U-^{14}C as substrates for aerated resting cells.   The results showed that for every 100 molecules 28 were oxidized to 2-ketogluconate, presumably by the particulate dehydrogenases, and of the remaining 72, 63 (corresponding to 88%) entered into the pentose cycle.   Essentially all of the CO_2 produced from glucose was estimated to arise *via* the pentose cycle.

It was concluded that such an active participation of the pentose cycle makes the absence of a TCA cycle reasonable in this organism.

The Entner-Doudoroff pathway   The operation of this pathway in the oxidative breakdown of glucose in *G. suboxydans* has been ruled out by Kitos *et al.*[476] from radiorespirometric experiments, because of the high yield of CO_2 from the carbon 2 of glucose (higher than from the carbon 4). However, a report recently presented by Kovachevich and Wood[479] indicates that a specific enzyme involved in this pathway, 2-keto-3-deoxy-6-phosphogluconate aldolase, which splits-phosphogluconate into pyruvate and glyceraldehyde-3-phosphate, is present in dried cells of *G. suboxydans* NRRL B-72.

Oxidative phosphorylation.   Oxidative phosphorylation in whole cells of *G. suboxydans* was investigated by Klungsöyr *et al.*[165] using glucose, glycerol, or fructose as the substrate and $^{32}PO_4^{3-}$ as a tracer.   The results showed that the P/O ratios were very low, averaging about 0.5; the amount of active nucleotides involved in phosphorylation was also very low, about 1.5 μ moles ATP per gram of respiring cells, and this could not be increased by the addition of acceptors.   The presence of relatively large amounts of inorganic pyrophosphate was also indicated, strongly suggesting the participation of inorganic pyrophosphate as an intermediate in oxidative phosphorylation.

The TCA Cycle

The normal TCA cycle does not function in *G. suboxydans*, or in other *Gluconobacter* strains, although *Acetobacter* strains have been shown to possess a complete enzyme system for this cycle, which actually functions in terminal oxidation. Even the catalase-negative *A. peroxydans* possesses a modified system for the TCA cycle.

Acetate, α-ketoglutarate, malate, succinate, fumarate, and citrate are oxidized by washed cells or cell-free extracts of *G. suboxydans* and pyruvate is only oxidized to the acetate stage (King and Cheldelin,[95, 96] Kondo and Ameyama[466]). Murooka *et al.*[234] obtained the same results, with the exception of oxaloacetate, which showed an oxygen uptake of 33.7 μl per 5 μmoles of substrate after 180 minutes, with dried cell preparations. Oxalosuccinate, glycollate, and tartrate were not oxidized. Stouthamer[350] also noted the non-oxidizability of succinate, fumarate, DL-malate, and α-ketoglutarate in two strains of *G. suboxydans*. *G. suboxydans* var. *biourgianum* strain 2 was observed to oxidize these C_4-dicarboxylic acids very weakly after a lag phase. Oxidative decarboxylation of α-ketoglutarate to succinate was shown not to occur by Rao.[343] According to King and Cheldelin,[161] resting cells of *G. suboxydans* failed to oxidize citrate, α-ketoglutarate, succinate, or fumarate even in the presence of glycerol as a potential "sparker." Soluble and insoluble fractions of the cells also failed to oxidize TCA cycle members even in the presence of a hot water extract of the organism or methylene blue, or both. Acetate and oxaloacetate were not condensed to citrate even in the presence of CoA and ADP.

From these experiments it appears that neither the TCA cycle nor the C_4-dicarboxylic acid cycle participates in the oxidative metabolism of this organism. It is interesting to note, however, that a considerable amount of α-ketoglutarate is produced from glucose by growing cells of *G. cerinus*, which is taxonomically closest to *G. suboxydans* (Asai *et al.*[290]). The pathway for the production of α-ketoglutarate is still obscure. Walker's hypothesis[480] of the conversion of 2-ketogluconate into α-ketoglutarate by oxidative decarboxylation and dehydration, and the conversion of 2,5-diketogluconate into α-ketoglutarate proposed by Datta and Katznelson[287] are both unlikely. The same is true for the conversion of L-arabinose to α-ketoglutarate *via* L-arabinolactone and L-arabonate, proposed by Weinberg and Doudoroff[481] for *Pseudomonas saccharophila*.

Recently, Sekizawa *et al.*[472] suggested two new routes for the synthesis of glutamic acid in *G. suboxydans* ATCC 621. In one of these reactions, glyoxylic acid condenses with oxaloacetic acid. The final product was identified as α-ketoglutaric acid by experiments using cell homogenates. Formation of the C_5 skeleton was also indicated by the isolation of γ-hydroxyglutamic acid when both substrates were incubated with α-alanine. In another experiment, α-ketoglutarate and γ-hydroxyglutamate in addition to both substrates always gave rise to glutamate in cell-free extracts. The conversion of hydroxyglutamate to glutamate appears to be NADH$_2$-linked. Another route of glutamate synthesis was thought to be initiated through condensation of pyruvate and acetate, the reverse of the fermentation of glutamate in *Clostridium tetanomorphum* (Munch-Peterson and Barker[482]). Both substrates and the presumed intermediates, DL-citramalate, mesaconate and DL-β-methyl aspartate, were converted to glutamate in the presence of alanine by cell-free extracts. These findings are noteworthy because there is no evidence in *G. suboxydans* of the TCA

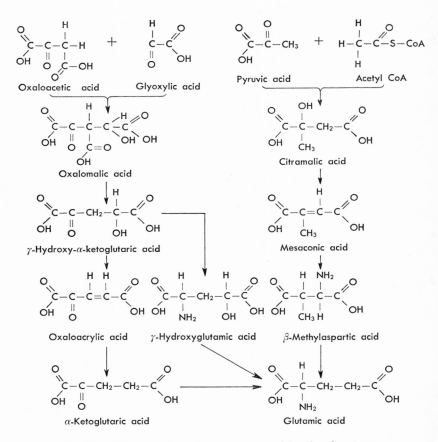

Fig. 30. Two condensation routes resulting in glutamate synthesis by *G. suboxydans* (after Sekizawa *et al.*).

cycle that causes the synthesis of glutamic, aspartic, and other amino acids and the formation of α-ketoglutaric acid. The presumed routes for the synthesis of glutamate are presented in Figure 30, although some speculative intermediates are involved in the reaction schema, and a question still remains as to the precursors of glyoxylate and oxaloacetate.

In a recent publication, Sekizawa et al[483] described the isolation and characterization of γ-hydroxyglutamic acid, and indicated that the γ-hydroxyglutamic acid obtained has the erythro-L-configuration, like that obtained from rats.

The formation of γ-hydroxy-α-ketoglutaric acid, considered to be the direct precursor of γ-hydroxyglutamic acid and an intermediate in glutamic acid formation, was also shown to occur in the reaction mixture of cell-free extracts of this organism with oxaloacetate and glyoxylate and in the absence of a nitrogen source.

The glyoxylate bypath.   Isocitrate lyase, which participates in the glyoxylate bypath, has been found in *A. aceti* grown previously on acetate medium (Smith and Gunsalus[484]). *A. dioxyacetonicus* strain A15 also possesses this enzyme, plus malate synthase, according to Kasai and Kida (unpublished). But the *Gluconobacter* generally lack these two enzymes, and hence the presence of a glyoxylate bypath is most unlikely.

Cheldelin,[485] however, suggested that isocitrate lyase might be active in *G. suboxydans*, and that it might aid isocitrate formation by acting in reverse to form isocitrate from glyoxylate plus succinate, or even to form oxalosuccinate from glyoxylate plus malate. Experiments to confirm this assumption, however, were unsuccessful because of the presence of active glyoxylate reductase (glycollate dehydrogenase) in the organism.

Evidence regarding the condensation of glyoxylate with oxaloacetate and acetate with pyruvate was discussed earlier.

The Cytochromes

According to Smith,[365] the cytochromes of *G. suboxydans* contain at least three components of reduced cytochromes of the b type, with an α-band at 554 mμ, β-band at 525 mμ, and γ-band at 442 mμ, in a preparation of intact cells, while the particulate fraction solubilized by deoxycholate showed three components of b-type with bands at 558–560 mμ, 528–532 mμ, and 426–428 mμ (King and Cheldelin[486]).

According to Yoshiie and Kameyama[487] three main absorption bands were observed: a well-marked reduction band at 554 mμ (α-band) and another weaker one at 520-530 mμ (β-band), and a faint band at about

560 mμ which can be ascribed to component b. The two prominent bands were regarded provisionally as those of component b_4. The intact cells showed one more absorption band at about 588 mμ, which was considered to be a_1 or a related compound. Neither the cytochrome band of component c, nor an indophenol oxidase system was detected.

Fewster[19] observed that the active pink cells of *G. suboxydans* ATCC 621, harvested after growth for 24, 48, or 72 hours, were rich in a cytochrome component having an α-band at 552 mμ and a β-band at 524 mμ. When the cells were grown on a medium containing glucose without $CaCO_2$ under conditions where acid production was allowed to proceed unchecked, they were shown to lack the characteristic cytochrome spectrum and to oxidize only glucose and 2-deoxy-D-glucose to the corresponding onic acids. In contrast, when the cells grown on glucose in the presence of $CaCO_3$ were incubated at pH 3.0 in phosphate buffer, the presence of the characteristic cytochrome, after reduction with dithionite, was demonstrated in apparently undiminished amounts and was shown to be oxidized by O_2 despite the fact that a loss of oxidizing ability and bleaching of the cells occurred.

Castor and Chance[368] had at one time suggested that no a-type cytochrome was involved in actual O_2 transport, but that some unknown b-type cytochrome might instead be responsible for the terminal oxidation. In subsequent experiments,[488] however, they came to doubt active participation of b-type cytochrome for these reasons: a) The reduced pigment does not show any sharp absorption band at 562 mμ. b) The protoheme, a prosthetic group of b-type cytochrome, is not detected in extracts of this organism. c) No evidence has been found for the presence of a b-type cytochrome which combines with CO and inhibits a terminal oxidase system. Cytochrome o instead of a b-type cytochrome was ultimately proposed as the main oxygen-transferring enzyme. A " CO-binding pigment " was first observed spectrophotometrically in *G. suboxydans* by Kubowitz and Haas.[369]

Meanwhile, Iwasaki[373] had isolated and purified cytochrome a_1-590, cytochrome b-560, cytochrome c_1-552, and cytochrome c_1-554 from a strain of *G. suboxydans* using sodium deoxycholate and digestive enzymes. The optical and physiological properties of these four cytochromes were investigated. The preparation of cytochrome c_1-554 was found capable of strongly reducing 2,6-dichlorophenolindophenol in the presence of lactate, succinate, or fumarate. It was emphasized that although various procedures were used to purify lactate dehydrogenase, cytochrome c_1-554 was always found associated with the enzyme activity, suggesting a certain functional link between lactate dehydrogenase and cytochrome c_1-554. The fact that the oxidized form of c_1-554 could be reduced readily by the addition of lactate strongly suggests the presence of the following electron-transfer system: lactate dehydrogenase→cytochrome c_1-554. A reconstruction of the oxygen-uptake system was accomplished when these four cytochromes and lactate dehydrogenase were combined.

Thus the components indispensable for lactate oxidation were specified to be a terminal oxidase, cytochrome c_1-554, and lactate dehydrogenase. The following scheme for the electron transfer system in lactate oxidation was presented:

Lactate $\longrightarrow$ Lactate dehydrogenase $\longrightarrow$ Cytochrome c_1^{554}

$\longrightarrow$ (Cytochrome c_1^{552}) $\longrightarrow$ Cytochrome oxidase $\longrightarrow$ O_2

(Cytochrome a_1)

Cytochrome c_1-552 was verified to be identical in function with the corresponding cytochrome found in a wide variety of bacteria, and the presence of Castor and Chance's cytochrome o was disproved, since no trace of the a-peak of cytochrome o at 567 mμ was found in the oxidase preparation.

The Fate of Acetate

Acetate is not oxidized by intact cells or cell-free extracts of *G. suboxydans*, either alone or in the presence of a " sparker." Kitos et al.[476] showed that a negligible amount of ^{14}C appeared in the respiratory CO_2, whereas 25% of the added ^{14}C was incorporated into the lipid fraction of the cells when $CH_3^{14}COOH$ was added to respiring cells. Another experiment (Kitos et al.[489]) showed that cell-free extracts or sonically disrupted cells could form acetyl CoA from acetate in the presence of CoA, ATP and Mg^{++} ion, yet the acetyl CoA formed could not be converted into citrate or acetyl-sulfanilamide, suggesting lack of suitable acceptor systems. Pyruvate did not form acetyl CoA during oxidation unless ATP was added.

Hromatka and Gsur[490] reported that *G. suboxydans* assimilated $^{14}CO_2$ when it was added in aeration gas. About 0.1% of the C content of the organism originated from gaseous CO_2. It was also noted that although most of the acetic acid in the fermentation originated from glucose, a small amount was derived from CO_2 by other, unknown metabolic pathways.

A tentative pattern of carbohydrate metabolism in *G. suboxydans* is shown in Figure 31.

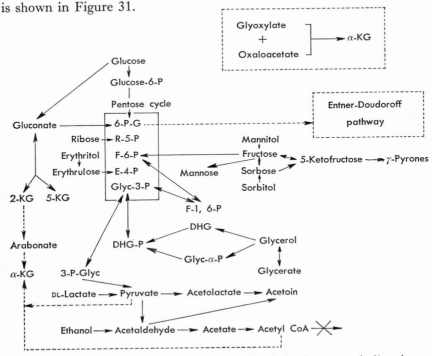

Fig. 31. A suggested metabolic map of carbohydrate metabolism by *G. suboxydans*.

Broken lines indicate reactions which are presumed to take place.

Gluconobacter melanogenus

This organism is characterized by brown colonies with soluble brown pigment when cultured in glucose agar medium in the presence of $CaCO_3$. Its capacity to produce 2,5-diketogluconate from glucose or gluconate also distinguishes it from other gluconobacters. This property, however, is not stable and mutation to a non-pigmented strain similar to *G. suboxydans* has been reported.

Non-Phosphorylative Oxidation

Oxidation of glucose, gluconate, and 2-ketogluconate   In experiments by Katznelson et al.,[254] aged intact cells and cell-free extracts of *G.*

melanogenus (MA 62) oxidized glucose, gluconate, and 2-ketogluconate to a single end product, with oxygen uptakes of 1.5, 1.0, and 0.5 moles respectively per mole of substrate without CO_2 liberation. The end product was isolated and identified as 2,5-diketogluconate. This indicates that the stepwise oxidation of glucose to 2,5-diketogluconate proceeds *via* gluconate and 2-ketogluconate, and not *via* 5-ketogluconate. Young intact cells of *G. melanogenus* carried out a more complete oxidation. In the presence of dinitrophenol, however, the metabolism resembled that of aged cells.

2,5-Diketogluconate is chemically unstable, especially at a pH above 4.5, and easily gives rise to brown decomposition products. The brown discoloration in glucose- or gluconate-containing media in the presence of $CaCO_3$ was attributed by Katznelson *et al.*[254] to spontaneous decomposition of 2,5-diketogluconate during cultivation. Stouthamer,[491] however, reported that a strain of *G. suboxydans* var. *biourgianum* (isol. 2) formed 2,5-diketogluconate as rapidly as *G. melanogenus*, but the brown coloration, on yeast extract-glucose-$CaCO_3$ slants, occurred only after a long incubation period (3–5 weeks).

When glucose-1-^{14}C was used as the substrate, aged cells of *G. melanogenus* formed 2,5-diketogluconate as the sole radioactive product in the medium and one half of the labeled carbon was found in the cells. Degradation of the radioactive 2,5-diketogluconate recovered showed that only the carboxyl group contained the carbon isotope. Katznelson *et al.* also confirmed the formation of 2,5-diketogluconate *via* 2-ketogluconate from maltose by this organism. No 5-ketogluconate was detected.

Stouthamer[350] reported that resting cells of *G. melanogenus* (Delft) oxidized gluconate mainly to 5-ketogluconate. A small amount of 2-ketogluconate formation was observed, although further oxidation of 2-ketogluconate to 2,5-diketogluconate was recognized. 5-Ketogluconate was not oxidized. NADP-linked glucose dehydrogenase and NADP-linked 2- and 5-ketogluconate reductases were demonstrated in the soluble fraction of the cells. The particulate fraction also oxidized glucose, gluconate and 2-ketogluconate.

The enzyme catalyzing the direct oxidation of 2-ketogluconate was investigated by Datta and Katznelson.[492] The purified enzyme responsible for the oxidation was obtained in the soluble fraction of sonically disrupted cells. This enzyme preparation did not react directly with oxygen but required a dye such as 2,6-dichlorophenolindophenol or phenazine methosulfate as a hydrogen carrier. Methylene blue and pyocyanine were not satisfactory as electron acceptors, as determined by the respirometric method. The enzyme preparation oxidized glucose and gluconate only very slightly. The enzyme itself is not NAD- or NADP-linked, nor is it inhibited by malonate, arsenite, or monoiodoacetate, indicating that it is probably not sulfhydryl active. Fluoride does not produce inhibition, suggesting that phosphorylation of the substrate is probably not involved. Cyanide inhibits enzyme activity to about 40% at a concentration of 10^{-3}M. Cytochrome c is reduced in the presence of the enzyme and 2-ketogluconate, but attempts to demonstrate the presence of cytochrome oxidase were not successful. The addition of Mg^{++} or Mn^{++} did not influence enzyme activity.

Oxidation of 2,5-diketogluconate.   Cell-free extracts of G. melanogenus (48-hour cells in shake culture) showed a marked increase in oxygen consumption with 2-ketogluconate when phenazine methosulfate was added to the reaction mixture (Datta and Katznelson[287]). Approximately 2.0 μmoles oxygen were consumed per μmole of 2-ketogluconate or 1.5 μmoles oxygen per μmole of 2,5-diketogluconate. At the same time 1 μmole of CO_2 was liberated per μmole of 2,5-diketogluconate and the formation of α-ketoglutarate was verified after the oxygen uptake had ceased. It was also demonstrated that the formation of a pentose-like substance gradually decreased, whereas that of α-ketoglutarate increased with time; at the end of the reaction approximately 50% of the 2,5-diketogluconate was accounted for as α-ketoglutarate. In another experiment cell-free extracts catalyzed the conversion of L-arabinose to α-ketoglutarate, suggesting that the pentose-like substance was most likely converted to α-ketoglutarate.

Datta et al.[493] made further investigations on 2,5-diketogluconate decarboxylase to discover whether or not the substrate is first decarboxylated to a five-carbon compound, and then oxidized to α-ketoglutarate. The decarboxylase preparation was obtained from the supernatant of centrifuged sonic extracts of cells grown for 48 hours in shaking culture. The purified enzyme decarboxylated 2,5-diketogluconate with the liberation of 1 μmole CO_2 per μmole of 2,5-diketogluconate. The optimum pH was found to be between 3.0 and 3.5. Dialysis for 24 hours or more caused considerable loss of activity which was not restored by the addition of thiamine pyrophosphate, Mg^{++} ions or Mn^{++} ions alone or in combination. The undialyzed enzyme preparation was not appreciably stimulated by ThPP but was considerably stimulated by Mn^{++}. NAD, NADP, ATP, ADP, AMP, pyridoxal phosphate, glutathione, and CoA had no stimulatory effect. The enzyme preparation decarboxylated α-ketoglutarate very slowly, but gluconate, 2-ketogluconate, acetate, malate, oxaloacetate, succinate, citrate or phenylalanine were not decarboxylated. Pyruvate was readily decarboxylated in the presence of Mn^{++}. The reaction product from 2,5-diketogluconate, though it gave a positive orcinol reaction, showed an independent R_f value differing from arabinose, ribulose, xylulose, sedoheptulose, and dihydroxyacetone. A crystalline o-nitrophenylhydrazone which gave a sharp melting point at 181°C was obtained from the reaction product, but could not be identified.

Another experiment by Stouthamer[350] showed that G. melanogenus (Delft) oxidized 2,5-diketogluconate slowly after an induction phase without gas exchange. The oxidation led to the formation of γ-pyrone derivatives, as reported by Aida et al.[281] These γ-pyrones, especially rubiginol, are unstable and decompose spontaneously, forming brownish-black pigment. Cell-free extracts did not produce α-ketoglutarate during oxidation of 2,5-diketogluconate, in contrast to the findings of Datta and Katznelson,[287] though an unknown acid was detected in culture filtrates and in the supernatants of resting cells which had been incubated with 2,5-diketogluconate.

Ameyama and Kondo[295] obtained a penturonic acid-like compound among the oxidation products from glucose in growing cultures of a strain of G. melanogenus. This acid, one of the main oxidative metabolites other

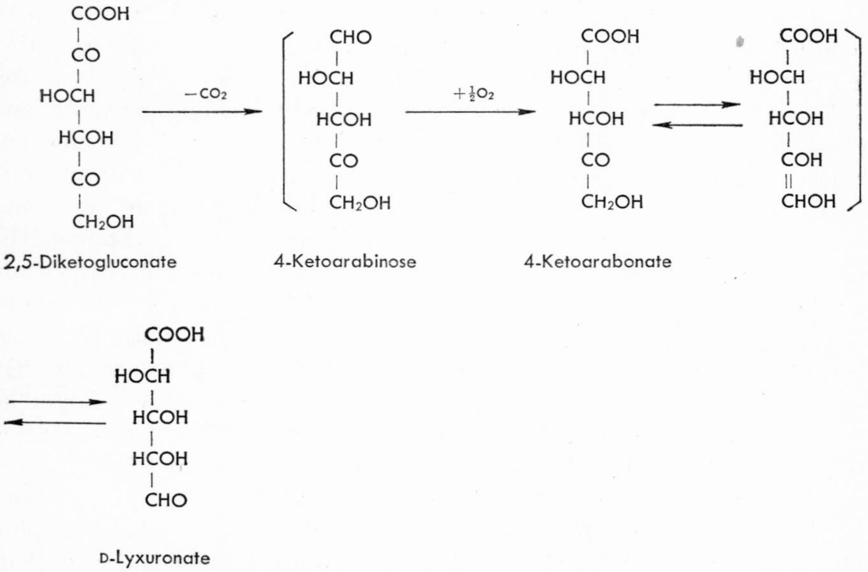

Fig. 32. A presumed pathway for the formation of D-lyxuronic acid from 2,5-diketogluconic acid by G. melanogenus
(after Ameyama and Kondo).

than glycollate, was isolated and identified as D-lyxuronic acid. It was presumed that this acid is produced from D-glucose *via* 2,5-diketogluconate and probably *via* " 4-ketoarabinose " and " 4-ketoarabonate." A substance with the same R_f value as 4-ketoarabinose was found in the nonacidic fraction of the products. Takahashi and Asai[288] observed the formation of a penturonic acid-like substance from glucose by a strain of *Gluconobacter*.

A possible metabolic pathway for the formation of D-lyxuronic acid, suggested by Ameyama and Kondo,[295] is shown in Figure 32.

De Ley and Stouthamer[316] found a mutant strain of *G. melanogenus* (Delft) which had lost the power to make brown pigment and produced a considerable amount of 5-ketogluconate. This behavior was attributed to the absence of 2-ketogluconate reductase and the low content of gluconate oxidase. 5-Ketogluconate was not metabolized by this strain, owing to the absence of gluconokinase.

Phosphorylative Breakdown

G. melanogenus, like *G. suboxydans*, possesses a system for hexose phosphate oxidation in addition to direct oxidation of glucose. Katznelson[494] reported that glucose could be phosphorylated and that the resulting hexose phosphate is oxidized predominantly to 6-phosphogluconate, followed by the pentose cycle or a C_3-C_3 split yielding pyruvate and glyceraldehyde-3-phosphate. Hexokinase, gluconokinase, 2-ketoglucono-kinase, glyceraldehyde-3-phosphate dehydrogenase, and aldolase were found. Sedoheptulose formation from 6-phosphogluconate was also verified by paperchromatography. The oxidative rather than the glycolytic pathway was thought to operate since phosphohexokinase activity could not be demonstrated. Under the conditions of Katznelson's experiments, a C_3-C_3 split (Entner-Doudoroff pathway) appeared more likely, as considerably more pyruvate was produced from 6-phosphogluconate than from ribose-5-phosphate, whereas approximately the same amount of pyruvate would be expected if 6-phosphogluconate was first decarboxylated to pentose phosphate. Pyruvate was thought to be formed from ribose-5-phosphate either *via* a transketolase-transaldolase series of reactions leading to fructose-6-phosphate, glucose-6-phosphate, and 6-phospho-gluconate followed by a C_3-C_3 split, or from glyceraldehyde-3-phosphate itself originating from the pentose phosphate through the transketolase reaction. Both pathways are probably operative.

Stouthamer[350] identified hexokinase and fructokinase in the soluble fraction of *G. melanogenus* (Delft) but failed to find gluconokinase. De Ley and Stouthamer[83] also could not find gluconokinase.

The Oxidation of TCA Cycle Intermediates and C_3 Compounds

Washed cell suspensions of a strain of *G. melanogenus* oxidized only pyruvate, according to Kondo and Ameyama;[125] no other intermediates of the TCA cycle were oxidized. Stouthamer[491] showed that resting cells of *G. melanogenus* (Delft) were unable to oxidize succinate, fumarate, DL-malate, and α-ketoglutarate, and only oxidized pyruvate and DL-lactate. Gluconate-grown cells of *G. melanogenus* (MA 62) oxidized pyruvate with an uptake of one atom of oxygen per mole of substrate, indicating the production of one mole of acetate and one mole of carbon dioxide (Bone and Hochster[495]). Sonically disrupted cell-free extracts of these gluconate-grown cells contained condensing enzyme and NAD-linked isocitrate dehydrogenase of low specific activity. α-Ketoglutarate dehydrogenase activity was demonstrated in the extracts when CoA, ThPP, ATP, Mg^{++}, lipoic acid and NAD were supplemented. Oxaloacetate decarboxylase, succinate dehydrogenase, fumarase, and malate dehydrogenase were also present. No evidence was obtained for the presence of the following enzymes: phosphotransacetylase, aconitase, NADP-linked isocitrate dehydrogenase, malate synthase, and phosphoenolpyruvate carboxylase. The presence of NAD-linked enzyme is unusual since the NADP-enzyme is usually more active and stable and the NAD-catalyzed system is quite labile in most tissues.

The pyruvate oxidase system was found to be more active. The maximal initial rate of pyruvate oxidation was dependent on the presence of Mg^{++}, ThPP, NAD or NADP, and GSH (reduced glutathione). Dimedone inhibited only oxygen uptake, suggesting that an aldehyde is probably an intermediate in pyruvate oxidation. Sulphathiazole and *p*-chloromercuribenzoate, which are inhibitors of yeast pyruvate decarboxylase, effectively inhibited the system. Arsenite inhibited oxygen uptake only slightly but inhibited α-ketoglutarate dehydrogenase completely. Acetaldehyde was actively oxidized by cell-free extracts, forming acetate as the main product.

Attempts to synthesize the four carbon compounds from pyruvate and from acetate were unsuccessful; thus it was concluded that the normal TCA cycle does not function in the usually accepted manner in extracts of this organism. Experiments with labeled acetate, bicarbonate, and pyruvate showed that acetate can be slowly incorporated into cell constitu-

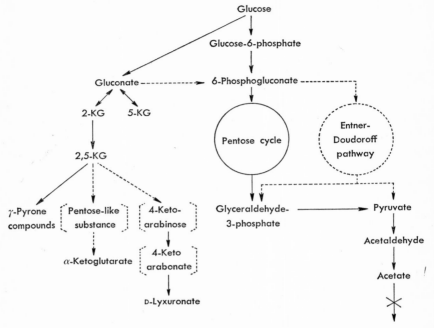

Fig. 33.   A suggested map of glucose metabolism by
G. *melanogenus*.

Abbreviations :   2-KG ; 2-ketogluconate, 5-KG ; 5-ketogluconate, 2,5-KG ;
2,5-Diketogluconate.

ents, though the mechanism is still obscure.

The pathway of glucose metabolism in *G. melanogenus* is shown in Figure 33.

Acetobacter xylinum

The metabolism of carbohydrates and related compounds by *A. xylinum* was touched upon in the chapter on biosynthesis of cellulose, a particular characteristic of this organism.   Here we will discuss the general aspects of the chemical pathways and enzyme systems by which glucose or other sugars are metabolized either completely or incompletely.

A. xylinum oxidizes glucose to gluconate and 2- and 5-ketogluconate as do the gluconobacters (Ameyama and Kondo[496]).   2,5-Diketogluconate is not formed.   Resting cells of *A. xylinum* (Davis) weakly oxidize 2-

ketogluconate but not 5-ketogluconate (Stouthamer[491]). However, cells of *A. xylinum* 8747 have been shown by De Ley[457] to oxidize ketogluconates: 0.6 mole O_2 per mole of 2-ketogluconate and >0.4 mole O_2 per mole of 5-ketogluconate were consumed.

Gromet *et al.*,[420] (see also p. 236 in the text) in a detailed study using cell-free extracts, found glucose-6-phosphate and 6-phosphogluconate dehydrogenases linked to both NAD and NADP, as well as the enzymes catalyzing the conversion of ribose-5-phosphate into hexose-6-phosphate in the extracts. The latter transformation indicates the presence of the following enzymes: phosphopentose isomerase, phosphopentose epimerase, transketolase, transaldolase and phosphohexose isomerase. Alkali-labile phosphate (presumably triose-P) was formed anaerobically from fructose-diphosphate in the presence of hydrazine added as a trapping agent. During the anaerobic split of fructose-1,6-diphosphate a rapid formation of Pi accompanied by release of a relatively small amount of reducing sugar was observed. The cell homogenate, by contrast, exhibited versatile phosphatase activity. In this case, Pi was rapidly released from glucose-6-phosphate, α-glycerol-1-phosphate, ribose-5-phosphate and ATP in an anaerobic system at pH 7.5.

The important discovery of the presence of the enzyme " fructose-6-phosphate phosphoketolase," responsible for the energy-yielding process, was made by Schramm *et al.*[497] in cell-free extracts of a celluloseless mutant of *A. xylinum*. This enzyme catalyzes the following reaction:

Fructose-6-phosphate + Inorganic phosphate $\longrightarrow$

Acetyl phosphate + Erythrose-4-phosphate + H_2O

The reaction was not inhibited by iodoacetate (3×10^{-3}M) and proceeded anaerobically. Crude cell-free extracts contained acetate kinase, hence ATP was formed from acetyl phosphate and ADP.

The over-all reactions involving fructose-6-phosphate phosphoketolase and other enzymes, which Schramm *et al.* named the fructose-6-phosphate pathway, can be visualized as follows (see next page).

The significance of this pathway is not apparent, but according to Schramm *et al.* it may serve as a " short circuit " for the production of acetate which is readily oxidizable by this organism. It yields 3 moles of ATP per mole of fructose-6-P or 2 moles of ATP per mole of glucose, a yield identical with that of glycolysis. On a molar basis it provides a rather low yield of energy as compared with acetate oxidation *via* the TCA cycle,

Reactions	Enzymes
Fructose-6-P + Pi ⟶ Acetyl-P + Erythrose-4-P	Fructose-6-P phosphoketolase
Erythrose-4-P + Fructose-6-P ⟶	
	Transaldolase
Sedoheptulose-7-P + Glyceraldehyde-3-P	
Sedoheptulose-7-P + Glyceraldehyde-3-P	
	Transketolase
⟶ Ribose-5-P + Xylulose-5-P	
Ribose-5-P ⟶ Xylulose-5-P	Ribose-5-P isomerase and
	Xylulose-5-P epimerase
2 Xylulose-5-P + 2 Pi ⟶	
	Phosphoketolase
2 Acetyl-P + 2 Glyceraldehyde-3-P	
2 Glyceraldehyde-3-P ⟶ Fructose-1,6-diP	Glyceraldehyde-3-P
	isomerase and Aldolase
Fructose-1,6-diP ⟶ Fructose-6-P + Pi	Fructose-1,6-diP Phosphatase

Sum : Fructose-6-P + 2Pi ⟶ 3Acetyl-P

and this may help to explain the lack of cellulose synthesis under anaerobic conditions. Partially purified preparations of fructose-6-phosphate phosphoketolase also split xylulose-5-phosphate into acetyl phosphate and glyceraldehyde-3-phosphate, but did not split ribulose-5-phosphate, sedoheptulose-7-phosphate, xylulose, fructose, hydroxypyruvate and pyruvate.

Prieur[498] reported that fructose-grown cells yielded cell-free extracts much more active in the oxidation of various substrates and phosphorylated intermediates, but deficient in 6-phosphofructo-1-kinase and phosphoglyceromutase which were prevalent in the extracts of glucose-grown cells. Phosphoglucomutase and UDPG dehydrogenase, found in the extracts of fructose-grown cells, were virtually absent in the extracts of glucose-grown cells.

Occurrence of the Entner-Doudoroff pathway in conjunction with the pentose cycle was confirmed recently by White and Wang[499] in *A. xylinum* ATCC 10821, using specially ^{14}C-labeled glucose and gluconate and active metabolizing cells. Synthesized alanine and pyruvate were degradated and the ^{14}C distribution determined. Both alanine and pyruvate formed from glucose-1-^{14}C maintained over 90% of the total activity at C-1 and 2–7% at C-3. Alanine formed from D-gluconate-1-^{14}C had approximately 100% of the total activity at C-1. The small fraction of activity at C-3 of alanine and pyruvate suggest that a weak glycolysis may operate in the metabolism, as do the facts that cells incubated under nitrogen evolved $^{14}CO_2$ from glucose-U-^{14}C and that cell extracts slowly converted fructose-6-phosphate into triose phosphate. The fact that the C-2 of glucose and gluconate appeared largely in the C-2 of pyruvate and alanine strongly indicates the participation of the Entner-Doudoroff pathway. C-6 of glucose was incorporated to the extent of 52–64% into the C-1, and 29–40% into the C–3 of pyruvate. Alanine from gluconate-6-^{14}C exhibited 84% of the total activity at C-1 and 12% at C-3. This was attributed to the preferred conversion of the C-6 of glucose and gluconate into a carboxyl group and thence to carbon dioxide *via* pyruvate, and a functional process that would introduce the C-6 of glucose into the C-1 of hexose phosphate and thence into the C-1 of pyruvate *via* the Entner-Doudoroff pathway. Aldolase, hexosediphosphatase and the enzymes responsible for this pathway were found in cell-free extracts. Phosphofructokinase was also detected, though the activity was extremely weak. The conversion of 3-phosphoglycerate into pyruvate and the synthesis of hexosephosphate from glyceraldehyde-3-phosphate were also demonstrated. Glyceraldehyde-3-phosphate generated *via* the Entner-Doudoroff pathway and the pentose cycle appeared to undergo extensive recombination at the triose phosphate-isomerase and aldolase level to yield fructose-1,6-diphosphate, and by reverse glycolysis fructose-1,6-diphosphate converted to 6-phosphogluconate *via* glucose-6-phosphate. The pyruvate formed from 6-phosphogluconate-1-^{14}C *via* the Entner-Doudoroff pathway was labeled at C-1.

The Entner-Doudoroff pathway appeared to be more active than the pentose cycle in hexose phosphate utilization. Subsequent studies by the same authors,[500] however, indicated that in young cells the pentose cycle appeared to be more active, and about 80-90% of the triose phosphate formed *via* the pentose cycle and the Entner-Doudoroff pathway appeared to recombine *via* fructose-1,6-diphosphate to yield hexose monophosphate. The various pathways for glucose and gluconate metabolism in *A. xylinum* are evaluated in Figure 34.

Enzymic studies in the degradation of glucose and other substrates using both the wild-type and celluloseless mutant were carried out by Leisinger.[501]

Enzymes of the Entner-Doudoroff pathway and CM-cellulose cellulase activity were about equal in a wild-type (M15/C$^+$) and celluloseless mutant (M91) of *A. xylinum*. *A. mesoxydans* had neither activity. Glucose oxidation of resting cells was ~33% higher in mutant M91 than in a wild type M15/C$^+$. Cell-free extracts of the above three bacteria showed a UDPG pyrophosphorylase activity of 0.22, 0.11, and 0.12 micromole

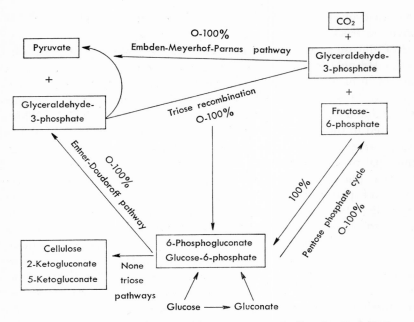

Fig. 34.   A model of glucose and gluconate metabolism in *A. xylinum*
(after White and Wang).

UDPG split/mg protein/hr, respectively. The enzyme system that transfers
glucose from UDPG to cellodextrin was present only in a wild-type
M15/C⁺.

Several papers have confirmed the activity of the TCA cycle in the
breakdown of pyruvate.   According to Schramm *et al.*,[419] acetate, pyruvate
and TCA cycle intermediates were oxidized to carbon dioxide.   Gromet-
Elhanan and Hestrin[431] reported that cells grown on ethanol, acetate or
succinate were able to transform TCA cycle intermediates into cellulose
whereas cells grown on glucose could not.   Oxidation of pyruvate and
α-ketoglutarate by resting cells of *A. xylinum* (Davis) was observed by
Stouthamer.[491]   Oxidation of α-ketoglutarate to succinate with cell suspensions of *A. xylinum* and oxidation of succinate, fumarate, malate, and
pyruvate by resting cells of *A. xylinum* were reported by Iwatsuru[502] and
Tanaka[503] respectively.   Benziman and Abeliovitz[504] observed the accumulation of oxaloacetate from fumarate or L-malate by washed succinate-
grown cells or cell-free extracts of *A. xylinum*, whereas no accumulation of
this acid was observed when succinate served as the substrate.   When
extracts preincubated with EDTA were allowed to oxidize malate, oxalo-

acetate accumulated at a constant rate, indicating that oxaloacetate does not inhibit malate oxidation. Malate oxidation by extracts with oxygen, ferricyanide, or dichlorophenolindophenol as electron acceptors showed that 1 atom of oxygen or 2 moles of ferricyanide were consumed, or 1 mole of dichlorophenolindophenol was reduced, per mole of oxaloacetate formed. Addition of NAD or NADP had no effect on malate oxidation, even when dialyzed extracts were used. This indicates that malate oxidation is irreversible and is catalyzed by an NAD- or NADP-independent enzyme. Benziman and Galanter,[505] in subsequent experiments, investigated the malate-oxidizing enzyme of this organism using supernatants obtained by high-speed centrifugation of sonic extracts. Ferricyanide, phenazine methosulfate and, to a lesser extent, dichlorophenolindophenol were found active as oxidants for malate oxidation. The enzyme lost its activity after acid ammonium sulfate precipitation and was reactivated by low concentrations of FAD but not by FMN or riboflavin. The activity was inhibited by hematin. Inhibition was prevented by imidazole or globin. o-Phenanthroline, 8-hydroxyquinoline, a,a-dipyridyl, and p-chloromercuribenzoate inhibited malate oxidation. Amytal markedly inhibited malate oxidation in the presence of oxygen, phenazine methosulfate, or dichlorophenolindophenol, but not in the presence of ferricyanide. It was deduced from these results that the malate-oxidizing enzyme of *A. xylinum* is an FAD-linked enzyme, containing an iron-binding site essential for its activity. Fumarate, succinate, and L-lactate were not oxidized by this enzyme. DL-Malate was oxidized at half the rate of L-malate, indicating a strict substrate specificity for L-malate. That the malic dehydrogenase is functionally linked to the cytochrome system was indicated by the cyanide and azide inhibition of malate oxidation in the crude extracts when oxygen served as the terminal electron acceptor.

Benziman and Abeliovitz[504] showed that succinate-grown cells of *A. xylinum* catalyzed a rapid exchange of $^{14}CO_2$ into oxaloacetate. The nature of the change reaction was studied subsequently by Benziman and Heller.[506] Extracts of succinate-grown cells, prepared by sonic treatment, were shown to catalyze the decarboxylation of oxaloacetate to pyruvate and CO_2, and the exchange of ^{14}C-carbon dioxide into the β-carboxyl of oxaloacetate. The ratio between the CO_2 exchange and oxaloacetate decarboxylation activities in the crude extracts was $1:100$. A purified preparation possessing specific activity 10-fold that of the enzyme system catalyzing the CO_2 exchange and oxaloacetate decarboxylation was obtained by fractionation of the extracts with ammonium sulfate. This preparation catalyzed the exchange of pyruvate-3-^{14}C into oxaloacetate. The CO_2 exchange and oxaloacetate decarboxylation activities had similar pH curves with maximal activity at pH 5.6. Mn^{++} or Mg^{++} ions were required for both reactions.   Dialysis or metal-chelating agents inhibited the oxaloacetate decarboxylation more strongly than the CO_2 exchange.   Avidin did not inhibit either reaction. ATP, ADP, GTP, GDP, pyrophosphate or inorganic phosphate stimulated neither oxaloacetate decarboxylation nor CO_2 exchange catalyzed by the purified preparation.   The purified preparation failed to catalyze the carboxylation of phosphoenolpyruvate in the presence of GDP, ADP, or inorganic phosphate, and that of pyruvate in the presence of ATP or GTP, even when supplemented with an oxaloacetate trapping system resulting from an excess of exogenous malate dehydrogenase and reduced NAD.

The following scheme was suggested for the mechanism of oxaloacetate (OAA) decarboxylation.

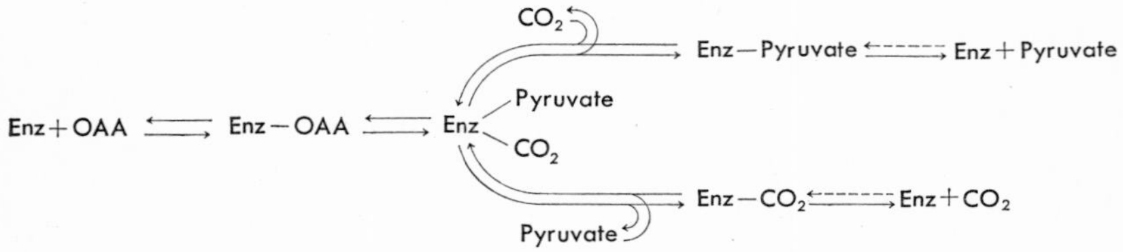

In this scheme, the enzyme complex with both the pyruvate and β-carboxyl moieties of oxaloacetate may be reversibly dissociated either to free pyruvate and enzyme-bound CO_2, or to free CO_2 and enzyme-bound pyruvate.   Formation of enzyme-bound pyruvate or enzyme-bound CO_2 occurs much faster from oxaloacetate than from free pyruvate or CO_2. Thus, an exchange of CO_2 or pyruvate with oxaloacetate would be expected, but the synthesis of oxaloacetate from these compounds would be very much slower.

Webb and Colvin[507] observed that cells of *A. xylinum* partially lysed by lysozyme (pH 8.1 in phosphate-citrate buffer) possessed the same oxidative capacity as whole cells with respect to glucose-6-phosphate dehydrogenase and to the oxidation of glucose and ethanol.   Tomlinson and Campbell[508] reported on oxidative assimilation of glucose with washed cell suspensions using glucose-U-^{14}C as the substrate. ^{14}C was incorporated into the nitrogenous fractions of the cell, but the level of assimilation was low due to the accumulation of extracellular cellulose.   Ammonia was formed as a product of endogenous respiration, and some fractions of ammonia were reincorporated into cellular material during glucose oxidation.

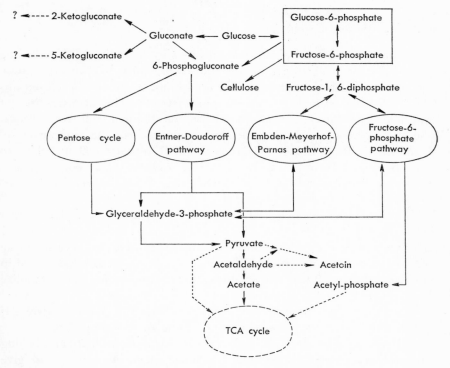

Fig. 35. Pathways proposed for glucose and fructose
dissimilation in *A. xylinum.*

Full lines indicate reactions which have been experimentally demonstrated.
Broken lines indicate reactions which are presumed to take place.

The metabolic pathways of glucose and fructose dissimilation in *A. xylinum* are schematized in Figure 35.

Acetobacters of the Oxydans and Mesoxydans Group

A. ascendens is unusual among the catalase-positive acetobacters in that it lacks the capacity to oxidize glucose. Neither gluconate nor keto-gluconate is oxidized by the resting cells and neither glucose oxidase nor glucose dehydrogenase can be detected. Other strains of *Acetobacter*, except the catalase-negative *Peroxydans* group of Frateur, metabolize glucose aerobically.

Frateur's *Oxydans* Group

This group of *Acetobacter* was divided by Stouthamer[491] according to the behavior of glucose oxidation into two major subgroups as follows.

a) Strains which oxidize glucose to gluconic acid with an oxygen uptake of 0.5 mole per mole of glucose: gluconate and ketogluconates are not oxidized. Hexokinase and gluconokinase are absent in cell-free extracts. The strains do not generally metabolize glucose *via* the pentose cycle. *A. rancens* Delft, *A. rancens* NCIB 4937 and *A. lovaniense* Delft belong to this subgroup. According to De Ley,[457] *A. rancens* 15, 23, and " Davis," *A. pasteurianus* 11 and 8856, *A. kützingianus* 3924 and *A. rancens* var. *turbidans* 8619 can also be included.

b) Strains which give irisation on yeast extract-glucose-$CaCO_3$ agar slants: the strains oxidize glucose with a high oxygen uptake ($>1.5 \rightarrow >3.2$ moles O_2 per mole of glucose) and also show an appreciable oxygen uptake with gluconate but not with ketogluconates. Gluconokinase is detected in cell-free extracts. By means of gluconokinase the strains can metabolize gluconate *via* the pentose cycle.

A. rancens var. *saccharovorans*, *A. rancens* (isol. 10), *A. rancens* (isol. 24) and *A. rancens* var. *turbidans* (isol. 16) belong to this subgroup.

A. rancens (isol. 29) and *A. rancens* var. *turbidans* (isol. 26) also belong here, but are intermediates, able to oxidize gluconate but not to give

Table 23. Enzymes found in cell-free extracts of *A. rancens* Delft and *A. rancens* (isol. 10) (after Stouthamer).

Enzyme	*A. rancens* Delft	*A. rancens* (isol. 10)
Glucose oxidase	+	+
Gluconate oxidase	−	−
Glycerol oxidase	−	−
Hexokinase	−	Not observed
Gluconokinase	−	+
Ketogluconate reductase	−	−
G-6-P dehydrogenase	+	+
6-P-G dehydrogenase	±	+
Isocitrate dehydrogenase	+	+

Notes: G-6-P and 6-P-G were oxidized by cell-free extracts in the presence of both NADP and NAD and phenazine methosulfate. Ribose-5-phosphate was oxidized very slowly under these conditions.

irisation on glucose-$CaCO_3$. They are able to form 2-ketogluconate from gluconate, but 2-ketogluconate is not metabolized further.

No glycolytic system seems to be present in strains of the *Oxydans* group, since none produce CO_2 from glucose under anaerobic conditions, whereas CO_2 is rapidly evolved from pyruvate.

Stouthamer[350] carried out a series of experiments on the enzymes involved in glucose oxidation; the results are shown in Table 23.

Stouthamer also showed that resting cells of strains of the *Oxydans* group oxidized pyruvate, succinate, fumarate, DL-malate, and α-ketoglutarate. Citrate and DL-isocitrate were oxidized by the cells when they were incubated in phosphate buffer at pH 6.0. The presence of α-ketoglutarate dehydrogenase, fumarase, and malate dehydrogenase was also confirmed.

Oxidation of pyruvate, succinate, fumarate, and malate by *A. ascendens* and *A. rancens* was observed by Kondo and Ameyama[509] and by Tanaka.[503] The presence of isocitrate dehydrogenase, as well as aconitase, α-ketoglutarate dehydrogenase, and fumarase was demonstrated in *A. pasteurianus* by King *et al.*[344] TCA cycle activity is evident in the oxidative metabolism of this bacterial group.

Mannitol is oxidized to fructose by *A. rancens* var. *turbidans* (isol. 26).

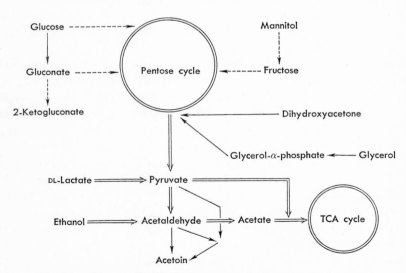

Fig. 36. A map of carbohydrate metabolism in *Acetobacter* belonging to Frateur's " *Oxydans* " group
(after Stouthamer).

Glycerol is not directly oxidized to dihydroxyacetone, but is slowly oxidized by cell-free extracts of *A. rancens* (isol. 10) in the presence of NAD, NADP, phenazine methosulfate, and ATP, suggesting that the oxidation occurs *via* glycerol-α-phosphate (Stouthamer[350]). Acetoin is produced by most strains. The pathways of carbohydrate metabolism in the *Oxydans* group are shown in Figure 36.

Frateur's *Mesoxydans* Group

This group includes *A. aceti,* *A. mesoxydans*, and *A. xylinum*. *A. xylinum*, however, differs from other acetobacters in certain characteristics, and details of its carbohydrate metabolism were discussed separately in the preceding section.

None of these strains produce CO_2 from glucose and fructose under anaerobic conditions. Pyruvate is not decarboxylated in an N_2 atmosphere. However, isotopic experiments by Bourne and Weigel[442] indicated the presence of the Embden-Meyerhof-Parnas pathway in the glucose metabolism of *A. acetigenus*, which is thought to be a variant of *A. aceti*. According to De Ley and Schell,[241] *A. aceti* Ch 31 possesses hexokinase and G-6-P isomerase, suggesting that a glycolytic breakdown might take place.

Stouthamer[491] divided this group into the two subgroups described below.

a) Strains which do not show irisation on yeast extract-glucose-$CaCO_3$ agar slants: these strains are able to oxidize glucose, partly to ketogluconate, and partly to enter into the pentose cycle by the hexokinase. Glucose is rapidly oxidized to gluconate which is slowly converted to 2-ketogluconate, but not to 5-ketogluconate. *A. aceti* Delft, *A. aceti* Ch 31, *A. mesoxydans* Delft and *A. xylinum* (Davis) belong to this subgroup.

A. aceti Ch 31, however, does not oxidize glucose to ketogluconates. The two strains of *Acetobacter* isolated from damaged apples and identified as *A. aceti* by Carr *et al.*[310] could produce 2,5-diketogluconate and 2-ketogluconate from glucose and also gave 2,5-D-threodiketohexose from fructose. Assignment of these two strains of *Acetobacter* to this subgroup is therefore questionable. Rao,[343] and Rao and Gunsalus[97] demonstrated the presence of certain enzymes of the TCA cycle in *A. aceti*, namely α-ketoglutarate dehydrogenase, condensing enzyme (acetyl CoA→oxaloacetate-transacetylase), aconitase, isocitrate and succinate dehydrogenase, fumarase, and malate dehydrogenase. Stouthamer[491] observed the oxidation of pyruvate, succinate, fumarate, DL-malate, and α-ketoglutarate by resting cells of *A. aceti* strains. Thus, the evidence is strong that the

TCA cycle operates in these strains.

It should be mentioned also that acetate-grown cells of *A. aceti* (F 4) can decompose isocitrate into glyoxylate and succinate. Isocitrate lyase and malate synthase have been demonstrated in cells of this organism by Smith and Gunsalus.[484]

De Ley and Schell[241] carried out an extensive study on the oxidation of various substrates by resting cells or cell-free extracts of *A. aceti* Ch 31, previously grown in a defined medium with DL-lactate as the carbon source. Gluconate and 2- and 5-ketogluconate were not oxidized by resting cells. Acetate, pyruvate, D- and L-lactate, succinate, fumarate, and citrate were oxidized. Cell-free extracts showed the presence of kinases for glucose, fructose, and glycerol. Mannitol, galactose, and pentoses were not phosphorylated. Cell-free extracts oxidized glucose-6-phosphate with phenazine methosulfate as a carrier. NADP and NAD increased the oxidation rate. The extracts also contained 6-phosphogluconate dehydrogenase. Fructose-6-phosphate was also oxidized by NADP, presumably after a phosphohexoisomerase action. Phosphofructokinase was not demonstrated. Fructose itself, or that formed from mannitol by mannitol dehydrogenase, was most probably metabolized by way of fructokinase and the pentose cycle. Since intact resting cells, as opposed to cell-free extracts, could not oxidize fructose, it was assumed that the first step in the metabolism of fructose was probably an inducible permease action.

The presence of the pentose cycle in organisms of this subgroup was demonstrated with a strain of *A. xylinum* by Gromet et al.[420]

A. aceti ATCC 8303, unlike *A. xylinum*, did not incorporate ammonia in the presence of glucose, α-ketoglutarate, or pyruvate, according to Tomlinson and Campbell.[508] The fact that ammonia is not utilized even in the presence of α-ketoglutarate agrees with the fact that this organism requires a complex nitrogen source for growth.

b) Strains showing irisation on yeast extract-glucose-CaCO$_3$ agar slants: only *A. mesoxydans* var. *saccharovorans* belongs to this subgroup. This strain oxidizes glucose, gluconate, 2-ketogluconate, and 2,5-diketogluconate with O$_2$ uptakes of about 70% of the amount needed for complete oxidation (Stouthamer[491]). 5-Ketogluconate was oxidized only when another oxidizable substance was added, as was shown in *G. suboxydans* by Asai and Murooka.[470]

Various enzymes related to glucose metabolism are present in cell-free extracts of this organism, according to Stouthamer.[350] Dehydrogenases found in the extracts are shown in Table 24.

Table 24.   Dehydrogenases in cell-free extracts of *A. mesoxydans* var. *saccharovorans* (centrifuged at 28,000×g, after Stouthamer).

Substrate	NAD-linked	NADP-linked
Gluconate	+(Weak)	+
2-Ketogluconate	+	−
5-Ketogluconate	+	+
2, 5-Diketogluconate	+	Not observed
Glucose-6-phosphate	+(Weak)	++
6-Phosphogluconate	Not observed	+
Glyceraldehyde-3-phosphate	+	Not observed
Sorbitol	−	+
Mannitol	+	+
DL-Isocitrate	++	+
Ethanol	++	+(Weak)

Pyridine nucleotide-linked dehydrogenases for DL-lactate, DL-malate, succinate, glucose, erythritol, and glycerol were not detected in cell-free extracts.   With resting cells, however, DL-malate, succinate, fumarate, oxaloacetate, and α-ketoglutarate were easily oxidized.   Citrate was found to reduce NAD with cell-free extracts.

Ribose-5-phosphate, fructose-1,6-diphosphate, and fructose-6-phosphate were oxidized in the presence of NADP and phenazine methosulfate by cell-free extracts in which were also detected kinases for glucose, fructose, gluconate, glycerol, and dihydroxyacetone (Stouthamer[350]). Epimerase (ketopentose-phosphate isomerase), pentose-phosphate isomerase, transketolase and transaldolase, as well as the other enzymes of the pentose cycle, were also identified.   Ketogluconate reductases, responsible for the conversion of ketogluconates to gluconate and for feeding them into the pentose cycle, were present.   Cell-free extracts were shown to be capable of reducing 2,5-diketogluconate in the presence of NADH$_2$.   Particulate fractions of the cells were able to oxidize glucose, D-galactose, D-mannose, D-xylose, and L-arabinose, but were not able to oxidize gluconate and ketogluconates.   DL-Lactate, pyruvate, DL-malate, glycerol, erythritol, sorbitol, and adonitol were oxidized.   These facts strongly support the presence and activity of the pentose cycle as well as the TCA cycle in this organism.

It is interesting to note that this organism is totally unable to oxidize acetate (Stouthamer[491]), whereas it completely oxidizes DL-lactate and

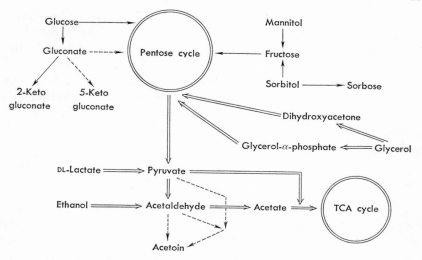

Fig. 37.   A map of carbohydrate metabolism in *A. aceti*
and *A. mesoxydans* Delft
(after Stouthamer).

⇒, Always present; →, Generally present; ⇢, Rarely present.

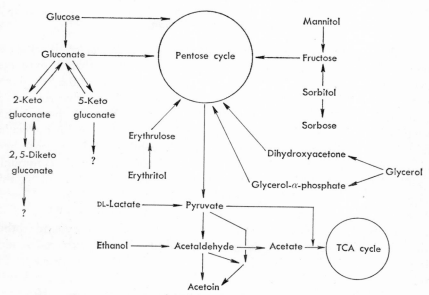

Fig. 38.   A map of carbohydrate metabolism in *A. mesoxydans*
var. *saccharovorans*
(after Stouthamer).

pyruvate. Indeed, the biochemical activities of *A. mesoxydans* var. *saccharovorans* are quite similar in many respects to *G. suboxydans* or other *Gluconobacter* strains, especially in the metabolic pattern of carbohydrates, except for the presence of the TCA cycle. In these respects it could reasonably be classified as *G. suboxydans*, or at least assigned to the genus *Gluconobacter*. The flagellation feature is an important criterion in deciding the generic assignment of acetic acid bacteria and therefore this strain should be reexamined to determine whether it is polarly or peritrichously flagellated when motile.

Figures 37 and 38 show the pathways through which carbohydrates are metabolized by Frateur's *Mesoxydans* group.

Polyalcohols are oxidized by most strains but oxygen consumption is generally lower than in *G. suboxydans* or *G. melanogenus*. Sorbitol is not oxidized by *A. aceti* Ch 31 or *A. mesoxydans* Delft. In general, the kinds of polyalcohols that can be fed into the pentose cycle are fewer than for the *G. suboxydans* group. Acetoin is produced by *A. mesoxydans* Delft (Stouthamer[350]).

The presence of a glyoxylate-bypath in *A. aceti* (F4) was suggested by Smith and Gunsalus.[484]   (See also glyoxylate-bypath, p. 296.)

Acetobacter peroxydans

Acetobacter peroxydans, unlike most of the *Acetobacter*, is catalase-negative and possesses peroxidase activity. It oxidizes ethanol, pyruvate, acetate and lactate, but does not oxidize glucose, gluconate, the ketogluconates, mannitol, fructose, mannose, galactose, xylose, glycerol, or *i*-erythritol. Earlier investigations by Wieland and Pistor[98] showed that acetaldehyde is the primary product of ethanol oxidation with molecular oxygen and hydrogen peroxide as hydrogen acceptors. These researchers also reported that the carbohydrates and related compounds commonly oxidized by most strains of *Acetobacter* were not oxidized at all by this organism. It can oxidize molecular hydrogen, according to Visser't Hooft,[115] Wieland and Pistor,[98] Atkinson,[100] and Tanenbaum.[102]

Frateur *et al.*[278] demonstrated that no reducing compounds other than the initial substrate are detectable in cultures in yeast extract-containing sugar media, indicating that the organism lacks the enzyme systems related to the oxidation of glucose and other carbohydrates. De Ley[457] detected neither glucose oxidase nor glucose dehydrogenase. However, several enzymes participating in glucose metabolism were demonstrated in cell-free extracts, glucose-6-phosphate dehydrogenase by Atkinson[100] and

De Ley,[510] and NAD-linked 6-phosphogluconate dehydrogenase and NADP-linked ribose-5-phosphate dehydrogenase by De Ley.[510] The latter suggested that the HMP cycle has only an anabolic function in the synthesis of hexoses and pentoses required for cell wall and RNA formation, *etc.*, since the organism is unable to metabolize any sugar. Investigations on this particular organism have centered largely on the oxidation systems of C_2 and C_3 compounds and related enzymes involving cytochromes and peroxidase.

Growth on Sugars and Related Substrates

De Ley[510] conducted growth tests on a number of carbon substrates with four strains of *A. peroxydans* (Strains 1,3, ATCC 838 and NCIB 8618) and found that they grew to an appreciable extent only on lactate and ethanol in synthetic medium, on 1% yeast extract or on beer. They were unable to grow on glucose, galactose, L-arabinose, D-arabinose, D-xylose, rhamnose, Ca gluconate, fructose, lactose, maltose, sucrose, ribose, raffinose, or starch, or on polyalcohols. Strain 1 showed good growth on Ca acetate. The fact that they could neither grow on, consume or oxidize glucose, a common carbon source for most other bacteria, is noteworthy.

Oxidation of Ethanol and Other Aliphatic Monoalcohols

Tanenbaum[101] reported the presence of NADP-linked alcohol dehydrogenase and NADP-linked aldehyde dehydrogenase in cell-free extracts of *A. peroxydans* NCIB 8618. Atkinson[99] found NAD-linked alcohol dehydrogenase in cell-free extracts of *A. peroxydans* ATCC 838. A washed cell or dried cell preparation of 24-hour cells of *A. peroxydans* 8618 showed an R. Q. value of 0.5, although the organism oxidizes ethanol to CO_2 and water in growing cultures. The amount of CO_2 evolved to O_2 consumed was in accord with the following equation:

$$CH_3 \cdot CH_2OH + 2O_2 \longrightarrow CO_2 + (CH_2O) + 2H_2O$$

The conversion of ethanol to acetaldehyde was not affected when washed cells in pathalate buffer were exposed to 1×10^{-3}M arsenate, whereas acetaldehyde oxidation was completely inhibited. Antimycin A did not influence the oxidation of ethanol or of acetaldehyde by intact cells or cell-free extracts. The participation of phosphate in aldehyde oxidation or high energy phosphate bond formation during oxidative transformations of acetaldehyde and ethanol was ruled out by the experiments of Tanenbaum.[101]

Resting cells of strains 1, ATCC 838, and NCIB 8618, after growth on media containing either ethanol or lactate, oxidized ethanol with an uptake of 2.3–2.6 moles O_2 per mole substrate (De Ley[510]). For complete oxidation 3 moles O_2 was required, showing that 13–23% of the ethanol was assimilated. These strains readily oxidized n-propanol, n-butanol, isobutanol, and n-amylalcohol with an uptake of nearly 1 mole O_2. Isopropanol and sec-butanol were oxidized to the corresponding ketones. Methanol was oxidized very slowly.

Oxidation of Acetate

Acetate is attacked by both fresh and aged cells, but experiments by Atkinson[99] showed that ethanol-grown cells do not oxidize acetate while they oxidize ethanol, pyruvate, succinate, malate, fumarate and oxaloacetate to CO_2 and water. Atkinson therefore suggested that free acetate is not the intermediate of ethanol oxidation. Cell-free extracts were devoid of enzyme activity to acetate, in contrast to sonic homogenates which retained this oxidative ability. The R.Q. value for acetate oxidation using 24-hour cells consistently approached 1, and indicated the following conversion:

$$CH_3 \cdot COOH + O_2 \longrightarrow CO_2 + (CH_2O) + H_2O$$

Oxidation was almost completely inhibited by $1 \times 10^{-4}M$ 2,4-dinitrophenol as well as by a similar concentration of monoiodoacetate in experiments with whole cells. De Ley's experiments[510] showed that resting cells of strains 1,3, ATCC 838, and NCIB 8618 grown on lactate medium oxidized acetate with an uptake of 1.3 moles of O_2 per mole substrate. However, crude cell-free extracts of strain NCIB 8618 grown on DL-lactate medium could not oxidize Na acetate, according to later experiments by De Ley and Schell.[337] NADP and NAD, added separately or together, did not affect any of the oxidation rates or the overall O_2 uptake.

Further investigations are needed to clarify the mechanism of acetate oxidation in this organism. Formate itself, which is considered the intermediate in the oxidation of ethanol or acetate with washed cells or sonically disrupted cell preparations, was quantitatively oxidized to CO_2 and water by intact fresh cells, according to Tanenbaum.[101]

Oxidation of Pyruvate and Lactate

Pyruvate is rapidly oxidatively decarboxylated by cell suspensions. The reaction links only to NADP reduction. Lactate behaves in the same manner. De Ley and Vervolet,[339] using *A. peroxydans* NCIB 8618, showed that there was no active participation of peroxidase in the oxidation of D-lactate. A comprehensive study by De Ley and Schell[337] has been described earlier (p. 206).

Oxidation of TCA Cycle Intermediates

Tanenbaum's experiments[101] with cell suspensions of *A. peroxydans* NCIB 8618 gave evidence that succinate, malate and fumarate are oxidatively decarboxylated without a lag period, while citrate, isocitrate, and α-ketoglutarate are not oxidized by intact or sonically disrupted cells which oxidize acetate, or by cell-free extracts. Attempts to add a sparker or cofactors such as CoA, NADP, NAD, lipoic acid or Mg^{++} were unsuccessful. Despite the presence of pyridinenucleotide-linked oxaloacetate, malate, lactate, ethanol and acetaldehyde dehydrogenases in cell-free extracts, attempts to reduce the coenzyme in the presence of isocitrate failed. It was concluded that the usual TCA cycle does not operate in this organism. Atkinson[99,100] detected aconitase, NADP-linked isocitrate dehydrogenase, succinate dehydrogenase, fumarase, malate dehydrogenase, and α-ketoglutarate dehydrogenase in *A. peroxydans* ATCC 838. De Ley[510] observed that resting cells of various strains of *A. peroxydans* could oxidize pyruvate, succinate, malate and acetate. Most strains oxidized fumarate but none could oxidize citrate.

These results indicate that *A. peroxydans* can oxidize C_2 and C_3 compounds, with the formation of CO_2 and water from lactate, pyruvate and ethanol, *via* acetaldehyde, acetate, and possibly the TCA cycle or a modification of it. The decisive answer to whether the usual TCA cycle operates in this organism or not must await results of suitable isotopic experiments.

Other Enzymes Participating in Oxidative Metabolism

$NADPH_2$-oxidase, $NADPH_2$-linked cytochrome *c* reductase, diaphorase, peroxidase, and the H_2O_2 peroxidations of reduced NADP and of reduced cytochrome *c* were demonstrated in cell-free extracts of *A. peroxy-*

dans NCIB 8618 by Tanenbaum.[101] Attempts to find cytochrome *c* oxidase in cell-free extracts, however, were unsuccessful. Chin[338] reported preliminary evidence that cytochrome a_2 is the autoxidizable component in *A. peroxydans*. Experiments with cell-free extracts showed that 2,6-dichlorophenolindophenol is rapidly reduced by the ethanol-NADPH$_2$ system. Reduction of methylene blue was rather sluggish, and ferricyanide acted as the hydrogen acceptor for NADPH$_2$. Indophenol reductase was found in aged cell extracts long after cytochrome *c* reductase had disappeared, suggesting that an independent diaphorase was present rather than cytochrome *c* reductase manifesting diaphorase activity.

Cytochromes. Chin[338] found cytochrome a_4 with an absorption band at 612 mμ. In aged cells this cytochrome was transformed to cytochrome a_3 (635 mμ), which was later displaced by the cytochrome *b* group (554 mμ) and cytochrome *c* (550 mμ). Atkinson[100] demonstrated the presence of cytochrome *b* in *A. peroxydans* ATCC 838. Tanenbaum[101] observed a single visible absorption band at 553±5 mμ when ethanol was added to cell suspensions of *A. peroxydans* NCIB 8618. Castor and Chance[488] detected cytochrome a_2. De Ley and Schell[337] reported that thick suspensions of resting cells of *A. peroxydans* NCIB 8618 showed three cytochrome bands: the darkest band had its centre at 555 mμ (extending from 546 to 565 mμ) and was probably identical to that described by Chin[338] (maximum at 554 mμ), Tanenbaum[101] (553±5 mμ), and Wieland and Pistor[98] (554 mμ). An α-band of cytochrome a_1 at 599 mμ (extending from 587 to 613 mμ), and a band at 525 mμ (extending from 518 to 536 mμ) were also observed though the latter was very weak. These bands disappeared when the suspension was oxidized by air. Addition of ethanol, Na D-lactate or pyruvate to an aerated suspension resulted in a simultaneous reduction of the three cytochrome bands.

Concerning the cytochromes' function in the oxidative metabolism of *A. peroxydans*, Tanenbaum[101] suggested that a cytochrome type of respiration was more probable than a direct flavoprotein to oxygen pathway, because of the cyanide and azide inhibitions and the presence of NADP-linked dehydrogenases and NADPH$_2$-cytochrome *c* reductase in cell-free extracts.

Peroxidases. A particular characteristic of *A. peroxydans* is its peroxidase activity. Wieland and Pistor[98] demonstrated the presence of peroxidases for D- and L-lactates, ethanol, formate, molecular hydrogen, and *p*-phenylenediamine in intact cells, and Tanenbaum[101] observed peroxidase activity for NADPH$_2$, cytochrome *c* and pyrogallol in cell-free extracts of this organism. De Ley and Vervolet[339] carried out an extensive study on the peroxidase activity for several physiological substrates.   Intact resting cells of *A. peroxydans* NCIB 8618 showed a marked peroxidase activity for ethanol, *n*-propanol, *n*-butanol, D-lactate, acetaldehyde, and aromatic diamines (*o*-, *p*- and dimethyl-*p*-phenylenediamine).   *n*-Amyl alcohol, isopropanol, *sec*-butanol, isobutanol, L-lactate, orcinol and cytochrome *c* were weak hydrogen donors.   Vitamin C, malate, methanol, *tert*-butanol, citrate, succinate, fumarate, formate, glucose, tyrosine, tryptophan, glucose-6-phosphate, fructose-6-phosphate and 3-phosphoglycerate were oxidized, very weakly or not at all.   The optimum pH for ethanol peroxidase was 5.5–6, the optimum temperature 30°C.   Peroxidase activity was destroyed by H_2O_2 and the complete stoichiometric conversions were never obtained

There was no evidence for the physiological function of peroxidase in the respiration of D-lactate, and only mechanism A in the following reactions was supported as a metabolic pathway for D-lactate oxidation.

Mechanism A

$$D\text{-Lactate} + 0.5 \ O_2 \longrightarrow Pyruvate + H_2O$$

$$Pyruvate \longrightarrow CH_3 \cdot CHO + CO_2$$

$$CH_3 \cdot CHO + 0.5 \ O_2 \longrightarrow Acetate$$

Sum : $D\text{-Lactate} + O_2 \longrightarrow Acetate + CO_2 + H_2O$

Mechanism B
(Eliminated)

$$D\text{-Lactate} + O_2 \xrightarrow{\text{(Flavin)}} Pyruvate + H_2O_2$$

$$Pyruvate + H_2O_2 \xrightarrow{\text{Peroxidase}} Acetate + CO_2 + H_2O$$

Sum : $D\text{-Lactate} + O_2 \longrightarrow Acetate + CO_2 + H_2O$

Mechanism C
(Eliminated)

$$D\text{-Lactate} + O_2 \xrightarrow{\text{(Flavin)}} Pyruvate + H_2O_2$$

$$D\text{-Lactate} + H_2O_2 \xrightarrow{\text{Peroxidase}} Pyruvate + 2H_2O$$

Sum : $2 \ D\text{-Lactate} + O_2 \longrightarrow 2 \ Pyruvate + 2 \ H_2O$

According to De Ley and Vervolet, at least four different peroxidases exist for *p*-phenylenediamine, lactate, acetaldehyde and primary alcohols. *p*-Phenylenediamine peroxidase was inhibited by 2-*n*-nonyl-4-hydroxy-quinoline N-oxide, sulfide, hydroxylamine, hydrazine, cyanide, and semicarbazide.   The participation of SH groups and metal ions was not apparent in enzyme activity.   D-Lactate peroxidase was inhibited by most of these inhibitors and azide.   CO had no effect on either peroxidase. Ethanol peroxidase was rapidly inactivated by ultrasonic treatment. Acetaldehyde peroxidase was also rapidly inactivated, but at a different rate, pointing to a separate enzyme. D-Lactate and p-phenylenediamine peroxidases were also inactivated but at the slowest rate. Both D-lactate and p-phenylenediamine peroxidases were shown to be located chiefly on the solid outer cell envelope, most probably on the cytoplasmic membrane.

p-Phenylenediamine peroxidase was detected in other acetic acid bacteria, including gluconobacters.

Hydrogenase. One of the more unusual properties of *A. peroxydans* is its ability to oxidize molecular hydrogen. It might be possible, on this basis, to class it in the *Hydrogenomonas*, with the additional property of oxidizing ethanol into acetic acid. De Ley,[510] however, showed that none of the strains of *A. peroxydans* was able to grow like the hydrogen bacteria, and favored keeping the organism in the genus *Acetobacter*.

According to Tanenbaum,[102] the hydrogenase system of *A. peroxydans* differs from that of other organisms. Experiments showed that it reduced hydrogen peroxide, dyes and oxygen with molecular hydrogen, and that it could reduce methylene blue but not methyl and benzylviologene and monomethyl hydrogen peroxide. No reduction was obtained for several coenzymes (NAD, NADP, FAD or FMN) with cell-free extracts and hydrogen, even when considerable amounts of methylene blue were present, though an increase in optical density at 340 mμ was observed when hydrogen was bubbled through a cuvette containing NAD or NADP and cell-free extracts. Reduction of cytochrome c by hydrogen was observed with cell-free extracts.

Cell suspensions which actively reduced methylene blue under hydrogen failed to catalyze the exchange reaction at 30°C. Some exchange activity was observed, however, at 37°C. Experiments also indicated that cell suspensions did not appreciably catalyze the conversion of para to normal hydrogen at 30°C.

A. peroxydans exhibits certain peculiarities in its metabolism of carbon substrates. A simplified diagram of this metabolism is shown in Figure 39.

"Intermediate" Strain IAM 1834* (Gluconobacter liquefaciens)

This strain was originally called *Gluconoacetobacter liquefaciens*, and was placed in the genus *Gluconobacter* by Asai.[246] The subgeneric name of *Gluconoacetobacter*, however, was withdrawn in 1958[135] and was re-

---

* Recently the strain was assigned by Asai to the genus *Acetobacter* on the basis of its peritrichous flagellation and acetate oxidizability, and was named *Acetobacter intermedius* nov.sp.

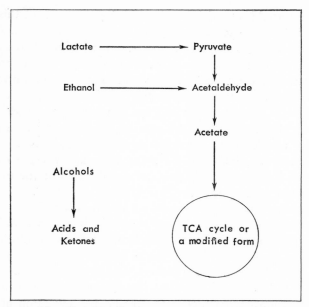

Fig. 39.   Metabolic map of *A. peroxydans*.

placed by *Gluconobacter liquefaciens*, which was used as the species name until 1964.[511]   When it was first isolated in 1935, the strain showed no motility, did not oxidize acetate, and produced a brown pigment from glucose in the presence of $CaCO_3$.   These properties are quite similar to those of *G. melanogenus*.   Long preservation in the laboratory, however, led to the appearance of an acetate-oxidizing and motile mutant with peritrichous flagella.   The other properties, except for gelatin liquefaction, remained the same as in the original isolate.

This strain, at present, is temporarily termed an " Intermediate " strain, assignable neither to the *Gluconobacter* nor to the *Acetobacter* on the basis of the taxonomic criteria described in Chapter 11 of Part I.

The oxidative metabolism of carbohydrates by this organism has been studied exclusively by Asai *et al.*, Stouthamer, and De Ley.

Non-Phosphorylative Oxidation

Oxidation of glucose, gluconate, and 2-ketogluconate.   This strain, like *G. melanogenus*, oxidizes glucose to gluconate, 2-ketogluconate and 2,5-diketogluconate.   5-Ketogluconate is also formed from glucose and gluconate.

According to Aida et al.,[280] an intact cell suspension of this organism showed an oxygen uptake of 1.5, 1.0 and 0.5 moles per mole of glucose, gluconate, and 2-ketogluconate, respectively, without liberation of CO_2 in the presence of 2,4-dinitrophenol; in the absence of 2,4-dinitrophenol these substrates were oxidized rapidly with a noticeable output of CO_2. When cell-free extracts were used instead of intact cells, the evolution of CO_2 did not occur and the oxygen uptake of each substrate was essentially the same as in the case of intact cells in the presence of 2,4-dinitrophenol. These results provided strong support for the theory that oxidation of glucose proceeds via gluconate and 2-ketogluconate without evolution of CO_2. NaF $(5 \times 10^{-2}M)$ did not affect either the rate or the total amount of oxygen uptake with glucose.

Stouthamer,[512] using glucose-grown resting cells, found that the O_2 uptake per mole of glucose, gluconate, 2-ketogluconate and 2,5–diketo-gluconate fell 0.5 mole for each substrate, while the evolution of CO_2 remained about the same. With 5-ketogluconate there was a much higher O_2 uptake and evolution of CO_2 than with the other substances. A similar phenomenon was observed by Murooka et al.[513] In this instance, 5-ketogluconate again gave a higher O_2 uptake and evolution of CO_2 than 2-ketogluconate, but there was a sharp decline of O_2 uptake after the uptake of 0.5 mole of O_2 per mole of substrate with 2-ketogluconate. This suggested that 5-ketogluconate is metabolized by a different pathway than 2-ketogluconate (Stouthamer,[512] Murooka et al.[513]). According to De Ley and Stouthamer,[83] the enzymes for the oxidation of glucose, gluconate and 2-ketogluconate to 2,5-diketogluconate are present in the particles (protoplasmic membranes) of the cells. 5-Ketogluconate, however, was not oxidized by the particles, and no coenzyme-linked glucose dehydro-genase could be detected in the particle-free supernatant (in contrast to G. suboxydans and G. melanogenus). 5-Ketogluconate was rapidly reduced by $NADPH_2$, indicating the presence of 5-ketogluconate reductase in the soluble fraction. Reoxidation of $NADPH_2$ by 2-ketogluconate also occurred, indicating the presence of 2-ketogluconate reductase; it is not clear as yet whether 2,5-diketogluconate is reduced by a specific enzyme (2,5-diketogluconate reductase) or by one of the ketogluconate reductases.

The enzymes that are known to catalyze glucose, gluconate and ketogluconate oxidation are listed in Table 25.

Oxidation of 2,5-diketogluconate. Glucose and gluconate are oxidized beyond the 2,5-diketogluconate stage by suspensions of cell particles.

Stouthamer[514] obtained a compound similar to rubiginol from such

Table 25. The enzymes responsible for the oxidation of glucose, gluconate and ketogluconates by "Intermediate" strain IAM 1834 (*G. liquefaciens*).

Enzyme	Soluble fraction	Particulate fraction
Glucose oxidase		+
Gluconate oxidase		+
2-Ketogluconate oxidase		+
5-Ketogluconate reductase	+	
2-Ketogluconate reductase	+	
2, 5-Diketogluconate reductase	+	

a reaction mixture.   Rubiginol and the other two γ-pyrone coumpounds, *i.e.* rubiginic acid and comenic acid, were first isolated and identified by Aida from a culture of the organism growing on glucose.[240, 285, 286]   Aida *et al.*[280] confirmed the formation of these three γ-pyrone compounds in reaction mixtures of 2,5-diketogluconate and dried cells or cell-free extracts. Formation of rubiginol from rubiginic acid was also demonstrated.   On the basis of these results and experiments using gluconate-1-^{14}C as substrate (Aida and Asai, unpublished) to check incorporation of radioactive carbon in the products, a pathway for the formation of these γ-pyrone compounds was presented as follows (Fig. 40).

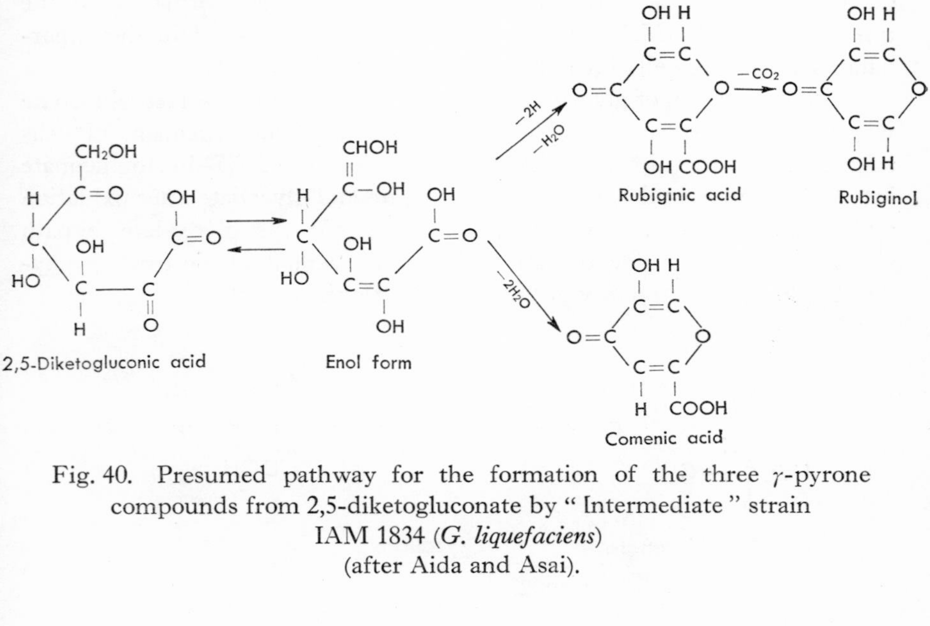

Fig. 40. Presumed pathway for the formation of the three γ-pyrone compounds from 2,5-diketogluconate by " Intermediate " strain
IAM 1834 (*G. liquefaciens*)
(after Aida and Asai).

Stouthamer[512] suggested that 3,4,5-trihydroxy-γ-pyran-6-carboxylic acid might be the intermediate in the formation of rubiginic acid. Datta et al.[493] demonstrated the presence of 2,5-diketogluconate decarboxylase in *G. melanogenus*. Stouthamer,[512] however, observed a slow decarboxylation of 2,5-diketogluconate under anaerobic conditions with soluble frac-

3, 4, 5-Trihydroxy-γ-pyran-6-carboxylic acid

tions of *G. liquefaciens*, which disagreed with the direct participation of 2,5-diketogluconate decarboxylase. Because a γ-pyrone compound ($R_f =$ 0.50) appeared in the reaction mixture, the decarboxylation was thought to take place during the later stage of multistepped conversion of 2,5-diketogluconate to γ-pyrone compounds. Aida et al.[280] in their early experiments supported the view that the decarboxylation occurs during the conversion of rubiginic acid to rubiginol. Stouthamer[350] noted the possibility of an arsenite-sensitive pathway for the oxidation of 2,5-diketogluconate. This pathway was thought to lead to the formation of the unidentified acid which was found in culture filtrates and in the supernatants of resting cells incubated with 2,5-diketogluconate.

The formation of glycolic and tartronic acids from 2,5-diketogluconate was confirmed by Aida et al.[280] with intact cells of the organism. It was suggested that the direct breakage of the molecules of 2,5-diketogluconate into C_3-C_3 compounds gave rise to tartronate and glycollate after oxidation and decarboxylation. However, the route via the glyoxylate bypath seems more likely. The formation of tartronic semialdehyde from glyoxylate by cell-free extracts was verified by Stouthamer et al.[514]

2, 5-Diketogluconate   Glycollate   Tartronate

Breakdown of 5-ketogluconate. 5-Ketogluconate is not oxidized by the particles but is reduced to gluconate by the soluble fraction of the cells. Accordingly, the formation of rubiginol, rubiginic and comenic acid from 5-ketogluconate demonstrated by Murooka *et al.*[513] is thought to be due to the degradation of 2,5-diketogluconate formed *via* the following paths:

5-Ketogluconate → Gluconate → 2-Ketogluconate → 2,5-Diketogluconate.

In fact, Murooka[467] found $NADPH_2$-linked 5-ketogluconate reductase in cell-free extracts of the organism.

The rapid oxidation by intact cells of 5-ketogluconate with a high degree of CO_2 evolution can be explained in the following terms: 5-ketogluconate enters the cytoplasm to be reduced to gluconate by 5-ketogluconate reductase; the latter is phosphorylated by gluconokinase, enters the pentose cycle, and is completely metabolized *via* a TCA cycle involving the glyoxylate bypath. Part of the gluconate is oxidized to 2,5-diketogluconate, from which γ-pyrone compounds can be produced without phosphorylative reaction.

According to Murooka *et al.*,[515] the oxygen uptake of intact cells with 5-ketogluconate was inhibited by arsenite, while the amount of keto acids (pyruvic and α-ketoglutaric acids) accumulated was increased by the addition of 10^{-3}M arsenite.

Location of oxidizing enzymes. De Ley and Dochy[375] studied the sites of various oxidizing enzymes in the cells of this organism. The cell débris and small particles oxidized glucose, gluconate and 2-ketogluconate to 2,5-diketogluconate and beyond. D-Lactate and ethanol were oxidized to the acetate stage. Mannose, xylose, and L-arabinose were oxidized to the corresponding acids. Sorbitol, glycerol, and *meso*-erythrytol were oxidized to the corresponding ketogenic compounds. Acetate, 5-ketogluconate, fructose, and mannitol were not attacked, whereas they were oxidized by intact cells.

The protoplasts oxidized most of the substrates at a rate about 60–80% that of intact cells. All the substrates which were oxidized by cell débris and small particles were oxidized to the same extent by " ghosts " and protoplast débris, with the exception of glycerol and *meso*-erythritol, the oxidation of which was thought to be inactivated (at 37°C) by heat-labile enzymes.

De Ley and Dochy suggested that the oxidase-bearing particles might be derived from the cell envelope, probably the cytoplasmic membrane. These particles adhere tightly to the cytoplasmic membrane and cannot be separated by ultrasonic treatment, and it was emphasized that all the particle-linked enzymes in acetic acid bacteria are most probably " ghost "-linked.

Phosphorylative Breakdown

The Embden-Meyerhof-Parnas pathway.   Cell-free extracts of the organism contain kinases for glucose, fructose, glycerol, and dihydroxyacetone, according to Stouthamer.[350] The presence of hexosephosphate isomerase and aldolase was demonstrated.   NAD-linked glyceraldehyde-3-phosphate dehydrogenase and pyruvate decarboxylase were also found in cell-free extracts.   Despite the presence of the enzymes necessary for the EMP pathway, the organism cannot decompose glucose under anaerobic conditions and neither CO_2 evolution from glucose and fructose, nor active decarboxylation of pyruvate in an N_2 atmosphere occurs.   This strongly suggests that the glycolysis pathway is absent.

The Entner-Doudoroff pathway.   Attempts to detect a dehydration of 6-phosphogluconate to 2-keto-3-deoxy-6-phosphogluconate, a key reaction in this pathway, were unsuccessful (Stouthamer[512]). This indicates that the Entner-Doudoroff pathway is not involved in the breakdown of glucose in this organism.   However, an entire system corresponding to the Entner-Doudoroff pathway has been detected by Stouthamer[512] in galactonate breakdown by galactose-grown cells of this organism; this will be discussed later.

The pentose cycle.   The enzymes related to this cycle were studied in detail by Stouthamer.[350] The presence of NADP-linked glucose-6-phosphate and 6-phosphogluconate dehydrogenase was demonstrated in the soluble fraction.   Hexosephosphate isomerase was indicated by rapid $NADPH_2$ production by the soluble fraction with fructose-6-phosphate. Transketolase, transaldolase, pentosephosphate isomerase and phosphoketopento-3-epimerase were shown to be present by the absorption spectra of reaction mixtures of ribose-5-phosphate with soluble fractions, to which the cysteine-H_2SO_4 reaction was applied after various incubation times.

The appearance of a peak with an absorption maximum at 505 mμ indicated the formation of sedoheptulose phosphate, and a peak at 410 mμ the formation of hexose phosphate.   Aldolase and fructose-1,6-diphosphatase were also detected.   The presence of gluconokinase, NADP-linked glucose-6-phosphate and 6-phosphogluconate dehydrogenase was verified by Murooka[467] in addition to active oxygen uptakes for fructose-6-phosphate, fructose-1,6-diphosphate and ribose-5-phosphate by cell-free extracts.

All the enzymes of the pentose cycle are thus present in this organism, and it metabolizes glucose mainly *via* the pentose cycle.

Oxidative phosphorylation.   Oxidative phosphorylation in cell-free extracts of the organism was investigated by Stouthamer[516] by measuring P/O ratios. The P/O ratios were found to be very low, ranging from 0.09 for $NADH_2$ to 0.40 for glucose-6-phosphate. This indicates the low efficiency of energy transfer in this organism; in growing cultures the molar growth yield is equally low. The low efficiency of energy transfer is partly explained by the presence of only one cytochrome, with absorption maxima at 418, 523, and 553 mμ, thought to be identical to the one found in *G. suboxydans* by Smith.[517] Cytochrome *c* oxidase is absent.

The TCA Cycle

In contrast to *G. melanogenus*, the operation of a TCA cycle involving the glyoxylate bypath is an important characteristic of this organism. Oxidation of acetate, pyruvate, succinate, fumarate, DL-malate, and α-ketoglutarate was observed with resting cells by Stouthamer[491] and this, plus oxidation of oxaloacetate, was confirmed by Murooka *et al.*[515] Murooka[467] also observed the oxidation of citrate, isocitrate, and *cis*-aconitate with cell-free extracts.

The enzymes related to the TCA cycle found in this organism (Stouthamer[350]) are listed below in Table 26.

Table 26.   Enzymes relating to the TCA cycle in *G. liquefaciens* (after Stouthamer).

Enzyme	Soluble fraction	Particulate fraction
Acetyl CoA synthetase	+	
Aconitase	+	
Isocitrate dehydrogenase (NADP-linked)	+	—
α-Ketoglutarate dehydrogenase (NAD-linked)	+	—
Succinate oxidase*	—	+
Fumarase	+	—
Malate oxidase*	—	+
Oxaloacetate decarboxylase	+	—
$NADH_2$ oxidase		+

* Pyridine nucleotide-linked succinate and malate dehydrogenases were not found; particle bound oxidase systems for both substrates were present.

The Glyoxylate bypath.   Isocitrate lyase was demonstrated in a strain of *A. aceti* by Smith and Gunsalus.[484]   An attempt to detect the glyoxylate-bypath in cell-free extracts of acetate-grown cells of *G. liquefaciens* in a complex medium was unsuccessful (Stouthamer[350]).   In later experiments by Stouthamer *et al.*[514] using cell-free extracts of this organism grown in mineral medium with acetate as the sole carbon source, isocitrate lyase, malate synthase, and a number of enzymes participating in the glyoxylate bypath were revealed.   Glyoxylate carboligase was detected.   The reaction proceeded as follows:

$$2C_2H_2O_3 \longrightarrow 1C_3H_4O_4 + 1CO_2$$

The reaction product was identified as tartronic semialdehyde. $NADH_2$ oxidation with cell-free extracts in the presence of enzymatically formed tartronic semialdehyde was positive, indicating the presence of tartronic semialdehyde reductase. Glycerate kinase was also detected. Cell-free extracts of ethyleneglycol-grown cells rapidly oxidized glycollate, while acetate-grown cells exhibited only a very weak oxidation, suggesting that glycollate oxidase is induced during growth on ethyleneglycol.

This organism can grow on ethanol or acetate as the sole carbon source.   Hence, it was inferred that a much higher activity of isocitrate lyase than malate synthase may lead to an accumulation of glyoxylate inside the cell, thus causing the induction of the pathway from glyoxylate to 3-phosphoglycerate, through which acetate can enter into the carbohydrate components of the cell.   An alternative pathway, however, consisting of phosphorylative decarboxylation of oxaloacetate to yield phosphoenol-pyruvate with subsequent synthesis of 3-phosphoglycerate was also presented.

Kasai *et al.*[372] found a glyoxylate oxidizing system in this organism. The enzyme responsible for glyoxylate oxidation, yielding oxalate, was found in the particles of cells grown in lactate medium.   Various strains of *Acetobacter*, such as *A. aceti*, *A. dioxyacetonicus*, *A. rancens*, *A. pasteurianus*, and *A. albuminosus* possess a glyoxylate oxidase system, whereas *Gluconobacter* strains do not.   Thus the glyoxylate formed by isocitrate lyase can yield oxalate without or as well as the reaction producing 3-phosphoglycerate *via* tartronic semialdehyde.   Oxalate is not metabolized further (Kasai *et al.*[141]).

Metabolism of Galactose

Early investigations by Takahashi and Asai[283] showed that growing cells of *G. liquefaciens* accumulated galactonic acid accompanied by a small amount of comenic acid (a γ-pyrone) from galactose. Recently the galactose metabolism of this organism was studied by Stouthamer,[512] who found that galactose and galactonate were only slowly oxidized by resting, glucose-grown cells. A small amount of 2-ketogalactonate was formed but no γ-pyrones were detected. The first step of this oxidation, the conversion of galactose to galactonate, occurred in the particles (protoplasmic membrane). No kinases for galactose and galactonate could be detected in the soluble fraction.

Table 27. Manometric experiments with resting glucose- or galactose-grown cells of *G. liquefaciens* on several substrates (after Stouthamer).

Substrate	Moles O_2/mole substrate		Moles CO_2/mole substrate	
	Glucose-grown cells	Galactose-grown cells	Glucose-grown cells	Galactose-grown cells
Glucose	>2.3	3.7	>1.6	3.4
Gluconate	>1.9	3.4	>1.6	3.8
2-Ketogluconate	>1.4	>0.7	>1.4	
5-Ketogluconate	3.6	3.6	4.1	
2, 5-Diketogluconate	>1.0		>1.4	
Galactose	>1.0	3.4	>0.5	3.4
Galactonate	>0.4	3.0	0	3.2

Galactose-grown cells, however, behaved differently. They oxidized glucose, gluconate, galactose and galactonate with much higher O_2 uptakes and CO_2 evolution than glucose-grown cells (see Table 27). No inflections in the respiration curves for glucose and gluconate took place with these cells and the CO_2 evolution did not lag behind the O_2 uptake. 2-Ketogluconate was oxidized only very slowly and no effects on 5-ketogluconate oxidation were observed. Neither a-ketoacids nor γ-pyrones were detected in the reaction mixture of galactose or galactonate.

The addition of DNP, however, changed the O_2 uptake with galactose by galactose-grown cells. The O_2 uptake dropped to 0.5 mole per mole of the substrate and no CO_2 was evolved. The reaction product was not galactonate, but an a-ketoacid, identified as 2-keto-3-deoxygalactonate. These results suggested that galactonate is converted to 2-keto-3-deoxygalactonate and that a phosphorylation is involved in its further metabolism. Phosphorylation of this compound with ATP by the soluble fraction was in fact shown. Galactonate dehydratase was also confirmed by incuba-

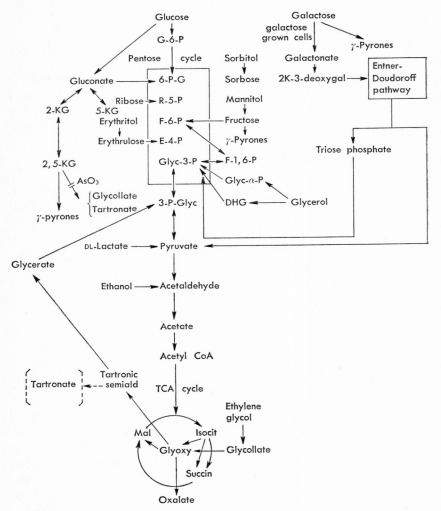

Fig. 41. A suggested map for glucose and galactose breakdown by
"Intermediate" strain IAM 1834
(*Gluconobacter liquefaciens*).

The broken line indicates a reaction which is assumed to take place.

Abbreviations : G-6-P, glucose-6-phosphate ; 6-P-G, 6-phosphogluconate ;
R-5-P, ribose-5-phosphate ; F-6-P, fructose-6-phosphate ; E-4-P, erythrulose-
4-phosphate ; Glyc-3-P, glyceraldehyde-3-phosphate ; 2-KG, 2-ketogluconate ;
5-KG, 5-ketogluconate ; 2.5-KG, 2,5-diketogluconate ; 2K-3-deoxygal, 2-
keto-3-deoxygalactonate ;   F-1, 6-P, fructose-1,6-diphosphate ;   Glyc-α-P,
glycerol-α-phosphate ; DHG, dihydroxyacetone ; 3-P-Glyc, 3-phosphoglyce-
rate.

Table 28. The enzymes related to carbohydrate metabolism
in "Intermediate" strain IAM 1834 (*G. liquefaciens*).

Enzymes	Soluble fraction	Particles
Hexokinase	+	
G-6-P dehydrogenase	+	
Gluconokinase	+	
6-P-G dehydrogenase	+	
Transketolase	+	
Transaldolase	+	
Aldolase	+	
Fructose-1, 6-diphosphatase	+	
Hexosephosphate isomerase	+	
Mannitol dehydrogenase	+	
Fructokinase	+	
Pentosephosphate isomerase	+	
Phosphoketopento-3-epimerase	+	
Glucose oxidase		+
Glucose dehydrogenase	−	
Gluconate oxidase		+
2KG oxidase		+
2KG oxidase (galactose-grown cells)		−
2KG reductase	+	
5KG reductase	+	
2, 5-KG reductase	+	
The enzymes producing γ-pyrones from 2, 5-KG	+	+
Sorbitol oxidase		+
Glycerol oxidase		+
Glycerol kinase	+	
DHA kinase	+	
G-3-P-dehydrogenase	+	
D- and L-Lactate dehydrogenase		+
Pyruvate carboxylase	+	−
Ethanol dehydrogenase	+	
Ethanol oxidase		+
Pyruvate dehydrogenase	+	
Acetyl CoA synthetase	+	
Aconitase	+	

Table 28 (Continued).

Enzymes	Soluble fraction	Particles
Isocitrate dehydrogenase	+	−
α-KG dehydrogenase	+	--
Succinate oxidase	−	+
Fumarase	+	−
Malate oxidase	−	+
Oxaloacetate decarboxylase	+	−
NADH$_2$ oxidase		+
Catalase		+
Isocitrate lyase	+	
Malate synthase	+	
Glyoxylate carboligase	+	
Glycerate kinase	+	
Tartonic semiald. reductase	+	
Galactose oxidase		+
Galactokinase	−	
Galactonokinase	−	
2-K-3-deoxygal. kinase (galactose-grown cells)	+ (adapt)	
Galactonate dehydratase (galactose-grown cells)	+ (adapt)	
Glycollate oxidase (glycol-grown cells)	+	

tion of the soluble fraction with galactonate. A further cleavage of 2-keto-3-deoxygalactonophosphate into pyruvate and triosephosphate was observed by trapping the formed pyruvate in the presence of quinone in a reaction mixture of galactonate and ATP. Thus it may be deduced that growth on galactose induced the adaptive formation of a mechanism for galactose breakdown corresponding to the Entner-Doudoroff pathway in *Pseudomonas saccharophila*.

Galactose-grown cells, in contrast to glucose-grown cells which formed a large amount of 2,5-diketogluconate, formed only small amounts of 2-ketogluconate from glucose, and neither 2,5-diketogluconate nor any γ-pyrone compound was detected. No 2-ketoglucono-oxidase was present in the protoplasmic membrane. No formation of brown pigments was observed, although this is characteristic of the organism when grown on glucose-CaCO$_3$ medium. These galactose grown cells are thought to oxidize glucose and gluconate after phosphorylation.

The possible pathways for glucose and galactose breakdown and the various enzymes actually involved in carbohydrate metabolism in this organism are shown in Figure 41 and Table 28.

Figure 42 shows the initial stages of glucose and ketogluconate metabolism, with the enzyme sites as determined by Stouthamer.[512]

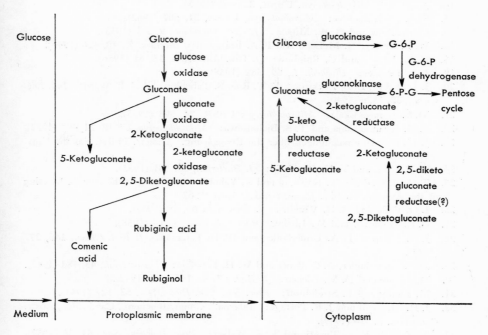

Fig. 42.   Relation between glucose metabolism and enzyme location in " Intermediate " strain IAM 1834 (*G. liquefaciens*).

REFERENCES

1. D. P. Hoyer: Dissert. Univ. Leiden, Waltmann, Delft (1898).
2. W. Henneberg: *Cent. Bakt.*, 2 Abt., **4**, 14, 67, 138 (1898).
3. M. W. Beijerinck: *Cent. Bakt.*, 2 Abt., **4**, 209 (1898).
4. A. Janke: *Cent. Bakt.*, 2 Abt., **45**, 1 (1916).
5. J. Frateur: *La Cellule*, **53**, fasc. 3, 333 (1950).
6. E. Leifson: *Antonie v. Leeuwenhoek*, **20**, 102 (1954).
7. A. N. Hall, L. Husain, K. S. Tiwari and T. K. Walker: *J. Appl. Bacteriol.*, **19**, 31 (1956).
8. J. L. Shimwell: *J. Inst. Brew.*, **63**, 44 (1957).

9. A. N. Hall, K. S. Tiwari, G. A. Thomas and T. K. Walker: *Arch. Biochem. Biophys.*, **46**, 485 (1953).

10. M. S. Dunn, S. Shankman, M. N. Camien and H. Block: *J. Biol. Chem.*, **168**, 1 (1947).

11. C. H. Gray and E. L. Tatum: *Proc. Natl. Acad. Sci.*, **30**, 404 (1944).

12. M. R. R. Rao and J. L. Stokes: *J. Bacteriol.*, **65**, 405 (1953).

13. M. R. R. Rao and J. L. Stokes: *J. Bacteriol.*, **66**, 634 (1953).

14. A. Alian: *Mikrobiologiya*, USSR, **32**, 869 (1963).

15. S. A. Shchelkunova: *Microbiologiya*, USSR, **31**, 507 (1962).

16. C. Rainbow and G. W. Mitson: *J. Gen. Microbiol.*, **9**, 371 (1953).

17. W. B. Emery, N. McLeod and F. A. Robinson: *Biochem. J.*, **40**, 426 (1946).

18. G. D. Brown and C. Rainbow: *J. Gen. Microbiol.*, **15**, 61 (1956).

19. J. A. Fewster: *Biochem. J.*, **69**, 582 (1958).

20. J. J. Stubbs, L. B. Lockwood, E. T. Roe, B. Tabenkin and G. E. Ward: *Ind. Eng. Chem.*, **32**, 1626 (1940).

21. M. S. Loitsyanskaja: *Mikrobiologiya*, USSR, **24**, 598 (1955).

22. S. G. Rasumofskaja and T. S. Bjelousowa: *Mikrobiologiya*, USSR, **21**, 403 (1952).

23. J. Gsur: Untersuchungen über die Essiggärung. Dissert. Philos. Fakult. Univ. Wien (1953).

24. T. E. King and V. H. Cheldelin: *J. Bacteriol.*, **66**, 581 (1953).

25. T. Asai, K. Aida, T. Kajiwara and A. Yahaba: Presented at the Annual Meeting of the Agr. Chem. Soc. Japan, Sendai, April (1962).

26. I. O. Foda and R. H. Vaughn: *J. Bacteriol.*, **65**, 79 (1953).

27. C. L. Goldman and W. Litsky: *Bacteriol. Proc.*, p. 15 (1958).

28. J. O. Lampen, L. A. Underkofler and W. H. Peterson: *J. Biol. Chem.*, **146**, 277 (1942).

29. L. A. Underkofler, A. C. Bantz and W. H. Peterson: *J. Bacteriol.*, **45**, 183 (1943).

30. M. Landy and D. M. Dicken: *J. Biol. Chem.*, **146**, 109 (1942).

31. M. Landy and F. Streightoff: *Proc. Soc. Exp. Biol. Med.*, **52**, 127 (1943).

32. J. H. Marshall and J. R. Postage: Internatl. Congr. Biochem., Cambridge, Abstr. Commun., p. 338 (1949).

33. A. N. Hall, K. S. Tiwari and T. K. Walker: *Proc. Biochem. Soc.*, **51**, 36 (1952).

34. W. Litsky, W. R. Esselen, B. S. Tepper and G. Miller: *J. Food Res.*, **18**, 250 (1953). W. Litsky and C. L. Goldman: *J. Food Res.*, **18**, 646 (1953).

35. B. W. Koft and J. H. Morrison: *Bacteriol. Proc.*, p. 99 (1954).

36. J. V. Karabinoos and M. Dicken: *Arch. Biochem.*, **4**, 211 (1944).

37. R. Baetsle: *Fermentatio*, **5**, 190 (1953).

38. M. Ameyama and K. Kondo: *Agr. Biol. Chem.*, **30**, 203 (1966).

39. M. Ameyama and K. Kondo: *Agr. Biol. Chem.*, **31**, 724 (1967).

40. H. U. Jlli, J. Mueller and L. Ettlinger: *Pathol. Microbiol.*, **28**, 1005 (1965).

41. C. L. Goldman, W. Litsky, M. Mandel and H. N. Little: *Can. J. Microbiol.*, **4**, 463 (1958).

42. H. P. Sarett and V. H. Cheldelin: *J. Biol. Chem.*, **159**, 311 (1945).

43. T. E. King, L. M. Locher and V. H. Cheldelin: *Arch. Biochem.*, **17**, 483 (1948).

44. T. E. King, I. G. Fels and V. H. Cheldelin: *J. Am. Chem. Soc.*, **71**, 131 (1949).

45. G. D. Novelli, R. M. Flymn and F. Lipmann: *J. Biol. Chem.*, **177**, 493 (1949).

46. G. D. Novelli: Phosphorus Metabolism, **1**, Johns Hopkins Press, Baltimore

(1951), p. 414.

47. J. Baddiley and E. M. Thain: *J. Chem. Soc.*, 3421 (1951).

48. T. E. King and V. H. Cheldelin: *Science*, **112**, 562 (1950).

49. T. E. King, A. L. Neal and F. M. Strong: *J. Biol. Chem.*, **189**, 309, 315 (1951).

50. T. E. King and F. M. Strong: *J. Biol. Chem.*, **191**, 515 (1951).

51. J. Baddiley, A. P. Mathias, G. D. Novelli and F. Lipmann: *Nature*, **171**, 76 (1953).

52. J. Baddiley and A. P. Mathias: *J. Chem. Soc.*, 2803 (1954).

53. G. M. Brown and E. E. Snell: *J. Am. Chem. Soc.*, **75**, 2782 (1953).

54. G. M. Brown and E. E. Snell: *J. Bacteriol.*, **67**, 465 (1954).

55. T. E. King and V. H. Cheldelin: *Proc. Soc. Exp. Biol. Med.*, **84**, 591 (1953).

56. G. M. Brown, M. Ikawa and E. E. Snell: *J. Biol. Chem.*, **213**, 855 (1955).

57. A. N. Hall, C. Russel and T. K. Walker: *J. Bacteriol.*, **68**, 279 (1954).

58. J. L. Stokes and A. Larsen: *J. Bacteriol.*, **49**, 495 (1945).

59. C. H. Gray and E. L. Tatum: *Proc. Nat. Acad. Sci.*, U.S.A., **30**, 404 (1944).

60. B. S. Tepper and W. Litsky: *Growth*, **17**, 193 (1953).

61. S. S. Kerwar, V. H. Cheldelin and L. W. Parks: *Bacteriol. Proc.*, p. 126 (1963), *J. Bacteriol.*, **88**, 179 (1964).

62. Y. Yamada, K. Tsuchiya, K. Aida and T. Uemura: Presented at the Annual Meeting of the Agr. Chem. Soc. Japan, Tokyo, April, 1965.

63. T. Asai, K. Aida, T. Kajiwara and Y. Kanehira: Unpublished.

64. H. B. Brown and F. W. Fabian: *Fruit Products*, **22**, 326, 347 (1943).

65. E. I. Fulmer, A. C. Bantz and L. A. Underkofler: *Iowa State Coll. J. Sci.*, **18**, 369 (1944).

66. R. Steel and T. K. Walker: *J. Gen. Microbiol.*, **17**, 12 (1957).

67. R. Steel and T. K. Walker: *J. Gen. Microbiol.* **17**, 445 (1957).

68. R. Steel and T. K. Walker: *Nature*, **180**, 812 (1957).

69. R. Steel and T. K. Walker: *J. Gen. Microbiol.*, **18**, 369 (1958).

70. R. B. Gilliland and J. P. Lacey: *Nature*, **202**, 727 (1963).

71. D. E. Bradley: *J. Gen. Microbiol.*, **41**, 233 (1965).

72. L. Pasteur: *Compt. Rend. Séan. Acad. Sci.*, **54**, 265 (1862), Etudes sur le vinaigre (1868).

73. W. Henneberg: Handbuch der Gärungsbakteriologie, Paul Parey, Berlin (1926), Bd. 2, p. 190–223.

74. A. Bertho: *Ergebnisse der Enzymforschung*, **1**, 231 (1932).

75. H. Wieland: *Ber. Chem. Ges.*, **46**, 3327 (1913).

76. H. Wieland and A. Bertho: *Liebig's Ann.*, **467**, 95 (1928).

77. H. Tamiya and K. Tanaka: *Acta Phytochimica*, **5**, 210 (1930).

78. K. Tanaka: *Acta phytochimica.* **5**, 265 (1930).

79. A Bertho: *Ergebnisse der Enzymforschung*, **1**, 254 (1932).

80. K. Tanaka: *J. Sci. Hiroshima Univ.*, Series B, Div. 2, Vol. 3, 125 (1938).

81. K. Tanaka: *J. Sci. Hiroshima Univ.*, Series B, Div. 2, Vol. 3, 101 (1938).

82. T. E. King and V. H. Cheldelin: *J. Biol. Chem.*, **198**, 131 (1952).

83. J. De Ley and A. H. Stouthamer: *Biochim. Biophys. Acta*, **34**, 171 (1959).

84. W. Henneberg: *Cent. Bakt.*, 2 Abt., **3**, 933 (1897).

85. C. Neuberg and E. F. Nord: *Biochem. Z.*, **96**, 158 (1919).

86. C. Neuberg and F. Windisch: *Biochem. Z.*, **166**, 454 (1925).

87. C. Neuberg: *Biochem. Z.*, **199**, 232 (1928).
88. C. Neuberg and E. Morinari: *Naturwiss.*, **14**, 758 (1926).
89. H. Wieland: *Ber. Chem. Ges.*, **46**, 3335 (1913).
90. E. Simon: *Biochem. Z.*, **224**, 253 (1930).
91. A. Bertho and K. P. Basu: *Liebig's Ann.*, **485**, 26 (1931).
92. M. Cozic: *Rev. gén. Bot.*, **48**, 209 (1936).
93. C. Lutwak-Mann: *Biochem. J.*, **32**, 1364 (1938).
94. T. E. King and V. H. Cheldelin: *J. Biol. Chem.*, **198**, 135 (1952).
95. T. E. King and V. H. Cheldelin: *Biochim. Biophys. Acta*, **14**, 108 (1954).
96. T. E. King and V. H. Cheldelin: *J. Biol. Chem.*, **220**, 177 (1956).
97. M. R. R. Rao and I. C. Gunsalus: *Feder. Proc.*, **14**, 267 (1955).
98. H. Wieland and H. J. Pistor: *Liebig's Ann.*, **522**, 116 (1936), **535**, 205 (1938).
99. D. E. Atkinson: *J. Bacteriol.*, **72**, 195 (1956).
100. D. E. Atkinson: *J. Bacteriool.*, **72**, 189 (1956).
101. S. W. Tanenbaum: *Biochim. Biophys. Acta*, **21**, 335 (1956).
102. S. W. Tanenbaum: *Biochim. Biophys. Acta*, **21**, 343 (1956).
103. N. Tamiya, K. Kameyama, Y. Kondo and S. Akabori: *J. Biochem.*, **42**, 613 (1955).
104. P. Prieur: *Compt. Rend. Séan. Acad. Sci.*, **244**, 253 (1957).
105. T. Nakayama: *J. Biochem.*, **46**, 1217 (1959).
106. T. Nakayama: *J. Biochem.*, **48**, 813 (1960).
107. M. Kida and T. Asai: Unpublished.
108. T. Nakayama: *J. Biochem.*, **49**, 158 (1961).
109. T. Nakayama: *J. Biochem.*, **49**, 240 (1961).
110. P. Dupuy and J. Maugenet: *Ann. Technol. Agr.*, **11**, 219 (1962).
111. J. De Ley and K. Kersters: *Bacteriol. Rev.*, **28**, 164 (1964).
112. T. E. King, M. K. Devlin and V. H. Cheldelin: *Feder. Proc.*, **14**, 236 (1955).
113. A. J. Brown: *J. Chem. Soc.*, 172 (1886).
114. W. Seifert: *Cent. Bakt.*, 2 Abt., **3**, 385 (1897).
115. F. Visser't Hooft: Dissert., Delft, W. D. Meinema (1925).
116. D. Müller: *Biochem. Z.*, **254**, 97 (1932).
117. M. Krehan: *Arch. Microbiol.*, **1**, 493 (1931).
118. J. Chauvet: *Enzymologia*, **11**, 57 (1943/1945).
119. K. Tanaka: *J. Sci. Hiroshima Univ.*, Series B, Div. 2, vol. 3, 127 (1938).
120. W. Henneberg: *Cent. Bakt.*, 2 Abt. **4**, 14 (1898), **14**, 681 (1905).
121. A. J. Brown: *J. Chem. Soc.*, 172, 182, 432, 435, 439 (1886), 638, 643 (1887).
122. H. Mosel: *Cent. Bakt.*, 2 Abt., **87**, 193 (1932).
123. S. Hermann and P. Neuschul: *Biochem. Z.*, **233**, 129 (1931).
124. T. Asai: *J. Agr. Chem. Soc. Japan*, **11**, 385 (1935).
125. K. Kondo and M. Ameyama: *Bull. Agr. Chem. Soc. Japan*, **22**, 371 (1958).
126. J. Tošić: *Biochem. J.*, **40**, 209 (1946).
127. W. Polesofsky: Dissert. philos. Fakult. Univ. Wien (1951).
128. A. Bertho: *Liebig's Ann.*, **474**, 1 (1929).
129. D. Müller: *Biochem. Z.*, **238**, 253 (1931).
130. T. Asai: *J. Agr. Chem. Soc. Japan*, **11**, 387 (1935).
131. T. Uemura and K. Kondo: *J. Agr. Chem. Soc. Japan*, **18**, 17 (1942).
132. D. Müller: *Biochem. Z.*, **254**, 102 (1932).

133. M. Cozic: Thesis, Nemours: André Lesot (1933).

134. T. Asai: *J. Agr. Chem. Soc. Japan*, **11**, 332 (1935).

135. T. Asai and K. Shoda: *J. Gen. Appl. Microbiol.*, **4**, 405 (1958).

136. K. Tanaka: *J. Sci. Hiroshima Univ.*, Series B, Div. 2, vol. 3, 121 (1938).

137. G. Bertrand: *Compt. Rend. Séan. Acad. Sci.*, **122**, 900 (1896).

138. A. J. Brown: *J. Chem. Soc.*, **51**, 638 (1887).

139. W. Henneberg: *Deutsch. Essigind.*, **2**, 145, 153, 161, 169, 177 (1898).   *Cent. Bakt.*, 2 Abt., **4**, 14, 67, 138 (1898).

140. F. Banning: *Cent. Bakt.*, 2 Abt., **8**, 453 (1902).

141. T. Kasai, I. Suzuki and T. Asai: *J. Gen. Appl. Microbiol.*, **9**, 49 (1963).

142. R. Kaushal and T. K. Walker: *Nature*, **160**, 572 (1947).

143. K. Kersters and J. De Ley: *Biochim. Biophys. Acta*, **71**, 311 (1962).

144. O. Hromatka and W. Polesofsky: *Enzymologia*, **24**, 372 (1962).

145. J. T. Cummins: Ph D. Thesis, Oregon State College, Corvallis (1957).

146. A. Kling: *Compt. Rend. Séan. Acad. Sci.*, **128**, 244, **129**, 1252 (1899), **133**, 231 (1901).

147. A. Copet, P. Fierens-Snoeck and H. Van Risseghem: *Bull. Soc. Chim. France*, 902 (1951).

148. H. Van Risseghem: *Bull. Soc. Chim. France*, 908 (1951).

149. A. Janke: Handbuch der Pflanzenphysiologie, Bd. 12, Springer-Verlag, Berlin (1960), p. 693.

150. K. R. Butlin and W. H. D. Wince: *J. Soc. Chem. Ind.*, **58**, 365 (1939).

151. J. Verloove: Ph. D. Thesis, State Univ., Ghent, (1960).

152. E. Grivsky: *Bull. Soc. Chem. Belg.*, **51**, 63 (1942).

153. E. I. Fulmer, L. A. Underkofler and A. C. Bantz: *J. Am. Chem. Soc.*, **65**, 1425 (1943).

154. A. Kling: *Ann. Chim. Phys.*, **5**, 471 (1905).

155. H. Van Risseghem: *Bull. Soc. Chem. Belg.*, **45**, 21 (1936).

156. E. P. Goldschmidt and L. O. Krampitz: *Bacteriol. Proc.*, p. 96 (1954).

157. G. Bertrand: *Compt. Rend. Séan. Acad. Sci.*, **126**, 762, 842 (1898).

158. T. Takahashi and T. Asai: *J. Agr. Chem. Soc. Japan*, **11**, 1008 (1935).

159. Y. Ikeda: *J. Agr. Chem. Soc. Japan*, **26**, 90 (1952).

160. K. Kondo and M. Ameyama: The Reports of the Dept. of Agr., Shizuoka Univ., **7**, 148 (1957).

161. T. E. King and V. H. Cheldelin: *J. Biol. Chem.*, **198**, 127 (1952).

162. V. H. Cheldelin, J. G. Hauge and T. E. King: *Proc. Soc. Exp. Biol. Med.*, **82**, 144 (1953).

163. J. G. Hauge, T. E. King and V. H. Cheldelin: *J. Biol. Chem.*, **214**, 1 (1955).

164. J. G. Hauge, T. E. King and V. H. Cheldelin: *J. Biol. Chem.*, **214**, 11 (1955).

165. L. Klungsöyr, T. E. King and V. H. Cheldelin: *J. Biol. Chem.*, **227**, 135 (1957).

166. L. Federico and L. Gobis: *Ann. Chim. Appl.*, (Roma), **39**, 278 (1949).

167. K. Ramamurti and C. P. Jackson: *Nature*, **204**, 402 (1964).

168. G. Bertrand: *Compt. Rend. Séan. Acad. Sci.*, **130**, 1472 (1900).

169. H. Müller, C. Montigel and T. Reichstein: *Helv. Chim. Acta*, **20**, 1468 (1937).

170. R. L. Whistler and L. A. Underkofler: *J. Am. Chem. Soc.*, **60**, 2507 (1938).

171. J. De Ley and R. Dochy: *Biochim. Biophys. Acta*, **40**, 277 (1960).

172. R. M. Hann, E. B. Tilden and C. S. Hudson: *J. Am. Chem. Soc.*, **60**, 1201 (1938).

173. A. C. Arcus and N. L. Edson: *Biochem. J.*, **64**, 385 (1956).

174. T. Reichstein: *Helv. Chim. Acta*, **17**, 996 (1934).

175. E. Votoček, F. Valentin and F. Rác: *Collect. Trav. Chim. Tschécosl.*, **2**, 402 (1930).

176. L. Anderson and H. A. Lardy: *J. Am. Chem. Soc.*, **70**, 594 (1948).

177. H. Müller and T. Reichstein: *Helv. Chim. Acta*, **21**, 271 (1938).

178. N. K. Richtmyer, L. C. Stewart and C. S. Hudson: *J. Am. Chem. Soc.*, **72**, 4934 (1950).

179. G. N. Bollenback and L. A. Underkofler: *J. Am. Chem. Soc.*, **72**, 741 (1950).

180. P. P. Regna: *J. Am. Chem. Soc.*, **69**, 246 (1947).

181. T. Asai and M. Kudaka: *J. Agr. Chem. Soc. Japan*, **20**, 83 (1944) ; G. Matsukura, M. Kudaka, S. Takahashi and T. Asai: *J. Agr. Chem. Soc. Japan*, **23**, 223 (1949).

182. T. Nehira: *J. Ferment. Technol.* (Japan), **21**, 534 (1943).

183. J. T. Cummins, T. E. King and V. H. Cheldelin: *J. Biol. Chem.*, **224**, 323 (1957) ; J. T. Cummins, V. H. Cheldelin and T. E. King: *J. Biol. Chem.* **226**, 301 (1957).

184. E. D. Mikhlin and M. G. Golysheva: *Biochimija*, **17**, 91 (1952).

185. E. D. Mikhlin and M. G. Golysheva: *Biochimija*, **19**, 549 (1954).

186. S. G. Rasumovskaya and S. M. Zhdan-Pushkina: *Mikrobiologiya*, **25**, 16 (1956).

187. N. M. Mitjushowa: *Mikrobiologiya*, **23**, 400 (1954).

188. S. M. Zhdan-Pushkina: *Mikrobiologiya*, **24**, 549 (1955).

189. B. Görlich: *Z. Naturforsch.*, **11**, 223 (1956).

190. S. A. Shchelkunova: *Microbiologiya*, **32**, 451 (1963).

191. S. M. Zhdan-Pushkina and R. A. Kreneva: *Mikrobiologiya*, **32**, 71 (1963).

192. M. Kulhánek and Z. Sevcikova: Tech. Publ. Stredisko Tech. Inform. Potravinar. Prumyslu., **139**, 161 (1963).

193. P. A. Wells, J. J. Stubbs, L. B. Lockwood and E. T. Roe: *Ind. Eng. Chem.*, **29**, 1385 (1937).

194. P. A. Wells, L. B. Lockwood, J. J. Stubbs, E. T Roe, N. Porges and F. A. Gastrock: *Ind. Eng. Chem.*, **31**, 1518 (1939).

195. E. I. Fulmer and L. A. Underkofler: *Iowa State Coll. Sci.*, **13**, 279 (1939).

196. H. S. Isbell and J. V. Karabinoos: *J. Res. Nat. Bur. Stand.*, **48**, 438 (1952).

197. H. L. Frush and L. J. Tregoning: *Science*, **128**, 597 (1958).

198. O. Terada, K. Tomizawa and S. Kinoshita: *J. Agr. Chem. Soc. Japan*, **35**, 127 (1961) ; O. Terada, K. Tomizawa, S. Suzuki and S. Kinoshita: *J. Agr. Chem. Soc. Japan.*, **35**, 131 (1961) ; O. Terada, S. Suzuki and S. Kinoshita: *J. Agr. Chem. Soc. Japan* **35**, 178 (1961).

199. T. Asai: *J. Agr. Chem. Soc. Japan*, **11**, 380 (1935).

200. M. Steiger and T. Reichstein: *Helv. Chim. Acta*, **18**, 790 (1935).

201. E. L. Totton and H. A. Lardy: *J. Am. Chem. Soc.*, **71**, 3076 (1949).

202. G. Bertrand: *Compt. Rend. Séan, Acad. Sci.*, **139**, 983 (1904).

203. G. Bertrand: *Compt. Rend. Séan. Acad. Sci.*, **149**, 225 (1909).

204. E. B. Tilden: *J. Bacteriol.*, **37**, 629 (1939).

205. L. C. Stewart, N. K. Richtmyer and C. S. Hudson: *J. Am. Chem. Soc.*, **71**, 3532 (1949).

206. V. Ettel, J. Liebster, M. Tatra and M. Kulhánek: *Chem. Listy*, **46**, 448 (1952).

207. G. Bertrand and C. Nitzberg: *Compt. Rend. Séan. Acad. Sci.*, **186**, 925, 1172 (1928).

208. M. Cozic: *Compt. Rend. Séan. Acad. Sci.*, **196**, 1740 (1933).

209. Y. Khouvine and G. Nitzberg: *Compt. Rend. Séan. Acad. Sci.*, **196**, 218 (1933).

210. W. B. Moore, A. C. Blackwood and A. Neish: *Can. J. Microbiol.*, **1**, 198 (1954).

211. D. Müller: *Enzymologia* **3**, 26 (1937).

212. C. Widmer, T. E. King and V. H. Cheldelin: *J. Bacteriol.*, **71**, 737 (1956).

213. Th. Elsaesser, J. Huber and H. Hirscher: *Z. Allgem. Mikrobiol.*, **2**, 249 (1962).

214. J. Huber, Th. Elsaesser and H. Hirscher: *Z. Allgem. Mikrobiol.*, **3**, 136 (1963).

215. K. Sasajima and M. Isono: Presented at the Annual Meeting of the Agr. Chem. Soc. of Japan, Tokyo, April, 1965.

216. D. French, R. J. Suhadolnik and L. A. Underkofler: *Science*, **117**, 100 (1953).

217. G. Bertrand: *Ann. Chim. Phys.* (8) **3**, 181, 195, 227 (1904).

218. L. A. Underkofler, E. I. Fulmer, A. C. Bantz and E. R. Kooi: *Iowa State Coll. J. Sci.*, **18**, 377 (1944).

219. E. I. Fulmer and L. A. Underkofler: *Iowa State Coll. J. Sci.*, **21**, 251 (1947).

220. J. W. Dunning, E. I. Fulmer, J. F. Guymon and L. A. Underkofler: *Science*, **87**, 72 (1938).

221. J. W. Dunning, E. I. Fulmer and L. A. Underkofler: *Iowa State Coll. J. Sci.*, **15**, 39 (1940).

222. A. J. Kluyver and A. G. J. Boezaardt: *Rec. Trav. chim. Pays-Bas*, **58**, 956 (1939).

223. T. Posternak: *Helv. Chim. Acta*, **24**, 1045 (1941).

224. E. Chargaff and B. Magasanik: *J. Biol. Chem.*, **165**, 379 (1946).

225. H. E. Carter, C. Belinsky, R. K. Clark Jr., E. H. Flynn, B. Lytle, G. E. McCasland and M. Robbins: *Feder. Proc.*, **6**, 243 (1946).

226. B. Magasanik and E. Chargaff: *J. Biol. Chem.*, **174**, 173 (1948).

227. B. Magasanik, R. E. Franzl and E. Chargaff: *J. Am. Chem. Soc.*, **74**, 2618 (1952).

228. T. Posternak and D. Reymond: *Helv. Chim. Acta*, **36**, 260 (1953).

229. L. Anderson, K. Tomita, P. Kussi and S. Kirkwood: *J. Biol. Chem.*, **204**, 769 (1953).

230. L. A. Anderson, R. Takeda. S. J. Angyal and D. J. McHugh: *Arch. Biochem. Biophys.*, **78**, 518 (1958).

231. T. Posternak, A. Rapin and A. L. Haenni: *Helv. Chim. Acta*, **40**, 1594 (1957).

232. T. Posternak: *Helv. Chim. Acta*, **33**, 350, 1954 (1950).

233. T. Posternak and F. Ravenna: *Helv. Chim. Acta*, **30**, 441 (1947).

234. H. Murooka, Y. Kobayashi and T. Asai: *Bull. Agr. Chem. Soc. Japan*, **24**, 196 (1960).

235. G. Bertrand: *Compt. Rend. Séan. Acad. Sci.*, **127**, 728 (1898).

236. J. Frateur: *La Cellule*, **53**, fasc. **3**, 334, 353, 358, 367, 382 (1950).

237. K. Bernhauer and E. Riedl-Tůmová: *Biochem. Z.*, **321**, 26 (1950).

238. J. Frateur: *Antonie v. Leeuwenhoek*, **20** 111 (1952).

239. J. Liebster, M. Kulhánek and M. Tadra: *Chem. Listy*, **47**, 1075 (1953).

240. K. Aida, T. Kojima and T. Asai: *J. Gen. Appl. Microbiol.*, **1**, 18 (1958).

241. J. De Ley and J. Schell: *J. Bacteriol.*, **77**, 445 (1959).

242. R. Kaushal, P. Jowett and T. K. Walker: *Nature*, **167**, 949 (1951).

243. E. B. Fred, W. H. Peterson and J. A. Underson: *J. Bacteriol.*, **8**, 277 (1923).

244. K. Yagi and Y. Hashitani: *J. Soc. Chem. Ind., Japan*, **26**, 1265 (1923).

245. S. Hermann: *Biochem. Z.*, **192**, 188 (1928), **205**, 297 (1929).

246. T. Asai: *J. Agr. Chem. Soc. Japan*, **10**, 621, 731, 932, 1124 (1934); **11**, 50, 331, 377, 499, 610, 674 (1935).

247. T. Asai: *J. Agr. Chem. Soc. Japan*, **10**, 1127, 1130 (1934).

248. K. Tanaka: *Acta Phytochimica*, **7**, 265 (1933).
249. T. Asai: *J. Agr. Chem. Soc. Japan*, **11**, 333 (1935).
250. T. Asai: *J. Agr. Chem. Soc. Japan*, **11**, 501 (1935).
251. D. Kulka and T. K. Walker: *Arch. Biochem. Biophys.*, **50**, 169 (1954).
252. K. Bernhauer and B. Görlich: *Biochem. Z.*, **280**, 367 (1935).
253. K. Kondo and M. Ameyama: The Reports of the Dept. of Agr., Shizuoka Univ., **7**, 137, 150 (1957).
254. H. Katznelson, S. W. Tanenbaum and E. L. Tatum: *J. Biol. Chem.*, **204**, 43 (1953).
255. T. Takahashi and T. Asai: *Cent. Bakt.*, 2 Abt., **84**, 13 (1931).
256. K. Bernhauer and K. Irrgang: *Biochem. Z.*, **280**, 360 (1935).
257. E. Riedl-Tůmová and K. Bernhauer: *Biochem. Z.*, **320**, 472 (1950).
258. A. Janke: Handbuch der Pflanzenphysiologie, Springer-Verlag, Berlin, **12**, (1960), p. 713.
259. L. Boutroux: *Compt. Rend. Séan. Acad. Sci.*, **102**, 924 (1866).
260. K. Bernhauer and H. Knobloch: *Biochem. Z.*, **303**, 308 (1939).
261. T. Takahashi and T. Asai: *J. Agr. Chem. Soc. Japan*, **6**, 407 (1930); *Cent. Bakt.*, 2 Abt., **82**, 390 (1930).
262. J. J. Stubbs, L. B. Lockwood, E. T. Roe, B. Tabenkin and G. E. Ward: *Ind. Eng. Chem.*, **32**, 1626 (1940).
263. K. Kondo and M. Ameyama: The Reports of the Dept. of Agr., Shizuoka Univ., **7**, 127 (1957).
264. A. J. Kluyver and F. J. G. De Leeuw: *Tijdschr. Verg. Geneesk.*, **10**, 170 (1924).
265. K. Bernhauer and K. Schön: *Z. Physiol. Chem.*, **180**, 232 (1929).
266. S. Hermann: *Biochem. Z.*, **214**, 357 (1929).
267. K. Bernhauer and H. Knobloch: *Biochem. Z.* **320**, 472 (1950).
268. S. Teramoto, R. Yagi and I. Hori: *J. Ferm. Technol.* (Japan), **24**, 22 (1946).
269. Y. Sumiki and Y. Hatsuta: *J. Agr. Chem. Soc. Japan*, **23**, 87 (1947).
270. S. Khesghi, H. R. Roberts and W. Bucek: *Appl. Microbiol.*, **2**, 183 (1954).
271. M. Yamazaki: *J. Agr. Chem. Soc. Japan*, **28**, 748 (1954).
272. M. Yamazaki and T. Asai: *J. Agr. Chem. Soc. Japan*, **31**, 818 (1957).
273. I. A. Foda and R. H. Vaughn: *J. Bacteriol.*, **65**, 233 (1953).
274. K. Bernhauer and H. Knobloch: *Naturwiss.*, **26**, 819 (1938), *Biochem. Z.*, **303**, 308 (1939).
275. H. Kondo and Z. Narita: *J. Pharm. Soc.*, Japan, **63**, 289, 301 (1943).
276. T. Asai and Y. Ikeda: *J. Agr. Chem. Soc. Japan*, **22**, 50 (1948).
277. H. Knobloch and H. Tietze: *Biochem. Z.*, **309**, 399 (1941).
278. J. Frateur, P. Simonart and T. Coulon: *Antonie v. Leeuwenhoek*, **20**, 111 (1954).
279. K. Kondo, M. Ameyama and T. Yamaguchi: *J. Agr. Chem. Soc. Japan*, **30**, 419 (1956).
280. K. Aida, M. Fujii and T. Asai: *Proc. Japan Academy*, **32**, 595 (1956).
281. K. Aida, M. Fujii and T. Asai: *Bull. Agr. Chem. Soc. Japan*, **21**, 30 (1957).
282. T. Takahashi and T. Asai: *J. Agr. Chem. Soc. Japan*, **9**, 55, 369 (1933); *Cent. Bakt.*, 2 Abt., **88**, 286 (1933).
283. T. Takahashi and T. Asai: *Cent. Bakt.*, 2 Abt., **93**, 248 (1936).
284. T. Yabuta: *J. Chem. Soc. Tokyo*, **37**, 1185, 1234 (1916).
285. K. Aida: *J. Gen. Appl. Microbiol.*, **1**, 30 (1955).

286. K. Aida: *Bull. Agr. Chem. Soc. Japan*, **19**, 97 (1955).

287. A. G. Datta and H. Katznelson: *Nature*, **179**, 153 (1957).

288. T. Takahashi and T. Asai: *J. Agr. Chem. Soc. Japan*, **9**, 351 (1933).

289. K. Aida, T. Kojima and T. Asai: *J. Agr. Chem. Soc. Japan*, **28**, 517 (1954).

290. T. Asai, Z. Sugisaki and K. Aida: *J. Agr. Chem. Soc. Japan*, **29**, 300 (1955).

291. D. Kulka, A. N. Hall and T. K. Walker: *Nature*, **167**, 905 (1951).

292. A. N. Hall, D. Kulka and T. K Walker: *Biochem. J.*, **60**, 271 (1955).

293 J Kamlet: U. S. Pat. 2,314,831 (1943).

294. K. Kondo, M. Ameyama and T. Yamaguchi: *J. Agr. Chem. Soc. Japan*, **30**, 423 (1956).

295. M. Ameyama and K. Kondo: *Bull. Agr. Chem. Soc. Japan*, **22**, 271, 380 (1958).

296. S. Hermann and P. Neuschul: *Biochem. Z.*, **270**, 62 (1934); *Bull. Soc. Chim. Biol.* **18**, 390 (1935).

297. J. Frateur: *La Cellule*, **53**, fasc. **3**, 341 (1950).

298. O. Terada, S. Suzuki and S. Kinoshita: *J. Agr. Chem. Soc. Japan*, **36**, 212, 854 (1962).

299. T. Komatsu: *J. Pharm. Soc., Japan*, **80**, 1387 (1960).

300. T. E. King and V. H. Cheldelin: *J. Biol. Chem.*, **224**, 579 (1957).

301. K. Takahashi and H. Kayamori: *Bull. Agr. Chem. Soc. Japan*, **24**, 231 (1960).

302. T. Komatsu: *J. Pharm. Soc., Japan*, **80**, 1392 (1960).

303. T. Komatsu: *J. Pharm. Soc., Japan*, **80**, 1395 (1960).

304. K. Kulhánek: *Chem. Listy*, **47**, 1075 (1953).

305. Y. Ikeda: *J. Agr. Chem. Soc. Japan*, **28**, 538 (1954).

306. K. Sakaguchi, T. Asai and Y. Ikeda: *J. Agr. Chem. Soc. Japan*, **20**, 155 (1944).

307. R. Weidenhagen and G. Bernsee: *Ber. Chem. Ges.*, **93**, 2924 (1960).

308. G. Avigad and S. England: *J. Biol. Chem.*, **240**, 2290, 2297 (1965).

309. O. Terada, S. Suzuki and S. Kinoshita: *J. Agr. Chem. Soc. Japan*, **35**, 1336 (1961).

310. J. G. Carr, R. A. Coggins and G. C. Whiting: *Chem. & Ind.*, **31**, 1279 (1963).

311. O. Terada, S. Suzuki and S. Kinoshita: *Agr. Biol. Chem.*, **25**, 802 (1961).

312. O. Terada, S. Suzuki and S. Kinoshita: *J. Agr. Chem. Soc. Japan*, **36**, 623 (1962).

313. K. Aida and Y. Yamada: *Agr. Biol. Chem.*, **28**, 74 (1963).

314. S. England and G. Avigad: *Proc. Israel Chem. Soc.*, **11A**, 72 (1962).

315. K. Aida, Y. Yamada and T. Uemura: Presented at the 16th Annual Meeting of Enzyme Chemistry Symposium Japan, Tokyo, June, 1964.

316. J. De Ley: *Biochim. Biophys. Acta*, **27**, 652 (1958), J. De Ley and A. H. Stouthamer: *Biochim. Biophys. Acta*, **34**, 171 (1959).

317. K. Okamoto: *J. Biochem.*, **53**, 448 (1963).

318. Y. Takagi: *Agr. Biol. Chem.*, **26**, 719 (1962).

319. Y. Yamada, K. Aida and T. Uemura: Presented at the Annual Meeting of the Agr. Chem. Soc. Japan, Sapporo, July, 1964.

320. Y. Yamada, K. Aida and T. Uemura: Presented at the 17th Annual Meeting of Enzyme Chemistry Symposium Japan, Tokushima, May, 1965, *Agr. Biol. Chem.*, **30**, 95 (1966).

321. Y. Yamada, K. Aida and T. Uemura: *J. Biochem.*, **61**, 636 (1967).

322. Y. Yamada, K. Aida and T. Uemura: Presented at the Annual Meeting of the Agr. Chem. Soc. Japan, Kyoto, April, 1966.

323. K. Kondo and S. Wada: *J. Ferment. Technol.* (Japan), **27**, 331 (1949).

324. K. Kondo and M. Ameyama: The Reports of the Dept. of Agr., Shizuoka Univ., 7, 139 (1957).

325. O. Terada, S. Suzuki and S. Kinoshita: *Agr. Biol. Chem.*, **25**, 870 (1961).

326. K. Sato, Y. Yamada, K. Aida and T. Uemura: *Agr. Biol. Chem.*, **31**, 877 (1967).

327. K. Sato, Y. Yamada, K. Aida and T. Uemura: *Agr. Biol. Chem.*, **31**, 640 (1967).

328. I. Isono, Z. Nakanishi, K. Sasajima, K. Mochizuki, T. Kanzaki, N. Okazaki and H. Yoshino: Presented at the Annual Meeting of the Agr. Chem. Soc. Japan, Tokyo, April, 1965.

329. W. B. Moore, A. C. Blackwood and A. C. Neish: *Can. J. Microbiol.*, **1**, 198 (1954).

330. T. Takahashi: *Bull. Coll. Agr. Tokyo Imp. Univ.*, **7**, 531 (1907), **10**, 103 (1909).

331. W. Henneberg: *Deutsch. Essigind.*, **2**, Nr. 19 (1898), **10**, 89 (1906).

332. S. W. Tanenbaum and H. Katznelson: *J. Bacteriol.*, **68**, 368 (1954).

333. T. Asai: *J. Agr. Chem. Soc. Japan*, **11**, 332 (1935).

334. G. A. Pitman and W. V. Cruess: *Ind. Eng. Chem.*, **21**, 1292 (1929).

335. W. Seifert: *Cent. Bakt.*, 2 Abt., **5**, 462 (1902).

336. S. Hermann and P. Neuschul: *Biochem. Z.*, **246**, 446 (1932).

337. J. De Ley and J. Schell: *Biochim. Biophys. Acta*, **35**, 154 (1959).

338. C. H. Chin: Abstr. Comm., Intern. Cong. Biochem. Biophys. Paris, p. 277 (1952).

339. J. De Ley and V. Vervolet: *Biochim. Biophys. Acta*, **50**, 1 (1961).

340. J. De Ley: *J. Appl. Bacteriol.*, **32**, 414 (1960).

341. J. De Ley and J. Schell: *J. Gen. Microbiol.*, **29**, 589 (1962).

342. T. E. King and V. H. Cheldelin: *J. Biol. Chem.*, **208**, 821 (1954), **220**, 177 (1956).

343. M. R. R. Rao: Thesis, Univ. Illinois, Urbana, 1955.

344. T. E. King, E. H. Kawasaki and V. H. Cheldelin: *J. Bacteriol.*, **72**, 418 (1956).

345. P. Dupuy and J. Maugenet: *Ann. Technol. Agr.*, **12**, 5 (1963).

346. T. Kitassato: *Biochem. Z.*, **195**, 118 (1928).

347. M. Lemoigne: *Ann. Inst. Pasteur*, **27**, 118 (1928).

348. J. Frateur: *La Cellule*, **53**, fasc. **3**, 382 (1950).

349. J. De Ley: *J. Gen. Microbiol.*, **21**, 352 (1959).

350. A. H. Stouthamer: Koohlhydraatstofwisseling van de Azijnzurbacteriën. Thesis, Univ. Utrecht (1960).

351. V. H. Cheldelin: Metabolic Pathways in Microorganisms, John Wiley & Sons, Inc., New York (1960), p. 19 ; T. E. King, M. K. Devlin and V. H. Cheldelin: *Feder. Proc.*, **14**, 236 (1955).

352. R. L. Wixom: *Biochem. J.*, **94**, 427 (1965).

353. K. Tanaka: *J. Sci. Hiroshima Univ.*, Series B, Div. 2, Vol. 3, 74 (1938) ; K. Tanaka and T. Kaibara: *J. Sci. Hiroshima Univ.*, Series B, Div. 2, Vol. 5, 45 (1942).

354. J. M. Wiame and R. Lambion: *Bull. Techn. Vinegar*, **7**, 195 (1951).

355. T. W. Dratwina: *Mikrobiologiya*, **6**, 468 (1937).

356. L. Federico and L. Gobis: *Bull. Soc. ital. Biol. sper.*, **25**, 236 (1949).

357. C. Antoniani, L. Federico and L. Gobis: *Bull. Soc. ital. Biol. sper.*, **25**, 526 (1949).

358. L. Federico: *Bull. Soc. ital. Biol. sper.*, **25**, 10 (1949).

359. C. Antoniani, L. Federico and L. Gobis: *Ann. Chim.* (Roma), **40**, 83 (1950).

360. K. R. Butlin: Dept. Sci. Ind. Res. Chemistry Research. Special Report No. 2, London (1936).

361. K. Tanaka: *Acta Phytochimica*, **8**, 285 (1935), *J. Sci. Hiroshima Univ.*, Series B,

Div. 2, Vol. 3, 70 (1938).

362. M. S. Loitsyanskaya and M. S. Mamkaeva: *Mikrobiologiya*, **33**, 344 (1964).

363. J. Banning: *Cent. Bakt.*, 2 Abt., **8**, 395, 425, 453, 520, 556 (1902).

364. G. C. Whiting and R. A. Coggins: *Biochem. J.*, **102**, 283 (1967).

365. L. Smith: *Arch. Biochem. Biophys.*, **50**, 299 (1954).

366. A. Fujita and T. Kodama: *Biochem. Z.*, **273**, 186 (1934).

367. B. Chance: *J. Biol. Chem.*, **202**, 383, 397, 407 (1953).

368. L. R. N. Castor and B. Chance: *J. Biol. Chem.*, **217**, 453 (1955).

369. F. Kubowitz and E. Haas: *Biochem. Z.*, **255**, 247 (1932).

370. O. Warburg and E. Negelin: *Biochem. Z.*, **262**, 237 (1933).

371. L. Smith: *Arch. Biochem. Biophys.*, **50**, 315 (1954).

372. T. Kasai, K. Aida and T. Uemura: Presented at the Annual Meeting of the Agr. Chem. Soc. Japan, Sapporo, July, 1964.

373. Y. Iwasaki: *Plant Cell Physiol.*, **1**, 207 (1960).

374. O. Warburg and W. Christian: *Biochem. Z.*, **262**, 237 (1933).

375. J. De Ley and R. Dochy: *Biochim. Biophys. Acta*, **42**, 538 (1960).

376. W. Zopf: *Ber. Deutsch. Bot. Ges.*, **18**, 32 (1900).

377. J. Frateur: *La Cellule*, **53**, fasc. **3**, 378, 381 (1950).

378. K. Tanaka and S. Masumoto: *J. Sci. Hiroshima Univ.*, **5**, 3, 61 (1942).

379. O. Hayaishi, H. Shimazono, M. Katagiri and Y. Saito: *J. Am. Chem. Soc.*, **78**, 5126 (1956).

380. F. Challenger, V. Subramaniam and T. K. Walker: *J. Chem. Soc.*, 200 (1927).

381. J. R. Quayle and G. A. Taylor: *Biochem. J.*, **78**, 611 (1961).

382. K. Kasai, K. Aida and T. Uemura: Presented at the 16th Annual Meeting of Enzyme Chemistry Symposium Japan, Tokyo, June, 1964.

383. K. Miyaji: *J. Chem. Soc. Japan*, **45**, 391 (1925).

384. C. Antoniani, L. Federico and L. Gobis: *Ann. Chim.* (Roma), **40**, 80 (1950).

385. L. Federico and L. Gobis: *Biol. Soc. ital. Biol. sper.*, **25**, 238 (1949).

386. J. J. Joubert, W. Bayens and J. De Ley: *Antonie v. Leeuwenhoek*, **27**, 151 (1961).

387. A. Janke, R. G. Janke and S. Perczel: *Arch. Mikrobiol.*, **45**, 7 (1963).

388. K. E. Cooksey and C. Rainbow: *J. Gen. Microbiol.*, **27**, 135 (1962).

389. M. Röhr: *Naturwiss.*, **48**, 478 (1961).

390. M. Röhr and D. Chiari: *Naturwiss.*, **51**, 519 (1964).

391. S. Klungsöyr and T. Aasheim: *Physiol. Plantarum*, **15**, 552 (1962).

392. H. Hibbert and J. Barsha: *Can. J. Research*, **5**, 580 (1931).

393. K. S. Barcley, E. J. Bourne, M. Stacey and M. Webb: *J. Chem. Soc.*, 1501 (1954).

394. H. L. A. Tarr and H. Hibbert: *Can. J. Research*, **4**, 372 (1930).

395. R. Kaushal and T. K. Walker: *Biochem. J.*, **48**, 618 (1951).

396. J. G. Carr: *Antonie v. Leeuwenhoek*, **24**, 157 (1958).

397. J. L. Shimwell and J. G. Carr: *J. Inst. Brew.*, **64**, 477 (1958).

398. H. Hibbert and J. Barsha: *Can. J. Research*, **10**, 170 (1934).

399. M. Aschner and S. Hestrin: *Nature*, **157**, 659 (1946).

400. K. Mühlethaler: *Biochim. Biophys. Acta*, **3**, 527 (1949).

401. E. Franz and E. Schiebold: *Makromol. Chem.*, **1**, 3 (1943).

402. B. G. Rånby: *Ark. Kemi* (Stockholm), **4**, 249 (1952).

403. K. Mühlethaler: *Makromol. Chem.*, **2**, 143 (1948).

404. E. Husemann and R. Werner: *Makromol. Chem.*, **59**, 43 (1963).

405. E. O. Dillingham, A. G. Jose and D. T. Kunth: *Bacteriol. Proc.*, **14**, 67 (1961).
406. Y. Khouvine: *Compt. Rend. Séan. Acad. Sci.*, **196**, 1144 (1933).
407. Y. Khouvine: *Compt. Rend. Séan. Acad. Sci.*, **198**, 1544 (1934).
408. S. Hestrin, M. Aschner and J. Mager: *Nature*, **159**, 64 (1947).
409. M. Schramm and S. Hestrin: *J. Gen. Microbiol.*, **11**, 123 (1954).
410. S. Hestrin and M. Schramm: *Biochem. J.* **58**, 345 (1954).
411. T. E. Webbs and J. R. Colvin: *Can. J. Biochem. Physiol.*, **41**, 1691 (1963).
412. T. E. Webbs and J. R. Colvin: *Can. J. Microbiol.*, **10**, 11 (1964).
413. F. W. Minor, G. A. Greathouse, H. G. Shirk, A. M. Schwartz and M. Harris: *J. Am. Chem. Soc.*, **76**, 1658 (1954).
414. F. W. Minor, G. A. Greathouse, H. G. Shirk, A. M. Schwartz and M. Harris: *J. Am. Chem. Soc.*, **76**, 5052 (1954).
415. F. W. Minor, G. A. Greathouse and H. G. Shirk: *J. Am. Chem. Soc.*, **77**, 1244 (1955).
416. G. A. Greathouse: *J. Am. Chem. Soc.*, **79**, 4507 (1957).
417. J. R. Colvin: *Biochem. J.*, **70**, 294 (1957).
418. R. G. Everson and J. R. Colvin: *Can. J. Biochem.*, **44**, 1567 (1966).
419. M. Schramm. Z. Gromet and S. Hestrin: *Biochem. J.*, **67**, 669 (1957).
420. Z. Gromet, M. Schramm and S. Hestrin: *Biochem. J.*, **67**, 679 (1957).
421. M. Schramm and E. Racker: *Nature*, **179**, 1349 (1957).
422. M. Schramm, Z. Gromet and S. Hestrin: *Nature*, **179**, 28 (1957).
423. G. Koshio and M. Katayama: Presented at the Annual Meeting of the Agr. Chem. Soc. Japan, Tokyo, April, 1961.
424. J. Weigl: *Arch. Mikrobiol.*, **38**, 350 (1961).
425. L. Glaser: *Biochim. Biophys. Acta*, **25**, 436 (1957) ; *J. Biol. Chem.*, **232**, 627 (1958).
426. S. Klungsöyr: *Nature*, **185**, 104 (1960).
427. H. Zeigler and J. Weigl: *Naturwiss.*, **46**, 20 (1959).
428. B. S. Enevoldsen: *Acta Chem. Scand.*, **18** (3), 834 (1964).
429. J. R. Colvin: *Nature*, **183**, 1135 (1959).
430. M. Benziman and H. Burger-Rachamimov: *J. Bacteriol.*, **84**, 625 (1962).
431. Z. Gromet-Elhanan and S. Hestrin: *J. Bacteriol.*, **85**, 284 (1963).
432. A. Janke: *Arch. Mikrobiol.*, **41**, 79 (1962).
433. J. A. Gascoigne: *Chem. & Ind.*, 1580 (1963).
434. D. T. Dennis and J. R. Colvin: Cellular Ultrastructure of Woody Plants, Proc. Advan. Sci. Seminar, Upper Saranac Lake, N. Y., 1964, 199 (1965).
435. J. R. Colvin and M. Beer: *Can. J. Microbiol.*, **6**, 631 (1960).
436. B. Millman and J. R. Colvin: *Can. J. Microbiol.*, **7**, 383 (1961).
437. J. R. Colvin: *Can. J. Microbiol.*, **11**, 641 (1965).
438. J. R. Colvin: *Can. J. Microbiol.*, **12**, 909 (1966).
439. J. H. Carson, L. C. Sowden and J. R. Colvin: *Can. J. Microbiol.*, **13**, 837 (1967).
440. R. Kaushal, T. K. Walker and D. G. Drummond: *Biochem. J.*, **50**, 128 (1951).
441. A. E. Creedy, P. Jowett and T. K. Walker: *Chem. & Ind.*, 1297 (1954).
442. E. J. Bourne and A. Weigel: *Chem & Ind.*, 132 (1954).
443. H. B. Wright and T. K. Walker: *Chem. & Ind.*, 18 (1955).
444. T. K. Walker and H. B. Wright: *Arch. Biochem. Biophys.*, **69**, 362 (1957).
445. R. Steel and T. K. Walker: *Nature*, **180**, 201 (1957).
446. K. Ramamurti and C. P. Jackson: *J. Biol. Chem.*, **237**, 2434 (1962).

447. W. F. Dudman: *J. Gen. Microbiol.*, **21**, 312, 327 (1959).
448. J. Tošić and T. K. Walker: *J. Gen. Microbiol.*, **4**, 192 (1950).
449. J. L. Baker, F. E. Day and H. F. E. Hulton: *J. Inst. Brew.*, **18**, 651 (1912).
450. J. L. Shimwell: *J. Inst. Brew.*, **42**, 585 (1936).
451. E. J. Hehre and D. M. Hamilton: *Proc. Soc. Exper. Biol. Med.*, **71**, 336 (1949); *J. Biol. Chem.*, **192**, 161 (1951).
452. O. Hoffmann-Ostenhof: Enzymologie, Wien, Springer (1954).
453. B. H. Arnold and A. N. Hall: *Nature*, **177**, 44 (1956).
454. S. A. Barker, E. J. Bourne, M. Stacey and R. B. Ward: *Nature*, **185**, 203 (1955).
455. T. K. Walker, D. N. Pellegrino and A. W. Khan: *Nature*, **183**, 682 (1959).
456. M. S. Loitsyanskaya: Tr. Petergof. Biol. Inst. Leningrad Gos. Univ., **19**, 20 (1965).
457. J. De Ley: *J. Gen. Microbiol.*, **24**, 31 (1961).
458. K. R. Butlin: *Biochem. J.*, **32**, 508, 1185 (1938).
459. Y. Iwasaki: *Plant Cell. Physiol.*, **1**, 195 (1960).
460. T. E. King and V. H. Cheldelin: *Biochem. J.*, **68**, 31 P (1958).
461. E. Galante, P. Scalaffa and G. A. Lanzani: *Enzymologia*, **26**, 23 (1963).
462. V. H. Cheldelin: Metabolic Pathways in Microorganisms, John Wiley & Sons, Inc., New York (1960), p. 5.
463. K. Takahashi: *J. Agr. Chem. Soc. Japan*, **36**, 524 (1962).
464. K. Okamoto: *J. Biochem.*, **53**, 348 (1963).
465. J. A. Fewster: *Biochem. J.*, **65**, 14 P (1957).
466. K. Kondo and M. Ameyama: The Reports of the Dept. of Agr., Shizuoka Univ., **7**, 118, 132, 136 (1957).
467. H. Murooka: Thesis, Univ. Tokyo (1962).
468. P. Scalaffa, E. Galante and G. A. Lanzani: *Bull. Soc. ital. Biol. sper.*, **39**, 667 (1963).
469. E. Galante, P. Scalaffa and G. A. Lanzani: *Enzymologia*, **27**, (3), 176 (1964).
470. T. Asai and H. Murooka: *J. Agr. Chem. Soc. Japan*, **30**, 1 (1956).
471. T. Asai, H. Murooka and Y. Kobayashi: *J. Agr. Chem. Japan*, **31**, 225 (1957).
472. Y. Sekizawa, M. E. Maragoudakis, S. S. Kerwar, M. Flikke, A. Baich, T. E. King and V. H. Cheldelin: *Biochem. Biophys. Res. Comm.*, **9**, 361 (1962).
473. J. A. Fewster: *Biochem. J.*, **66**, 9 P (1957).
474. P. A. Kitos, T. E. King and V. H. Cheldelin: *J. Bacteriol.*, **74**, 565 (1957).
475. V. H. Cheldelin: Metabolic Pathways in Microorganisms, John Wiley & Sons, Inc., New York (1960), p. 22.
476. P. A. Kitos, C. H. Wang, B. A. Mohler, T. E. King and V. H. Cheldelin: *J. Biol. Chem.*, **233**, 1295 (1958).
477. C. H. Wang, I. Stern, C. M. Gilmour, S. Klungsöyr, D. J. Reed, J. J. Bialy, B. E. Christensen and V. H. Cheldelin: *J. Bacteriol.*, **76**, 207 (1958).
478. T. E. King and V. H. Cheldelin: *Feder. Proc.*, **16**, 204 (1957).
479. R. Kovachevich and W. A. Wood: *J. Biol. Chem.*, **213**, 745, 757 (1955).
480. T. K. Walker: Advances in Enzymology, **9**, 575 (1949).
481. R. Weinberg and M. Doudoroff: *Feder. Proc.*, **14** (1), 302 (1955).
482. A. Munch-Peterson and H. A. Barker: *J. Biol. Chem.*, **230**, 649 (1958).
483. Y. Sekizawa, M. E. Maragoudakis, T. E. King and V. H. Cheldelin: *Chimica Chronika*, **30A**, 131 (1965).

484. R. A. Smith and I. C. Gunsalsus: *Nature*, **175**, 774 (1955); *J. Am. Chem. Soc.*, **76**, 5002 (1954).

485. V. H. Cheldelin: Metabolic Pathways in Microorganisms, John Wiley & Sons, Inc., New York, 1960. p. 24.

486. T. E. King and V. H. Cheldelin: *J. Biol. Chem.*, **224**, 579 (1957).

487. Y. Yoshiie and K. Kameyama: *Bot. Mag. Tokyo*, **70**, 19 (1957).

488. L. R. N. Castor and B. Chance: *J. Biol. Chem.*, **234**, 1587 (1959).

489. P. A. Kitos, T. E. King, J. A. Ambrose and V. H. Cheldelin: *Feder. Proc.*, **14**, 236 (1955).

490. O. Hromatka and H. Gsur: *Enzymologia*, **25**, 81 (1962).

491. A. H. Stouthamer: *Antonie v. Leeuwenhoek*, **25**, 241 (1959).

492. A. G. Datta and H. Katznelson: *Arch. Biochem. Biophys.*, **65**, 576 (1956).

493. A. G. Datta, R. M. Hochster and H. Katznelson: *Can. J. Biochem. Physiol.*, **36**, 327 (1958).

494. H. Katznelson: *Can. J. Microbiol.*, **4**, 25 (1958).

495. D. H. Bone and R. M. Hochster: *Ccn. J. Biochem. Physiol.*, **38**, 193 (1960).

496. M. Ameyama and K. Kondo: *Bull. Agr. Chem. Soc. Japan*, **22**, 373 (1958).

497. M. Schramm, V. Klybas and E. Racker: *J. Biol. Chem.*, **233**, 1283 (1958).

498. P. Prieur: *Compt. Rend. Séan. Acad. Sci.*, **256**, 5660 (1963).

499. G. A. White and C. H. Wang: *Biochem. J.*, **90**, 408 (1964).

500. G. A. White and C. H. Wang: *Biochem. J.*, **90**, 424 (1964).

501. Th. Leisinger: *Pathol. Microbiol.*, **29**, 756 (1966); *Arch. Mikrobiol.*, **54**, 21 (1966).

502. R. Iwatsuru: *Biochem. Z.*, **168**, 34 (1925).

503. K. Tanaka: *J. Sci. Hiroshima Univ.*, Series B, Div. 2, Vol. 3, 74 (1938).

504. M. Benziman and A. Abeliovitz: *J. Bacteriol.*, **87**, 270 (1964).

505. M. Benziman and Y. Galanter: *J. Bacteriol.*, **88**, 1010 (1964).

506. M. Benziman and N. Heller: *J. Bacteriol.*, **88**, 1678 (1964).

507. T. E. Webb and J. R. Colvin: *Can. J. Microbiol.*, **8**, 841 (1962).

508. J. J. R. Tomlinson and J. Campbell: *J. Bacteriol.*, **86**, 1165 (1963).

509. K. Kondo and M. Ameyama: *Bull. Agr. Chem. Soc. Japan*, **22**, 369 (1958).

510. J. De Ley: *Antonie v. Leeuwenhoek*, **24**, 281 (1958).

511. T. Asai, H. Iizuka and K. Komagata: *J. Gen. Appl. Microbiol.*, **10**, 95 (1964).

512. A. H. Stouthamer: *Biochim. Biophys. Acta*, **48**, 484 (1961).

513. H. Murooka, Y. Kobayashi and T. Asai: *Agr. Biol. Chem.*, **26**, 135 (1962).

514. A. H. Stouthamer, J. H. van Boom and A. J. Bastiaanse: *Antonie v. Leeuwenhoek*, **29**, 393 (1963).

515. H. Murooka, Y. Kobayashi and T. Asai: *Agr. Biol. Chem.*, **26**, 142 (1962).

516. A. H. Stouthamer: *Biochim. Biophys. Acta*, **56**, 19 (1962).

517. L. Smith: *Bacteriol. Rev.*, **18**, 106 (1954).

ADDENDUM

A biologically active substance produced by A. aceti

Antoniani *et al.*[1] and Yamamoto *et al.*[2] reported the presence of a biologically active substance in vinegar that markedly inhibits the growth of human tumor cells cultivated *in vitro*. The substance was obtained by a purification process similar to that used for cytochrome isolation. It had a marked oxidizing ability and promoted the activity of some catalase systems.

Further studies on this substance, especially on its chemical characteristics and biological activities, were carried out by Lanzani and Pecile.[3] When the substance was added to the medium in very low concentration in *in vitro* respiration experiments with Ehrlich's ascites tumor cells, an increase of respiration occurred, while at higher concentrations the respiration was blocked.

The substance was isolated in pure state as a white powder with a melting point of 184°C. Its probable structural formula is:

It may be in equilibrium with a γ-lactone form or may exhibit an enolic form. Its toxicity is very low: 2.5 g/kg intravenously is not lethal for mice. It induced a digitalis-like pattern without producing heart block. When the pure substance in saline solution adjusted to pH 7.4 was added in doses of 3 mg (equivalent to 2.9 mg nitrogen) to 2.8 ml fluid containing the ascites tumor cells, it produced an increase of respiration for the first 20 min and then a nearly complete inhibition of respiration. The results of *in vivo* experiments using Ehrlich's ascites tumor in white mice showed that a daily injection dose of 3 mg of substance/animal, beginning from the second day after the intraperitoneal inoculation of the ascitic fluid containing the tumor cells, completely inhibited tumor growth and caused a very marked decrease in the number of cells per ml of ascitic fluid.

Polyol dehydrogenases of G. suboxydans

Bygrave and Shaw[4] reported the presence of soluble NAD-linked mannitol dehydrogenase in cell-free extracts of *G. suboxydans* ATCC 621. They also found a soluble NADP-linked D-mannitol dehydogenase which oxidizes D-glucitol (to L-sorbose) and D-arabitol (to D-xylulose) in addition to D-mannitol. These two dehydrogenases were later investigated.[5] The reduction of NAD by mannitol or sorbitol proceeded at a constant rate, while the reduction of NADP by mannitol or sorbitol was not linear. NAD was reduced only to a slight extent by mannitol or sorbitol, unless KCN was added to the reaction mixture. By contrast, the addition of KCN was not required for NADP-linked activity. NAD-linked mannitol dehydrogenase appears to be a constitutive enzyme since cell-free extracts from cells grown on glucose, sorbitol, or mannitol in place of glycerol showed about the same specific activity. The purified enzyme preparation freed from $NADH_2$-oxidase activity was obtained by addition of protamine sulfate to cell-free extracts and column chromatography of the supernatant solution on DEAE-cellulose (equilibrated with glutathione and phosphate buffer, pH 7.0), eluting with a linear NaCl gradient from 0 to 0.25 M. By this procedure NAD-linked mannitol dehydrogenase was clearly separated from NADP-linked mannitol and NAD-linked sorbitol dehydrogenases.

NAD-linked mannitol dehydrogenase was less stable in stroage than the other two enzymes, but its activity was restored by incubating it with cysteine or glutathione.

In the absence of cysteine or glutathione, column chromatography on DEAE-cellulose resulted in the complete disappearnace of NAD-linked mannitol dehydrogenase activity, which accounts for the fact that this enzyme has been overlooked by a number of researchers.

An extensive investigation of polyol dehydrogenases was carried out recently by Kersters *et al.*[6] using *G. suboxydans* strain SU. The supernatant of cell-free extracts was treated with ammonium sulphate and the precipitates obtained were subjected to column chromatography on DEAE-cellulose after dialysis. This yielded at least four different soluble polyol dehydrogenases and two different gluconate dehygrogenases, each with distinct substrate and cofactor specificities.

These dehydrogenases are NAD-linked xylitol dehydrogenase (I) which oxidizes xylitol to L-xylulose, NAD-linked D-*erythro* dehydrogenase (II) which oxidizes pentitols only when the secondary hydroxyl group at C-2 possesses a D-*erythro* configuration, NAD-linked D-*xylo* dehydrogenase (III) which oxidizes pentitols, hexitols and heptitoles with the D-*xylo* configuration (and D-mannitol weakly), NADP-linked D-*lyxo* dehydrogenase

(IV) which oxidizes pentitols and hexitols with the D-*lyxo* configuration, and two different NADP-linked dehydrogenases (V), (VI) which oxidize D-gluconate.

The substrate specificities and the products are listed in Figures 1–4.

The particulate fraction of cell-free extracts oxidized only polyols with the L-*erythro* configuration, *i.e.* polyols with the Bertrand-Hudson

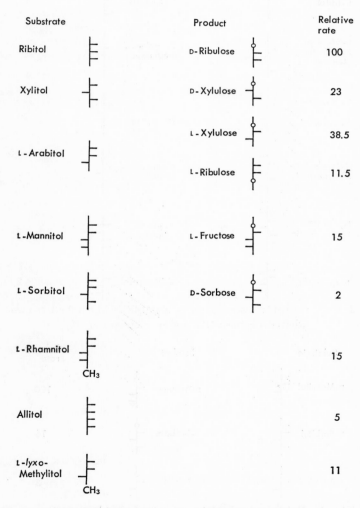

Fig. 1. Substrate specificity of soluble NAD-linked D-*erythro* dehydrogenase (II) (after Kerters *et al.*).

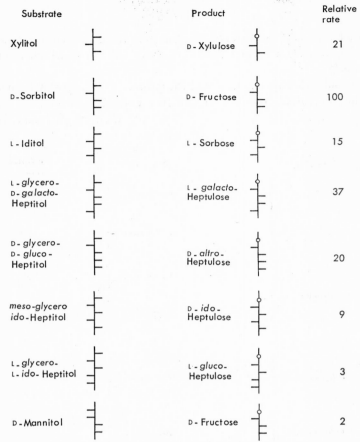

Fig. 2. Substrate specificity of soluble NAD-linked D-*xylo* dehydrogenase (III) (after Kersters *et al.*).

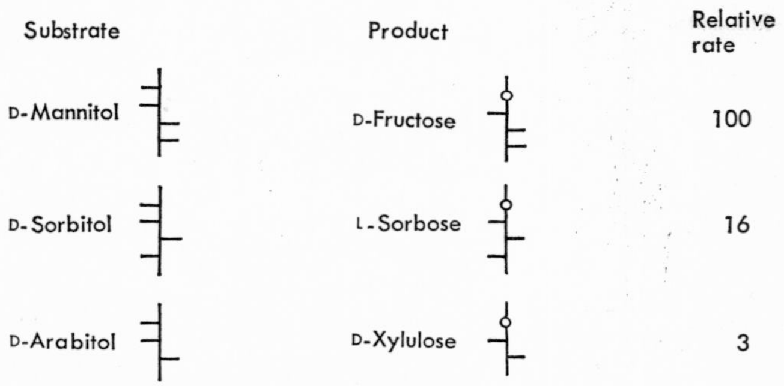

Fig. 3. Substrate specificity of soluble NAD-linked D-*lyxo* dehydrogenase (IV) (after Kersters *et al.*).

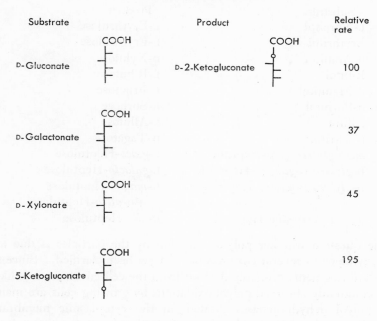

2-Ketogluconate reductase

Substrate	Product	Relative rate
D-Gluconate	D-2-Ketogluconate	100
D-Galactonate		37
D-Xylonate		45
5-Ketogluconate		195

5-Ketogluconate reductase

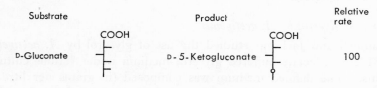

Substrate	Product	Relative rate
D-Gluconate	D-5-Ketogluconate	100
D-Galactonate		

Fig. 4. Substrate specificities of NADPH₂-linked 2-keto-gluconate reductase (V) and NADPH₂-linked 5-keto-gluconate reductase (VI) (after Kersters *et al.*)

configuration (with the exception of L-threitol). The substrates and the products were as follows:

Substrate	Product
L-Threitol	L-Erythrulose
Erythritol	L-Erythrulose
D-Arabitol	D-Xylulose
Ribitol	L-Ribulose
D-Mannitol	D-Fructose
D-Sorbitol	L-Sorbose
Allitol	L-Allulose
D-Altritol	D-Tagatose
meso-glycero-gulo-Heptitol	L-*gluco*-Heptulose
D-*glycero*-D-*galacto*-Heptitol	L-*galacto*-Heptulose
D-*glycero*-D-*manno*-Heptitol	D-*manno*-Heptulose +D-*altro*-Heptulose
meso-glycero-allo-Heptitol	L-*allo*-Heptulose

The question whether polyol oxidation by the particles is due to a single enzyme or to several enzymes has not yet been clarified. Since the particles are fragments probably derived from the cell envelope, it is likely that the commonly observed polyol oxidations by growing cells are mainly due to polyol dehydrogenases located on the cytoplasmic membrane; probably Bertrand and Hudson's rule was formulated with growing cultures of acetic acid bacteria.

Utilization of glycerol by A. acetigenus

Ramamurti and Jackson[7] studied the use of glycerol by *A. acetigenus* NCIB 8132 in a lactate-buffered glycerol medium under various cultural conditions. The defined medium was composed (in grams per litre of distilled water) of glycerol, 30.0 ; lactic acid, 10.3 ; KH_2PO_4, 2.0 ; $MgSO_4 \cdot 7H_2O$, 1.0 ; $FeSO_4 \cdot 7H_2O$, 0.01 ; $(NH_4)_2 HPO_4$, 2.5 ; Ca D-pantothenate, 0.002 ; riboflavin, 0.002 ; biotin, 0.001. pH=4.6.

The results of the experiments indicated a more complete use of glycerol in shaken cultures than under static conditions. Cellulose formation seemed to be best in shaken cultures, whereas the yield of it considerably decreased in static cultures.

5-Keto-D-fructose reductase

The oxidation of D-fructose to 5-keto-D-fructose by growing cells of *G. cerinus* IFO 3267 has been investigated recently by Englard and Avigad.[8]

The maximum rate of 5-keto-D-fructose production and accumulation was shown when the pH in culture medium dropped to 4.8 or less and the growth nearly ceased. From the fact that ^{14}C-fructose, in the presence of other utilizable carbon sources, was oxidized to 5-keto-D-fructose without any decrease in specific activity, and from the finding that 5-^{3}H-fructose lost its tritium when oxidized by the cells to 5-keto-D-fructose, it is obvious that a direct oxidition of the hydroxyl at C-5 position of fructose took place. In addition to the $NADH_2$-linked 5-ketofructose reductase activity, the crude cell-free extracts of *G. cerinus* IFO 3267 exhibited two additional dehydrogenase activities *viz.*, NADP-linked mannitol dehydragenase which reversibly oxidizes D-fructose and NADP-linked D-aldose (glucose, mannose and 2-deoxyglucose) dehydrogenase. A procedure of separation and partial purification of these enzymatic activities was described. D-Mannose isomerase activity was also detected in the cell-free extracts.

The subsequent experiments of England *et al.*[9] presentend evidence that $NADH_2$-linked 5-keto-D-fructose reductase catalyzes a stereospecific transfer of hydrogen from the A-side of the reduced pyridine nucleotide ring to the substrate. Most recently Avigad *et al.*[10] obtained a purified $NADPH_2$-linked 5-keto-D-fructose reductase preparation exhibiting approximately 1200-fold activity from the cells. The details of the purification procedure were described. Using this preparation, the following reaction was evidenced.

$$5\text{-Keto-D-fructose} + NADPH_2 \rightleftharpoons \text{D-Fructose} + NADP$$

(with an equilibrium far towards the right, $K_{eq} = 2 \times 10^{10}$.)

The *Km* values for $NADPH_2$, NADP, 5-ketofructose, and fructose were found to be 1.8×10^{-6} M; 1.3×10^{-3} M; 4.5×10^{-3} M; and 7.0×10^{-2} M, respectively, at pH 7.4. The enzyme appeared to be highly specific for $NADPH_2$ and 5-ketofructose. It was thought that under normal physiological conditions of growth, *G. cerinus* cells utilize 5-ketofructose as a carbon source *via* the reduction to fructose, and that a separate enzyme responsible for fructose oxidation to 5-ketofructose may exist.

2-*Ketogluconate and* 5-*ketogluconate reductases*

Galante and Scalafa[11] and Scalafa and Galante[12] reported some experiments on the factors which influence the enzymes forming 2-keto and 5-ketogluconic acids from gluconate in the cells of *A. suboxydans* NRRL B 223. The results seemed to indicate that the pH of the growth medium is a factor in determining the quantities of these enzymes, while the substitution of sorbitol or mannose for glucose or gluconate as carbon sources seems to have little effect. Serial transfer of the growing culture on glycerol medium (ten times in all) had no influence on the ketogenic activities of the cells, suggesting that 6-carbon carbohydrates or shorter chain derivatives do not act as inducers. Amino acids, vitamins and nitrogeneous bases also had no influence on the induction of these enzymes.

Galante et al.[13] further observed that cells grown in unfavorable conditions show high ketogenic activities. Experiments on the influence of glucose concentration in the medium showed that 1% glucose produced maximum exponential growth for 24 hours, but very low ketogenic activity, while 0.5% and 16% glucose gave very poor growth, but much higher ketogenic activity. There was evidence that the enzymatic formation of 2-keto and 5-keto-gluconates from gluconate in G. suboxydans depends quantitatively on the particular vegetative state of the cells, rather than on external chemical inducers.

Lethal effect of Acetobacter on yeasts

Gilliland and Lacey[14] confirmed the lethal effect of an *Acetobacter* strain on yeasts in bottled beer under anaerobic conditions. This activity appeared when *Acetobacter* AY (a strain belonging to *A. rancens*) was mixed with yeasts in the ratio of 3 : 1. The characteristic was not exhibited by other *Acetobacter* species, e.g. *A. xylinum*, *A. lovaniense*, *A. rancens*, *A. mesoxydans*, or *A. ascendens* or by *G. suboxydans*.

The active principle could pass through a dialysis membrane but was very quickly destroyed and could not be isolated. The filtrate of a growing culture of *Acetobacter* AY had no effect on yeasts. The activity was lost when the bacteria were killed by heat or chemicals, but it was maintained when they were killed by streptomycin.

The activity increased with rising temperatures (from 7° to 24°C), was enhanced by ethanol and was strongest at pH 3.5–4.5 ; it was inhibited by oxygen. The strain was active not only against *Saccharomyces cerevisiae* and *S. carlsbergensis* but also against *S. diastaticus*, *S. pastorianus*, *Hansenula anomala*, *Pichia fermentans*, *Schizosaccharomyces pombe*, *Zygosaccharomyces acidifaciens*, *Torula utilis*, *Candida mycoderma* and *Brettanomyces bruxellensis*. *Lactobacillus pastorianus* was not entirely susceptive, although growth was inhibited. *Pediococci* were not affected.

Participation of vitamin K in malate oxidation

Benziman and Perz[15] showed that in the malate oxidase system of *A. xylinum* a vitamin K-like compound may participate in the transfer of electrons from malate to molecular oxygen. In experiments with sonic extracts of succinate-grown cells, malate oxidase activity decreased significantly when the sonication time was increased. The activity of such extracts could be restored by the addition of low concentrations of vitamin K_3 (2-methyl-1, 4-naphthoquinone) or higher concentrations of vitamin K_1 (2-methyl-3-phytyl-1, 4-naphtoquinone). Inactivation of this quinone by irradiation with light at 360 mμ resulted in a loss of malate oxidase activity, which could be restored by replacing the inactivated quinone with exogenous vitamin K_3 or vitamin K_1. That a vitamin K-like compound is involved in the malate oxidase system is further corroborated by the fact that dicoumarol, a competitive inhibitor of vitamin K, inhibited malate oxidation, even in a system not supplemented with vitamin K_3. When present, vitamin K_3 seems to act directly as an intermediary electron carrier in the malate oxidase reaction.

Further experiments with various electron-transport inhibitors on the different oxidatvie activities, using amytal, 2-n-hepty-4-hydrazyquinoline-N-oxide, cyanide, azide or p-hydroxymercuribenzoate, suggested that K_3 acts before cytochrome c and possibly before cytochrome b. The pattern of inhibition places K_3 and probably the natural quinone in $A.$ $xylinum$ between the primary flavoprotein dehydrogenase and cytochrome c.

Phosphorylation coupled with malate oxidation

Benziman and Levy,[16] using succinate-grown cells of $A.$ $xylinum$, showed that malate oxidation by particles or crude extracts couples with phosphorylation. Substitution of oxaloacetate or pyruvate for malate gave no oxidation or phosphorylation, suggesting that the oxidation of malate is a one-step oxidation reaction and that oxaloacetate is the sole carbon compound formed from malate. Phosphorylation but not oxidation was observed in the presence of an ATP-trapping system. Omission of ADP or its substitution by AMP greatly decreased phosphorylation without affecting oxidation. Thus the reaction systems in oxidation and phosphorylation were regarded as loosely coupled.

Crude extracts catalyzed oxidative phosphorylation with $NADH_2$, $NADPH_2$ and isocitrate in addition to malate. In no cases was any phosphorylation observed under anaerobic conditions. Oxidation of $NADH_2$ proceeded at a rate more than two times greater than that of $NADPH_2$, and these oxidations were accompanied by phosphorylation. Since the ability to form ATP, estimated by the P/O ratio observed with various substrates, was significantly lower than that of mammalian systems, it was suggested that the respiratory chains in $A.$ $xylinum$ have fewer phosphorylating sites than those of animal tissues, or that microbial system is

much more readily damaged or uncoupled.   It was also suggested that in
A. xylinum the respiratory chain from malate to oxygen and the phosphorylation associated with it is to a large extent separate from that catalyzing electron transport from $NADH_2$ to oxygen.

Direct phosphorylation of pyruvate by A. xylinum

According to Benziman[17] succinate-grown cells of *A. xylinum* possess
an enzyme system which carries out a direct net synthesis of phosphoenolpyruvate from pyruvate and ATP with the concomitant formation of AMP
and Pi.   This synthesis was absent in glucose-grown cells.   The reaction
probably proceeds in the following way:

$$\text{Pyruvate} + \text{ATP} \xrightarrow{Mg^{++}} \text{PEP} + \text{AMP} + \text{Pi}$$

The enzyme activity could be separated from the pyruvate kinase
activity.   Other nucleotide triphosphates could not be substituted for
ATP.

It was concluded that the physiological role of PEP synthase in
succinate-grown cells is to catalyze the formation of PEP from pyruvate,
thus enabling them to synthesize cellulose which is essential for growth
in a static liquid medium.   This is consistent with the earlier finding
(Benziman and Burger-Rachamimov[18]) that the anhydroglucose carbon
chain of cellulose arises from pyruvate *via* PEP in *Acetobacter*.

Lactate oxidation system in G. suboxydans

The lactate oxidation system of *G. suboxydans* IAM 1828, with special
reference to the carbon monoxide-binding pigment, has recently been
examined by H. Iwaski[19].   The absence of *c*-type cytochromes in this
strain was noted, in contrast to the strain described in a report by Y.
Iwasaki[20], which possesses two *c-type* cytochromes (cytochrome 552 and
cytochrome 554) in addition to cytochrome a_1.   The present strain was
found to have cytochrome *a* and *b*.   Haemoprotein 558, (whose *a*-maximum
in the ruduced form is at 558 mμ) a CO-binding, was also present.

The purified haemoprotein 558, obtained from cell-free extracts by
column chromatography and ammonium sulfate fractionation, has pro-
tohaem as its prosthetic group and showed a peroxidase-like absorption
spectrum.   It formed compounds with cyanide and azide, and showed a
spectrum similar to horseradish peroxidase at pH 7.0.   CO formed a
complex with reduced haemoprotein 558 which is spectroscopically similar
to the CO compound of ferro-peroxidase, but haemoprotein 558 differs
somewhat from horseradish peroxidase.   It has no specific peroxidase

activity : the azide compound was formed even at pH 7.0, while peroxidase formed the corresponding compound at a pH below 4.5, and the cyanide compound of ferro-haemoprotein 558 had a Soret peak at 423 mμ, while that of ferro-peroxidase had a peak at 432 mμ. When oxygen was passed through the solution of reduced haemoprotein 558, a complex with a Soret peak at 423 mμ was formed, but peroxidase did not form such a complex. The absorption maximum then shifted to 417 mμ, indicating a transformation to another compound. One of these two products was probably oxygenated ferrohaemoprotein 558.

Inhibition of lactate oxidation by carbon monoxide in the dark, the recovery with light, inhibition by cyanide and azide, and the fact that Warburg's partition coefficient was estimated as 7, indicates the presence of some haemin oxidase in the reaction system, haemoprotein plus lactate dehydrogenase. These facts strongly suggest that haemoprotein 558 is the terminal oxidase in this organism.

The physiological function of cytochrome a in this organism is not clear.

Later H. Iwasaki[21] obtained haemoprotein 558 as needle-like crystals. Ultracentrifugal analysis showed that the crystallized sample had a simple and symmetrical peak. The sedimentation coefficient, S_{20}, was estimated to be 2.9 S, independent of protein concentration. Molecular weight was estimated by the gel filtration method to be 48,000 in $M/20$ Tris buffer, pH 7.5.

It does not make a complex with H_2O_2. At high pH's, the absorption spectrum of haemoprotein 558 changed to the true ferrohaemochrome type spectrum, reminiscent of cytochrome b. This transformation was found to be reversible.

Levan synthesis by A. suboxydans var. muciparum

Loitsyankaya[22] reported that strains of *A. suboxydans* var. *muciparum*, grown in media containing sucrose as the sole carbon source, inverted sucrose and synthesized levan. Liberated glucose was oxidized to gluconic acid and to 5-ketogluconic acid. When raffinose or melibiose was supplied instead of sucrose, growth of the organism and levan synthesis required the presence of 0.2–0.5% glycerol or ethanol.

Effect of phosphates on the activity of acetic acid bacteria

The influence of phosphate concentration on the growth of *A. suboxydans*, *A. melanogenus* and *A. aceti* was investigated by Shchelkunova[23]. The optimum concentration was found to be dependent on the carbon source: for media containing glucose and glycerol it was 10–20 μg P/ml, for media containging sorbitol, 10–50 μg P/ml. *A. suboxydans* did not grow in ethanol-containing medium unless 0.1% B-complex vitamins was added, and the optimum concentration of phosphate for growth was 10–12 μg P/ml.

Later the same author[24] investigated the effect of phosphate on the growth and dihydroxyacetone (DHA) formation of *A. suboxydans* and *A. melanogenus* in a medium containing glycerol, $(NH_4)_2SO_4$, and B-complex vitamins. With 0.1–0.6% KH_2PO_4, the growth of *A. suboxydans* increased 1.5–1.7-fold and the formation of DHA increased about 2-fold compared with the control values. With 0.1% K_2HPO_4, the growth was 1.5-fold greater than that of the control. The formation of DHA remained almost the same when the concentration of KH_2PO_4 varied from 0.1 to 0.8%. With Na_2HPO_4, growth and the formation of DHA were good only at a 0.1% concenration, higher concentrations being inhibitory. The growth and formation of DHA by *A. melanogenus* were good only at a 0.1% concentration, higher concentrations of the three salts were inhibitory. The effect of acetate- and citrate-buffer was also investigated. The organism could grow and oxidize glycerol in the pH range of 3.2–5.8.

Formation of α-isopropylmalic and citric acids in G. suboxydans

Two condensation reactions, catalyzed by cell-free extracts of *G. suboxydans* ATCC 621, were demonstrated by Maragoudakis and Strassman[25]. One is the formation of ^{14}C-labeled α-isopropylmalate by the condensation of α-ketoisovalerate with ^{14}C-labeled acetyl CoA, which indicates that the pathway for leucine biosynthesis *via* α-isopropylamalate which has been shown to exist in yeast and *Salmonella typhimurium* also occurs in this organism. The reaction product was isolated by paper and column chromatography and the identification was established by recrystallization with synthetic α-isopropylmalic acid to constant specific radioactivity.

The other reaction is the formation of citrate by the condensation of oxaloacetate with ^{14}C-labeled acetyl CoA, indicating the existence of the classic condensing enzyme, which has hitherto not been demonstrated in this organism. This reaction was further confirmed by means of spectrophotometric assy at 232 mμ, which measured the decrease of acetyl CoA due to condensation with oxaloacetate.

Nutrition and metabolism of acetic acid bacteria

A review of the acetic acid bacteria, with particular emphasis on morphological, nutritional, biochemical, and metabolic characters, was briefly introduced by Rainbow[26] who concluded that these bacteria comprise two genera.

REFERENCES

1. C. Antoniani, E. Cernuschl, G. A. Lanzani and A. Pecile: *La Clin. Ost. Ginecol.*, **60**, 225 (1958).
2. T. Yamamoto, G. A. Lanzani and E. Cernuschl: *La Clin. Ost. Ginecol.*, **61**, 35 (1959).
3. G. A. Lanzani and A. Pecile: *Nature*, **185**, 175 (1960).
4. F. L. Bygrave and D. R. D. Shaw: *Proc. Univ. Otago, Med. School*, **39**, 15 (1961).
5. D. R. D. Shaw and F. L. Bygrave: *Biochim. Biophys. Acta*, **113**, 608 (1966).
6. K. Kersters, W. A. Wood and J. De Ley: *J. Biol. Chem.*, **240**, 965 (1965).
7. K. Ramamurti and C. P. Jackson: *Can. J. Microbiol.*, **12**, 75 (1966). *J. Biol. Chem.*, **237**, 2434 (1962).
8. S. Englard and G. Avigad: *J. Biol. Chem.*, **240**, 2297 (1965).
9. S. Englard, G. Avigad and L. Prosky: *J. Biol. Chem.*, **240**, 2302 (1965).
10. G. Avigad, S. Englard and S. Pifko: *J. Biol. Chem.*, **241**, 373 (1966).
11. E. Galante and P. Scalaffa: *Boll. Soc. Ital. Biol. sper.*, **40**, 1265 (1964).
12. P. Scalaffa and E. Galante: *Boll. Soc. Ital. Biol. sper.*, **40**, 1576 (1964).
13. E. Galante, G. A. Lanzani and P. Sequi: *Enzymologia*, **30**, 257 (1966).
14. R. B. Gilliland and J. P. Lacey: *J. Inst. Brew.*, **72**, 291 (1966).
15. M. Benziman and L. Perez: *Biochem. Biopys. Res. Commun.*, **19**, 127 (1965).
16. M. Benziman and L. Levy: *Biochem. Biophys. Res. Commun.*, **24**, 214 (1966).
17. M. Benziman: *J. Bacteriol.*, **24**, 391 (1966).
18. M. Benziman and H. Burger-Rachamimov: *J. Bacteriol.*, **84**, 625 (1962).
19. H. Iwasaki: *Plant Cell Physiol.*, **7**, 199 (1966).
20. Y. Iwasaki: *Plant Cell Physiol.*, **1**, 207 (1960).
21. H. Iwasaki: Abstracts V. p. 881, Seventh International Congress of Biochemistry, Tokyo, Aug. 19–25 (1967).
22. M. S. Loitsyanskaya: *Tr. Petergofsk. Biol. Inst., Leningr. Gos. Univ.*, **19**, 20 (1965).
23. S. A. Shchelkunova: *Tr. Petergofsk. Biol. Inst., Leningr. Gos. Univ.*, **19**, 46 (1966).
24. S. A. Shchelkunova: *Tr. Petergofsk. Biol. Inst., Leningr. Gos. Univ.*, **19**, 57 (1966).
25. M. E. Maragoudakis and M. Strassman: *J. Bacteriol.*, **94**, 512 (1967).
26. C. Rainbow: *Wallerstien Lab. Commun.*, **29**, No. 98/99, 5 (1966).

Index of Bacteria

A

Acetobacter, 7
—— *aceti, 17, 54, 57,* **90, 91**
—— *aceti* var. *agile,* **92**
—— *aceti* var. *albuminosum,* **91**
—— *aceti* var. *friabile,* **91**
—— *aceti* var. *muciparum, 39,* **96**
—— *aceti viscosum,* **93**
—— *aceticum rosaceum,* **93**
—— *aceticum* var., **94**
—— *acetigenoideum,* **94**
—— *acetigenoideus,* **24**
—— *acetigenum,* **91**
—— *acetigenus, 19,* **58**
—— *acetosum,* **91**
—— *acetosum* var. *nairobiense,* **98**
—— *acetosus, 17*
—— *acetum,* **98**
—— *acetum-mucosum, 39*
—— *acetum* var. *nariobiense, 39*
—— *acidi oxalici,* **92**
—— *acidophilum,* **98**
—— *acidum-mucosum,* **97**
—— *acidum-polymyxa,* **96**
—— *albidus, 37,* **98**
—— *albuminosum,* **91**
—— *alcoholophilus,* **98**
—— *ascendens, 11, 19, 57,* **91**
—— *aurantium,* **98**
—— *aurantius, 36,* **57, 71**
—— *capsulatum,* **95**
—— *capsulatus, 19*
—— *curvum,* **92**
—— *curvus, 24, 26*
—— *dihydroxyacetonicum,* **94**
—— *dihydroxyacetonicus, 25*
—— *dioxyacetonicus, 36,* **95**
—— *estunense, 63,* **98**
—— *estunensis, 54*
—— *friabile,* **91**
—— *gluconicum,* **94**

Acetobacter
—— *gluconicus, 17*
—— *industrium,* **91**
—— *industrius, 19, 58*
—— *intermedius, 83*
—— *ketogenum, 95, 98*
—— *kützingianum,* **91**
—— *lafarinum,* **96**
—— *lafarinus, 24*
—— *lafarinum* var. *vindovonense,* **96**
—— *lermae, 98*
—— *lindneri,* **91**
—— *lovaniense, 22,* **96**
—— *melanogenum, 14,* **93**
—— *melanogenum* var. *malto-saccharovo-rans,* **97**
—— *melanogenum* var. *maltovorans,* **97**
—— *melanogenus, 17, 57, 58*
—— *mesoxydans, 22, 57,* **97**
—— *mesoxydans* var. *lentum,* **97**
—— *mesoxydans* var. *lentum-saccharovo-rans,* **97**
—— *mesoxydans* var. *saccharovorans, 39,* **97**
—— *mesoxydans vini, 39*
—— *mobile, 19,* **96**
—— *nikitinsky, 98*
—— *operans,* **97**
—— *orleanense, 17,* **92**
—— *oxydans, 19,* **91**
—— *paradoxum,* **96**
—— *paradoxus, 22*
—— *pasteurianum,* **90**
—— *pasteurianum* var. *agile,* **92**
—— *pasteurianum* var. *colorium,* **91**
—— *pasteurianum* var. *variabie,* **92**
—— *pasteurianus, 54, 58*
—— *peroxydans, 17, 57,* **94**
—— *plicatum,* **92**
—— *rancens, 17, 57,* **92**
—— *rancens* var. *agile,* **92**
—— *rancens* var. *celiae,* **92**

329

I

*"Intermediate" strain, **71**, 84*
"Intermediate" strain IMA 1834, 288

M

Micrococcus oblongus, 4
Mycoderma, 3
—— aceti, 3
*—— aceti gélatineux, **90***

P

Pseudomonadaceae, 8
Pseudomonas, 84, 85, 86
—— fluorescens, 20
—— geniculata, 60

S

Saccharomyces
—— carlsbergensis, 322
—— cerevisiae, 322
—— diastaticus, 322
—— pastorianus, 322
Schizosaccharomyces pombe, 322

T

Termobacterium, 7
*—— aceti, 9, **91***

U

Ulvina, 7
—— aceti, 3
Umbina, 7

General Index

A

48203